SECOND EDITION

Pspice and Circuit Analysis

SECOND EDITION

Pspice and Circuit Analysis

JOHN KEOWN
Southern College of Technology

PRENTICE HALL, Englewood Cliffs, NJ 07632

Library of Congress Cataloging-in Publication Data
Keown, John (John L.)
    PSpice and circuit analysis / by John Keown.—2nd ed.
      p. cm.
    Includes index
    ISBN 0-02-363526-6
    1. Electric circuit analysis—Data processing. 2. PSpice.
  I. Title.
  TK454.K46 1994
  621.3815'0285'5369—dc20

                                         92-42067
                                               CIP

Editor: David Garza
Production Editor: Mary M. Irvin
Art Coordinator: Ruth A. Kimpel
Cover Designer: Robert Vega
Production Buyer: Patricia A. Tonneman

This book was set in Times Roman by Macmillan Publishing Company and was printed and
bound by Book Press, Inc. The cover was printed by Phoenix Color Corp.

©1993 by Prentice-Hall, Inc.
A Simon & Schuster Company
Englewood Cliffs, New Jersey 07632

Printed in the United States of America

10 9 8 7 6 5 4 3

ISBN 0-02-363526-6

Prentice-Hall International (UK) Limited, *London*
Prentice-Hall of Australia Pty. Limited, *Sydney*
Prentice-Hall Canada Inc., *Toronto*
Prentice-Hall Hispanoamericana, S.A., *Mexico*
Prentice-Hall of India Private Limited, *New Delhi*
Prentice-Hall of Japan, Inc., *Tokyo*
Simon & Schuster Asia Pte. Ltd., *Singapore*
Editora Prentice-Hall do Brasil, Ltda., *Rio de Janeiro*

Preface

Methods of circuit analysis vary widely, depending on the complexity of the problem. While some circuits require nothing more complicated than the writing of a single equation for their solution, others require that several equations be solved simultaneously. When response of a circuit is to be performed over a wide range of frequencies, the work is often both tedious and time consuming. Various tools ranging from trig tables and slide rules to calculators and computers have been used by those anxious to ease the burden of lengthy computations.

In many cases the problem to be solved requires that the student have an understanding of which basic laws and principles are involved in the solution. In some cases, if the topology of a network is known, along with complete descriptions of the elements that are connected among the various nodes, computer programs can be used to perform the analyses.

Such programs have been under development for several decades. If you have access to a computer language such as BASIC, Pascal, or FORTRAN, you can devise your own programs to readily solve certain types of problems. More powerful programs, capable of solving many types of electrical networks under a variety of conditions, require years to develop and update.

Such a program is SPICE, which stands for *Simulation Program with Integrated Circuit Emphasis*. The version of SPICE used in this book is PSpice, a commercial product developed by the MicroSim Corporation. The evaluation version of the program, which is available at no cost, is sufficient to perform all of the exercises and programs in this book.

The SPICE program is both powerful and flexible. At the same time, it can be intimidating and bewildering to the beginner, who might well ask, "How do I use this mighty tool in the most elementary way?"

Although it might appear foolish to use a powerful hammer to drive a tack, if the novice can solve problems with SPICE *for which he already knows the answers*, he will gain confidence to move ahead. Thus the book begins with dc circuit analysis, proceeds with ac circuit analysis, then goes into the various topics involving semiconductors.

PSpice is used in industry for the main purpose of allowing the designer to investigate the behavior of a circuit without having to actually breadboard the circuit in the laboratory. This allows for a considerable savings in materials and labor. If the design needs to be modified or tweaked, changes can easily be submitted to the computer for another look at the results. The designer is familiar with the components he will eventually use in the actual circuit. He understands their electrical properties and behavior. How large numbers of these components will interact, however, is sometimes difficult to predict. This is where the computer program takes over, going through the tedious solutions much more quickly and with far less chance for mistakes than the human approach.

Should every electrical student, practitioner, and designer learn SPICE and use it? I believe the answer is an unqualified *yes*. It has become a standard in both the academic and professional worlds. Your education will not be complete without an exposure to this valuable tool.

Will SPICE teach you what you need to know in order to perform both circuit analysis and design? I believe the answer is an unqualified *no*. A study of the basic laws that govern circuit behavior is just as important today as it ever was. SPICE and other computer aids of the same nature will merely free you of the drudgery of lengthy and repetitive computations. You will surely gain some additional knowledge in the process, which you might otherwise overlook. You will also enjoy using Probe, a feature of PSpice that allows you to plot circuit response involving functions of frequency and time, among other things.

The motivation for this book comes from a desire to present a simple, easy-to-follow guide to PSpice to all those who want to learn more about computer aids to circuit analysis. The material is presented in such a way that anyone who is studying or has studied the various electrical topics will be able to immediately put PSpice to practical use.

An important feature of the book is the development of models for such devices as the bipolar junction transistor, the field-effect transistor, and the operational amplifier. The models need be no more complicated than necessary for the problem at hand. For example, if you are interested in bias voltages and currents for the BJT, there is no need for a model of the transistor that takes ac quantities into account. It is hoped that the reader will be able to develop his own models for other devices, especially those where linear approximations are all that is needed.

When reading this book be aware that you will learn much more by going through each example on the computer. It is important that you produce the required input files, submit them to the PSpice program, then look at the output files and/or Probe to see the results. Only by actual experience with the computer will you begin to appreciate the power at your disposal and the satisfaction which comes from seeing the solutions appear on your monitor and printer.

WHAT'S NEW IN THE SECOND EDITION

There have been numerous improvements and changes in the capabilities of PSpice since the material for the first edition was developed. New versions of PSpice are being released on a regular basis. The material for the second edition was developed using the January, 1992, version of the various products from MicroSim. An IBM compatible 486 machine was used for testing the Windows version of PSpice.

The second edition of *PSpice and Circuit Analysis* contains a new Preview chapter. Its purpose is to give the reader an overview of some of the main features of PSpice.

Chapter 1, DC Circuit Analysis, contains more material of an introductory nature. It has an expanded coverage of Norton's theorem and introduces non-planar circuits.

Chapter 2, AC Circuit Analysis, contains more examples of three-phase circuits, including power-factor improvement, voltage regulation, three-phase rectifiers, and two-phase systems.

Chapter 3, Transistor Circuits, contains a new section on dc sensitivity.

Chapter 6, Transients and the Time Domain, contains several new examples using switch-opening and initial conditions.

Chapter 8, Stability and Oscillators, has been rewritten to include examples using the initial conditions necessary to produce oscillation.

Chapter 9, An Introduction to PSpice Devices, contains an added example of a logic circuit.

Chapter 10 is a new chapter on the BJT with emphasis on the library models for transistors.

Chapter 11 is a new chapter on the Field-effect Transistor. The emphasis is on the library models for various types on FETs.

Chapter 12 is a new chapter on Parts and User-defined Libraries. It deals primarily with diodes, since they can be created and modified with the evaluation version of PSpice.

Chapter 13 is a new chapter on Two-port Networks. Examples include the y, z, and *ABCD* parameters.

Chapter 14 is a new chapter on the Filter Designer, one of the new products from the MicroSim corporation.

Chapter 15 is a new chapter on Genesis and Schematics. It introduces the concept of the Design Center and the Windows version of the PSpice set of programs.

Thanks to the colleagues and students of Southern College of Technology for their encouragement in developing this material. Thanks also to the following reviewers for their suggestions for additions and modifications to the first and second editions: Peter Aronhime, University of Louisville; William Barnes, County College of Morris; Robert Cox, Wentworth Institute of Technology; George Fredericks, Tri Cities State Tech; Frank Gergelyi, Metropolitan Technical Institute; David Hata, Portland Community College; Sayed Akhavi, Jefferson Technical College; Tom Brewer, Georgia Institute of Technology; Victor Gerez,

Montana State University; Arzind Karnik, Springfield Technical Community College; Nick Massa, Springfield Technical Community College; John Polus, Purdue University; Russ Puckett, Texas A&M University; and Stephen Titcomb, University of Vermont.

Contents

1

DC Circuit Analysis   25

2

AC Circuit Analysis (for Sinusoidal Steady-State Conditions)   69

3

4

5
The Operational Amplifier   191

8

Stability and Oscillators

9

An Introduction to PSpice Devices

10
The BJT and Its Model

11
The Field-Effect Transistor

12
Parts and User-Defined Libraries   367

13
Two-Port Networks and Passive Filters   381

14

Filter Synthesis and the Filter Designer

15

Genesis and Schematics

16
Nonlinear Devices

Appendix A

483

Brief Summary of PSpice Statements

Appendix B

487

PSpice Devices and Statements as Listed in PSPICE.HLP

Appendix C

505

Making PSpice Work

Appendix D

519

PSpice Devices and Model Parameters

Appendix E

531

Sample Standard Device Library

Introduction

PSpice and Circuit Analysis, 2nd edition, will reinforce basic circuit analysis principles using the version of SPICE called PSpice. It is not a complete text on circuit analysis; it is intended to serve as a supplement or reference to a more detailed, conventional textbook. Here the reader can take an active part in learning new ideas through the use of PSpice on the IBM PC/PS2 (or clone) or the Apple Macintosh computer.

The examples shown in this book were run using the January, 1990, and January, 1992, evaluation versions of PSpice. The latest version is available to faculty members of electrical departments in technical institutes, colleges, and universities free of charge directly from MicroSim. All of the examples given here can be used with the evaluation version of PSpice.

PSpice version 5.1 was released early in 1992. This is the full-fledged version of the program which is widely used in industry. Educational site-licensing is now available from MicroSim. This allows for the deluxe version to be used at a substantial cost reduction. Platforms include the PC-DOS and PC-OS/2, Sun-3, and VAX.

Further information is available from

MicroSim Corporation
20 Fairbanks
Irvine, CA 92718 USA
General Offices:
(714)770-3022
Orders:
(800)245-3022

The material in this book may be used in a variety of settings, beginning with elementary dc circuit analysis, progressing through ac analysis including polyphase circuits, electronic devices and circuits, as well as more advanced topics such as operational amplifiers, frequency response, transients, Fourier analysis, nonlinear devices, and active filters. As you move through the material, it is important that the exercises and problems be actually worked on the computer. This gives you an immediate feel for the ease or difficulty of a particular problem. It also allows you to build the confidence that is needed to move ahead with a thorough understanding of basic principles and concepts.

Whether your main interest is in dc/ac circuits or electronics, we advise you to start at the beginning. The mechanics of using the computer as a learning device are best applied to simple models in the first examples. As the mechanics are better known, you will find that the more difficult concepts require only a little more effort.

As you move to more advanced topics, you should feel more comfortable jumping about from chapter to chapter as your interests and studies dictate. The sequence of topics is not critical beyond the introductory chapters dealing with circuit analysis.

The examples in this text were run on a Zenith AT–compatible machine with numeric coprocessor (model ZW-248-82) using the latest release of the evaluation version of PSpice obtained by the author directly from MicroSim. The numerical results that you obtain when running the example program may agree exactly with those shown in the text. If they do not, differences in equipment or software versions may be involved.

A BIT OF BACKGROUND

SPICE is widely used in the academic and industrial worlds to simulate the operation of various electric circuits and devices. It was developed at the University of California and used in the beginning on main-frame computers. The successor to the original, SPICE2, is more powerful. Later versions, such as PSpice by the MicroSim Corporation, are designed to operate on PCs, Macintoshes, and minicomputers.

The material in this book is designed to be used with PCs and Macs. MicroSim has made the educational version of PSpice available to the academic community at no cost. Schools may obtain a master version of the program from MicroSim and make it available to students for use in the computer labs.

SPICE requires a text editor; suitable editors include EDLIN (a part of DOS), EZ (a bulletin-board editor available at nominal cost), the Turbo Pascal editor (from Borland), and TED (published in *PC Magazine* for free use). You may also use a word processor, but it is not recommended because of its unnecessary complexity.

Another possibility is to use the editor that is included with the interactive shell as a part of the PSpice software package. Details on using this shell are included in Appendix C. This method of using PSpice is more difficult than using an independent editor, especially for the type and length of input files that are included in the examples given in this text. For this reason, you are urged to find a simple full-screen editor—one that saves files in ASCII format—and use it.

GETTING STARTED

You will need the educational version of PSpice obtained from MicroSim or elsewhere. The minimum computer requirements for the personal computer are a PC/XT/AT machine with at least 512 kbytes of RAM, a fixed disk, MS-DOS version 3.0 or later, and a monochrome or color graphics monitor. We recommend that you have a machine in at least the AT class, with a 20Mb hard disk and EGA graphics. A coprocessor will allow for much faster speed. Install PSpice on the hard disk. We recommend that the program along with its various files be simply copied to the hard disk in a SPICE subdirectory.

Follow the directions to create a subdirectory from the root directory. At the C:\> prompt type

```
md spice
```

This (make directory) command creates your SPICE directory. Then type

```
cd spice
```

and you will enter the spice directory. (cd is for change directory.) We strongly recommend that your autoexec.bat file contain the statement

```
$p$g
```

This statement will allow the DOS prompt to be C:\SPICE> when you are in the SPICE directory. You may now unpack all the files from the floppy disk containing the PSpice files by typing

```
a:install
```

assuming that the floppy disk is in drive A. Full details of both the installation and set-up procedures for various machines are given in Appendix C.

When you have decided on a circuit to analyze, you must create an input file in a text editor. We assume that you know, or will learn, how to do this before you continue. A no-frills text editor is all that you will need. If nothing else is available, use the DOS editor, EDLIN. Its use is explained in a chapter of the DOS manual. There is also an elementary treatment of EDLIN in Appendix C. There is a brief summary of PSpice statements in Appendix A and a listing of PSpice devices in Appendix B. As needed, the statements and devices are introduced and explained in the examples. Appendix D is a summary of PSpice devices and model parameters.

That is all you will need except for a few formatted floppy disks if you are using an institutional computer and are not allowed to save your own files on that machine. Always back up your work onto floppies even if you have your own machine.

A FEW HELPFUL POINTS

When you create an input file for the circuit under investigation, always begin with a complete sketch of the circuit. Label the nodes using a distinct marking, such as red or blue ink. There must always be a zero (0) node, which will be the *reference* node. The other nodes can have either numerical or alphabetical designations (numbers are usually easier to use). Decide on a name for the input file, for example, *DC12.CIR*. The extension CIR designates an input file (although some other extension, such as *BAK*, will work).

Include a statement in your input file for every element in the circuit. The order of the statements for the elements is of no consequence; however, the first statement in your input file must be a title or description. If it describes an element, it will be ignored. The last statement must be the *.END* statement. Any statement that begins with a period is called a control statement.

Upper- and lower-case alphabetic characters may be used interchangeably; use whichever is more suitable for the particular circuit.

When dealing with either large or small values, observe the following SPICE conventions:

Value	Symbolic Form	Exponential Form
10^{-15}	F	1E−15
10^{-12}	P	1E−12
10^{-9}	N	1E−9
10^{-6}	U	1E−6
10^{-3}	M	1E−3
25.4×10^{-6}	MIL	25.4E-6
10^{3}	K	1E3
10^{6}	MEG	1E6
10^{9}	G	1E9
10^{12}	T	1E12

Remember that the symbolic form may be shown as either upper- or lower-case letters. For example, *M* or *m* will be correct for the prefix *milli*.

In describing a capacitor, a statement such as this might be used:

```
C 4 5 25NF
```

This means that a capacitor is connected between nodes *4* and *5*. The value of the capacitor is 25 nanofarads. The statement could have been given simply as

```
C 4 5 25n
```

Thus the unit symbol is optional.

Pay particular attention to the fact that the symbolic form for the prefix is immediately adjacent to the numeric value. Do not put a space between the number and the prefix. This also should be done when using the exponential form of the prefix, as in

```
C 4 5 25E-9
```

As another example,

```
R3  2  3  33kilohms
```

obviously describes a 33-kΩ resistor connected between nodes *2* and *3*. It would also be correct to use

```
R3  2  3  33k
```

An independent-voltage source is shown, for example, as

```
V  1  0  DC  40V
```

The *DC* designation is optional as is the final *V* (after the 40). Thus an alternate form for this statement might be

```
V  1  0  40
```

Some readers of the first edition have asked for a more detailed introduction to the mechanics of creating an input file for PSpice on the PC. Before you begin working on the material in Chapter 1, you may want to look at this complete example.

HERE'S HOW IT'S DONE

A dc circuit with a voltage source and four resistors is shown in Fig. I.1. The nodes have been given numbers ranging from zero to three. A SPICE analysis requires that all nodes be numbered (or given alphabetic designation). There must also be a reference node, which is called node zero. Assuming that you have PSpice on your computer in a directory called SPICE, begin with the prompt C:\ and type

```
cd spice
```

to go to the PSpice directory. The prompt will now be C:\SPICE.

Creating the Input File

If you are using DOS 5.0 (or a later version), you may use the editor EDIT. If you are using an earlier version of DOS, choose an appropriate editor instead. To use

Fig. I.1 First circuit for PSpice.

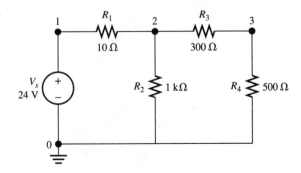

EDIT, type

```
edit ex1.cir
```

where *ex1.cir* is to be the name of the PSpice file that you desire to create. You will see a screen which indicates by the blinking cursor that it is ready for your input. Type the following:

```
First Circuit for PSpice
VS 1 0 24V
R1 1 2 10
R2 2 0 1k
R3 2 3 300
R4 3 0 500
.OPT nopage
.OP
.END
```

When you type the last line of this input file (containing the word .END), do not press the [*Rtn*] key at the end of the line. If you do, PSpice will consider this to be a blank line at the end of the file, which should be avoided. Refer to Fig. I.2, which shows the screen version of what your file should look like.

Saving Your Work

Leave the editor by pressing *Alt-f*, followed by either *s* or *x* (to Save or eXit).

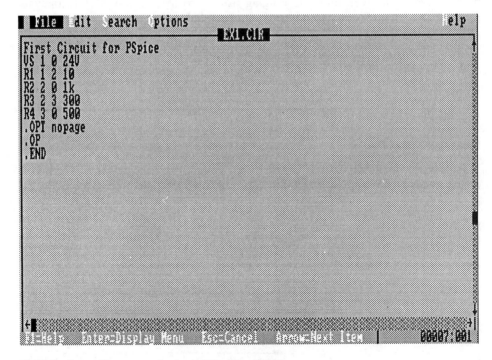

Fig. I.2   Input file as seen in the DOS editor EDIT.

Running the PSpice Analysis

Now you are ready to run the analysis. At the C:\SPICE prompt type

`pspice ex1`

This calls on the PSpice program to analyze your input file ex1.cir. The PSpice screen contains the information shown in Fig. I.3. When the analysis is completed, the message "No errors" should appear beneath the last line shown. Then this will be replaced by a line stating "Bias point calculated."

Notice the line that says "Writing results to EX1.OUT." This means that your analysis created a file containing the results of the PSpice investigation.

Examining the Output File

You may look at this in the editor by typing

`edit ex1.out`

If your output contains no errors, you may want to print the results. The output file EX1.OUT is shown in Fig. I.4. The only editing of this file has been the removal of some blank lines in order to make the output more compact.

Conventional circuit analysis will confirm that $V_{20} = 23.472$ V and $V_{30} = 14.67$ V. The current I is the (negative of the) current shown in the output file as

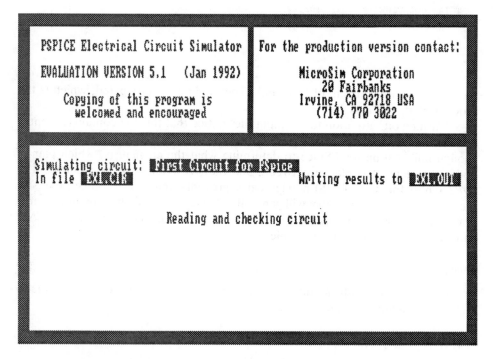

Fig. I.3   The PSpice screen during the analysis of ex1.cir.

```
First Circuit for PSpice

****       CIRCUIT DESCRIPTION

****************************************************************************

VS 1 0 24V
R1 1 2 10
R2 2 0 1k
R3 2 3 300
R4 3 0 500
.OPT nopage
.OP
.END

****       SMALL SIGNAL BIAS SOLUTION          TEMPERATURE =    27.000 DEG C

NODE    VOLTAGE        NODE    VOLTAGE      NODE    VOLTAGE      NODE     VOLTAGE

(    1)   24.0000  (    2)    23.4720  (    3)    14.6700


    VOLTAGE SOURCE CURRENTS
    NAME            CURRENT

    VS           -5.281E-02

    TOTAL POWER DISSIPATION    1.27E+00   WATTS

****       OPERATING POINT INFORMATION        TEMPERATURE =    27.000 DEG C

    JOB CONCLUDED

    TOTAL JOB TIME            102.22
```

Fig. I.4

the voltage-source current of 52.81 mA. The total power dissipated, which is the product of V_s and I, is 1.27 W.

Turning our attention to the input file, notice that this file contains an entry for every element in the circuit. Each element is shown on a separate line with enough information for PSpice to determine what the element is, which pair of nodes the element is placed between, and the size (value) of the element. The *.OPT* (short for .OPTIONS) nopage entry prevents unnecessary page breaks in the output file. The *.OP* entry will generally be used to supply bias-point information in transistor-circuit analysis. For the dc circuit it may be omitted. The *.END* entry is required for every input file.

Changing the Input File

In order to get more information from the PSpice analysis, revise the input file to include two additional lines as follows:

```
.DC VS 24V 24V 24V
.PRINT DC I(R1) I(R2) I(R3)
```

The placement of these lines is not critical, but you may refer to Fig. I.5 for their suggested location in the file. Save the new version of the input file as

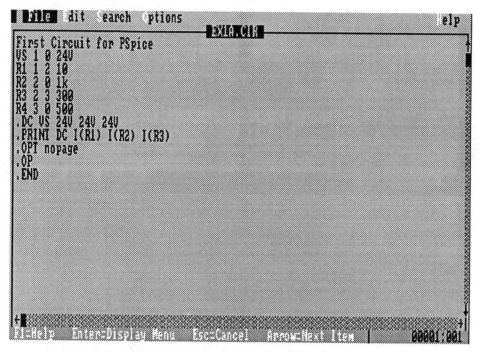

Fig. I.5 Insertion of new lines in the input file.

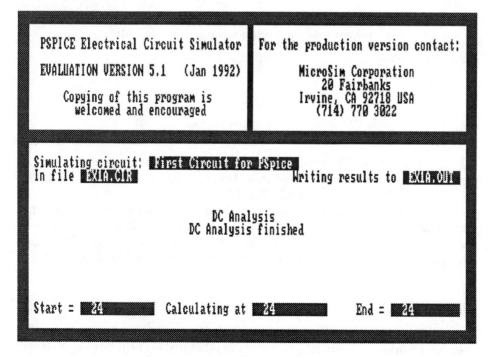

Fig. I.6 The PSpice screen during the analysis of ex1a.cir.

```
First Circuit for PSpice

****      CIRCUIT DESCRIPTION

*************************************************************************

VS 1 0 24V
R1 1 2 10
R2 2 0 1k
R3 2 3 300
R4 3 0 500
.DC VS 24V 24V 24V
.PRINT DC I(R1) I(R2) I(R3)
.OPT nopage
.OP
.END

****      DC TRANSFER CURVES              TEMPERATURE =   27.000 DEG C

   VS          I(R1)        I(R2)        I(R3)

   2.400E+01   5.281E-02   2.347E-02   2.934E-02

****      SMALL SIGNAL BIAS SOLUTION      TEMPERATURE =   27.000 DEG C

NODE   VOLTAGE       NODE   VOLTAGE     NODE   VOLTAGE     NODE   VOLTAGE

(    1)   24.0000  (    2)   23.4720  (    3)   14.6700

   VOLTAGE SOURCE CURRENTS
   NAME          CURRENT

   VS            -5.281E-02

   TOTAL POWER DISSIPATION   1.27E+00  WATTS

****      OPERATING POINT INFORMATION     TEMPERATURE =   27.000 DEG C

   JOB CONCLUDED

   TOTAL JOB TIME          117.48
```

Fig. I.7

ex1a.cir; then run the PSpice analysis, following the same procedure as before. At the conclusion of the job, the screen should contain the information shown in Fig. I.6. The new output file is shown in Fig. I.7.

The output file contains some new information. It shows the currents that were called for in the .PRINT statement. Verify by conventional circuit analysis that the branch currents are $I_{R1} = 52.81$ mA, $I_{R2} = 23.47$ mA, and $I_{R3} = 29.34$ mA. The explanation of the two new lines in the input file will be found in Chapter 1.

FURTHER READING

Many of the example problems in this text are references to similar examples that have been worked by conventional means in other texts. The following texts will serve as resource material for further study and explanation of the topics briefly covered in this book:

Introduction to Electric Circuit Analysis
Robert C. Carter, HRW, 1966

Electric Networks
Hugh Skilling, Wiley, 1974

Integrated Electronics
Millman and Halkais, McGraw-Hill, 1972

Elementary Linear Circuit Analysis, 2nd edition
Leonard S. Bobrow, HRW, 1987

Communication Circuits, 3rd edition
Ware and Reed, Wiley, 1949

Microelectronics
Jacob Millman, McGraw-Hill, 1979

Electronic Devices
William D. Stanley, Prentice-Hall, 1989

Electronic Circuits
Mohammed S. Ghausi, D. Van Nostrand, 1971

Electronic Devices and Circuits, 2nd edition
Theodore F. Bogart, Jr., Merrill/Macmillan, 1990

Circuit Analysis
Irving L. Kosow, Wiley, 1988

PSpice Overview

This section is one of the new features of the second edition. It will give the reader an overview of some of the things that can be done using PSpice. Explanations will be given in detail in later sections of the book. After you see a feature that is of special interest, you may want to move ahead to the chapter that covers the topic in more detail.

DC CIRCUIT ANALYSIS

Figure P.1 represents a dc circuit with a voltage source and three resistors. The PSpice solution for the various currents and voltages in the circuit can easily be found. If you have read the "Getting Started" section of the introduction, you will be ready to construct an input file. In the beginning, it should look like this:

```
Resistive Circuit with Voltage Source
Vs 1 0 dc 12V
R1 1 2 50ohms
R2 2 0 100ohms
R3 2 0 200ohms
.END
```

Do not insert a *Return* (*Enter*) keystroke after the .END. If you do, PSpice will think there is more to come in the input file. Do not put a space between the numerical values of the elements and their units. Now save the file, using the name *preview.cir*, and begin the PSpice analysis by typing

```
pspice preview
```

Fig. P.1 A dc circuit for
SPICE analysis.

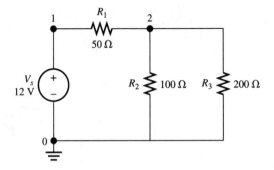

The analysis produces what is called the small-signal bias solution. You can view
the results by entering your text editor using the file name preview.out. This file is
shown in Fig. P.2. Before this file was printed, some slight editing took place to
delete the extra page-feed characters (after the .END line and at the end of the
output file) and unnecessary blank lines.

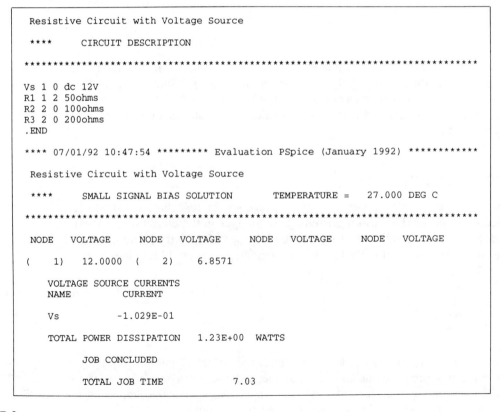

```
      Resistive Circuit with Voltage Source

      ****      CIRCUIT DESCRIPTION

      ***********************************************************************

      Vs 1 0 dc 12V
      R1 1 2 50ohms
      R2 2 0 100ohms
      R3 2 0 200ohms
      .END

      **** 07/01/92 10:47:54 ********* Evaluation PSpice (January 1992) ************

      Resistive Circuit with Voltage Source

      ****      SMALL SIGNAL BIAS SOLUTION      TEMPERATURE =   27.000 DEG C

      ***********************************************************************

      NODE   VOLTAGE      NODE   VOLTAGE      NODE   VOLTAGE      NODE   VOLTAGE

      (    1)   12.0000  (    2)    6.8571

         VOLTAGE SOURCE CURRENTS
         NAME           CURRENT

         Vs             -1.029E-01

         TOTAL POWER DISSIPATION   1.23E+00  WATTS

            JOB CONCLUDED

            TOTAL JOB TIME            7.03
```

Fig. P.2

The output file gives three items of interest: (1) the voltage at node *2*, (2) the current furnished by the voltage source, and (3) the total power dissipation. Identify each of these and verify the values from your pencil-and-paper solution.

It would be helpful to obtain more information about the circuit. The various currents can be shown if the input file contains two additional lines: a .dc voltage statement and a .PRINT dc statement. The extra page feeds are easily eliminated by including an option (.OPT) statement. The new input file is

```
Resistive Circuit with Voltage Source
Vs 1 0 12V
R1 1 2 50
R2 2 0 100
R3 2 0 200
.OPT nopage
.dc Vs 12V 12V 12V
.PRINT dc I(R1) I(R2) I(R3)
.END
```

The first four lines of the file have been modified slightly. The dc description of V_s is not required. The ohms designation for each resistor is not required; in fact, it makes the value more difficult to read. The .PRINT statement will not produce the desired results unless a dc sweep statement is included. The sweep begins at 12 V, ends at 12 V, and uses a 12-V step. This means that there is no actual sweep of voltages, involving several different voltage values.

Run the PSpice analysis and examine the output file. There is a section called DC TRANSFER CURVES, which shows the current through each resistor. Note, however, that the node voltages are missing. They can be restored to the output by the use of the .OP statement. This gives the operating-point information of node voltages and source currents.

In order to obtain still more information from the analysis, a transfer function statement will be included in the input file. The final version of this file is

```
Resistive Circuit with Voltage Source
Vs 1 0 12V
R1 1 2 50
R2 2 0 100
R3 2 0 200
.OPT nopage
.OP
.PRINT dc I(R1) I(R2) I(R3)
.dc Vs 12V 12V 12V
.TF V(2) Vs
.END
```

The output file is shown in Fig. P.3. The small-signal bias solution has been restored with the .OP statement. The .TF statement gives the ratio V(2)/Vs along with the input resistance at Vs and the output resistance at V(2). Check the results against your own solution. What should be the value of the input resistance? Note that the output resistance is found with Vs shorted, placing the three resistors in parallel.

```
     Resistive Circuit with Voltage Source

      ****      CIRCUIT DESCRIPTION

 ********************************************************************
 *
 Vs 1 0 12V
 R1 1 2 50
 R2 2 0 100
 R3 2 0 200
 .OPT nopage
 .op
 .PRINT dc I(R1) I(R2) I(R3)
 .dc Vs 12V 12V 12V
 .TF V(2) Vs
 .END

      ****      DC TRANSFER CURVES              TEMPERATURE =   27.000 DEG C

     Vs        I(R1)        I(R2)        I(R3)

    1.200E+01  1.029E-01  6.857E-02  3.429E-02

      ****      SMALL SIGNAL BIAS SOLUTION     TEMPERATURE =   27.000 DEG C

 NODE   VOLTAGE       NODE   VOLTAGE     NODE   VOLTAGE      NODE   VOLTAGE

 (    1)  12.0000  (    2)   6.8571

     VOLTAGE SOURCE CURRENTS
     NAME           CURRENT

     Vs            -1.029E-01

     TOTAL POWER DISSIPATION   1.23E+00   WATTS

      ****      OPERATING POINT INFORMATION    TEMPERATURE =   27.000 DEG C

      ****      SMALL-SIGNAL CHARACTERISTICS

         V(2)/Vs =  5.714E-01

     INPUT RESISTANCE AT Vs =  1.167E+02

     OUTPUT RESISTANCE AT V(2) =  2.857E+01

         JOB CONCLUDED

         TOTAL JOB TIME          7.41
```

Fig. P.3

AC CIRCUIT ANALYSIS

An ac circuit example will show some of the features available for steady-state sinusoidal circuits.

Figure P.4 shows a circuit with a source voltage of 100 V at a frequency of 100 Hz. This could be the effective (rms) value or the peak value; all other voltages and currents will be shown accordingly. The circuit has resistance, ca-

Fig. P.4 An ac circuit for
SPICE analysis.

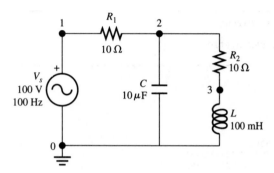

pacitance, and inductance. The input file is named *acpre.cir* and contains the
following:

```
Series-Parallel ac Circuit
Vs 1 0 ac 100V
R1 1 2 10
R2 2 3 10
L 3 0 100mH
C 2 0 10uF
.ac LIN 1 100Hz 100Hz
.PRINT ac I(R1) IP(R1) V(2) VP(2)
.PRINT ac I(C) IP(C) I(R2) IP(R2)
.OPT nopage
.END
```

Since the source voltage is ac rather than dc, it must be shown as such. The
inductance is 100 mH; the *H* is optional but is used for appearance sake. The .ac
statement provides an ac sweep. It is a linear sweep (LIN), although either an
octave or a decade sweep could be chosen. The sweep involves only a single value
of frequency, but without it the desired results could not be printed. The *.OPT*
statement prevents unnecessary title banners and page feeds. In most of the
examples in the rest of the book, this statement will not be used; it can always be
included if desired.

 The results of the PSpice analysis are shown in Fig. P.5. The small-signal
bias solution is of no interest. It would apply only if there were dc sources
involved. This portion of the output file is often edited out before the results are
printed. The current I(C) is the magnitude of the current in the *C* branch; IP(C) is
the phase angle in degrees of this current. The current I(R2) is the magnitude of
the current in the branch with R_2 and *L*; IP(R2) is the phase angle of this current.

 Using a calculator, verify that the sum of these currents is the same as the
current shown through R_1. In rectangular form these are

$$I_C + I_{R_2} = (0.0548, 0.60823) + (0.32, -0.873)$$
$$= 0.9298\underline{/-69.87°}$$

Note that the statement for R_1 is given as

```
R1 1 2 10
```

```
    Series-Parallel ac Circuit

    ****       CIRCUIT DESCRIPTION

    *********************************************************************************

    Vs 1 0 ac 100V
    R1 1 2 10
    R2 2 3 10
    L 3 0 100mH
    C 2 0 10uF
    .ac LIN 1 100Hz 100Hz
    .PRINT ac I(R1)IP(R1) V(2) VP(2)
    .PRINT ac I(C) IP(C) I(R2) IP(R2)
    .OPT nopage
    .END

    ****       SMALL SIGNAL BIAS SOLUTION       TEMPERATURE =    27.000 DEG C

    NODE    VOLTAGE      NODE    VOLTAGE      NODE    VOLTAGE      NODE    VOLTAGE

    (    1)    0.0000  (    2)    0.0000  (    3)    0.0000

       VOLTAGE  SOURCE  CURRENTS
       NAME            CURRENT

       Vs             0.000E+00

       TOTAL  POWER  DISSIPATION     0.00E+00   WATTS

    ****       AC ANALYSIS                       TEMPERATURE =    27.000 DEG C

     FREQ         I(R1)        IP(R1)       V(2)         VP(2)

      1.000E+02    9.295E-01   -6.988E+01    9.719E+01    5.152E+00

    ****       AC ANALYSIS                       TEMPERATURE =    27.000 DEG C

     FREQ         I(C)         IP(C)        I(R2)        IP(R2)

      1.000E+02    6.107E-01    9.515E+01    1.528E+00   -7.580E+01

             JOB CONCLUDED

             TOTAL JOB TIME            8.57
```

Fig. P.5

The nodes are given in the order $1, 2$. This will reference the current in the direction away from the source. When an attempt is made to add currents, the reference directions should be carefully noted and placed on the circuit diagram.

A more interesting ac analysis can be obtained when the frequency is swept between two chosen limits. In our example, a rough approximation of the resonant frequency shows it to be near $f_o = 160$ Hz. The input file will be changed as follows:

```
Series-Parallel ac Circuit
Vs 1 0 ac 100V
R1 1 2 10
```

```
R2 2 3 10
L 3 0 100mH
C 2 0 10uF
.ac LIN 151 50Hz 200Hz
.probe
.end
```

The .ac statement will give a linear sweep of 151 values of frequency in the range from 50 Hz to 200 Hz. This means that calculations will be performed for each integer frequency in the chosen range. The *.probe* statement will place the results in a probe data file, which in this case is called acpre.dat. When the PSpice analysis is run, the results will automatically be available to you in the Probe post processor. The screen will show the axes for a graph with frequency along the horizontal axis.

Choose *Add_trace (a);* then type

```
IP(R1)
```

This produces a plot, called a trace, of the phase of the current through R_1, which is the total circuit current. The plot may now be changed as follows:

Select *X_axis* (press *x*), *Linear* (press *l*), *Set_range* (press *s*); then type

```
50 200 (Rtn)
```

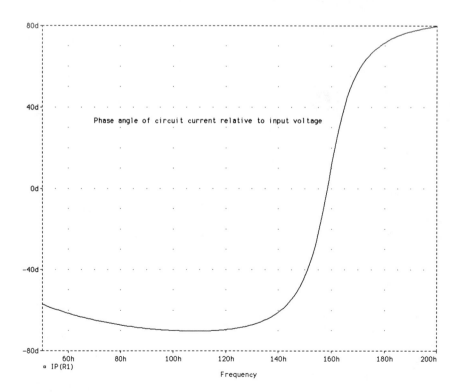

Fig. P.6   Series-Parallel ac Circuit.

Exit (press *e*), *Label* (press *l*), *Text* (press *t*), Now type in this description:

```
Phase angle of circuit current relative to input voltage
```

and press [*Rtn*]. Use the mouse to move the text to a convenient location on the graph. Next *Exit* (press *e*), and obtain *Hard_copy* (press *h*), 1_page_long [*Rtn*].

Before leaving Probe, choose *Cursor* (press *c*), and use the arrows to move the cursor to the location where the phase shift is near zero degrees. The box at the bottom right of the screen should show

```
C1 = 158.355, 107.1E-18
C2 = 50.000, -57.073
```

The *C* represents *cursor*; the value *158.355* is the resonant frequency. At this frequency, the phase shift is 107.1×10^{-18} degrees, which is virtually no phase shift. The cursor values shown on the second line are the beginning frequency and phase shift, when the cursor is at the left side of the graph. This means that when *f* = 50 Hz, the circuit current has an angle of −57.073°. Refer to Fig. P.6 for this graph.

Many interesting plots can be obtained using Probe. You may want to experiment with this before going to the next topic. See if you can produce a graph like the one shown in Fig. P.7, showing the branch currents as well as the total current. A second plot has been added on the same graph in order to show the

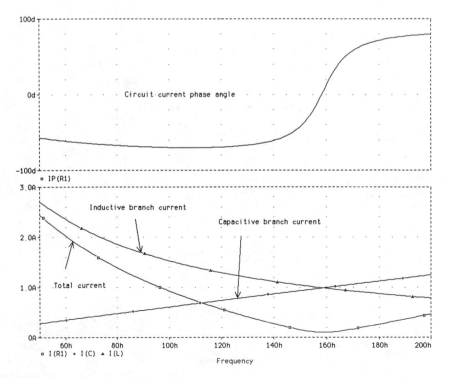

Fig. P.7   Series-Parallel ac Circuit.

phase angle. Note that the magnitude of the total current is less than the magnitude of the inductive-branch current. It is also sometimes less than the magnitude of the capacitive-branch current.

TRANSISTOR CIRCUIT ANALYSIS

The next preview circuit is that of a bipolar-junction transistor (BJT) with a typical biasing set of resistors. This circuit is shown in Fig. P.8. PSpice allows the use of "built-in" models for BJTs and other devices. Assume that the transistor has a large-signal current gain, $h_{FE} = 80$, and that under typical forward-bias conditions $V_{BE} = 0.8$ V.

Before going to the PSpice solution, we will consider the usual method for finding bias currents and voltages. It is assumed that from your studies you are familiar with these methods, and this will serve as a brief review. When the circuit is opened at the base, the Thevenin voltage is found using voltage division:

$$V_{Th} = \frac{V_{CC}R_2}{R_1 + R_2} = \frac{(12)(5)}{45} = 1.333 \text{ V}$$

To find the Thevenin resistance, V_{CC} is shorted, placing R_1 and R_2 in parallel. The Thevenin resistance is

$$R_{Th} = R_1 \parallel R_2 = 40 \text{ k}\Omega \parallel 5 \text{ k}\Omega = 4.444 \text{ k}\Omega$$

Applying KVL to the loop containing R_{Th} and R_E, we have

$$V_{Th} = R_{Th}I_B + V_{BE} + R_E(h_{FE} + 1)$$
$$1.333 \text{ V} = (4.444 \text{ k}\Omega)I_B + 0.8 \text{ V} + (100 \text{ }\Omega)(80 + 1)$$

Solving for I_B yields

$$I_B = 42.5 \text{ }\mu\text{A}$$

Since $I_C = h_{FE}I_B$, this gives a collector current of 3.4 mA. The emitter current is the sum of the collector and base currents and is 3.44 mA. The voltages at nodes *3*, *4*, and finally node *1* are now found using the known currents.

Fig. P.8 A BJT biasing circuit.

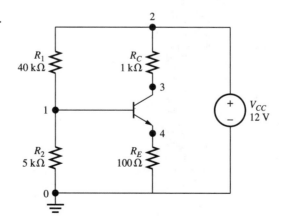

```
   BJT Biasing Circuit

    ****        CIRCUIT DESCRIPTION
VCC 2 0 12V
R1 2 1 40k
R2 1 0 5k
RC 2 3 1k
RE 4 0 100
Q1 3 1 4 QN
.MODEL QN NPN(BF=80)
.dc vcc 12V 12V 12V
.OP
.OPT nopage
.PRINT dc I(R1) I(R2) I(RC) I(RE)
.END

    ****        BJT MODEL PARAMETERS
                QN
                NPN
         IS  100.000000E-18
         BF   80
         NF    1
         BR    1
         NR    1

    ****        DC TRANSFER CURVES              TEMPERATURE =    27.000 DEG C
   VCC         I(R1)        I(R2)       I(RC)        I(RE)
    1.200E+01   2.713E-04    2.293E-04   3.366E-03    3.408E-03

    ****        SMALL SIGNAL BIAS SOLUTION       TEMPERATURE =    27.000 DEG C

  NODE    VOLTAGE       NODE    VOLTAGE      NODE    VOLTAGE      NODE     VOLTAGE
 (    1)   1.1464   (     2)   12.0000   (    3)    8.6345   (    4)     .3408

    VOLTAGE SOURCE CURRENTS
    NAME          CURRENT
    VCC          -3.637E-03

    TOTAL POWER DISSIPATION    4.36E-02   WATTS

    ****        OPERATING POINT INFORMATION      TEMPERATURE =    27.000 DEG C

**** BIPOLAR JUNCTION TRANSISTORS
NAME          Q1
MODEL         QN
IB            4.21E-05
IC            3.37E-03
VBE           8.06E-01
VBC          -7.49E+00
VCE           8.29E+00
BETADC        8.00E+01
GM            1.30E-01
RPI           6.15E+02
RX            0.00E+00
RO            1.00E+12
CBE           0.00E+00
CBC           0.00E+00
CBX           0.00E+00
CJS           0.00E+00
BETAAC        8.00E+01
FT            2.07E+18
```

Fig. P.9

The voltage at the collector is

$$V_3 = V_{CC} - R_C I_C = 12 - (1 \text{ k}\Omega)(3.4 \text{ mA}) = 8.6 \text{ V}$$

The voltage at the emitter is

$$V_4 = R_E I_E = (100 \ \Omega)(3.44 \text{ mA}) = 0.344 \text{ V}$$

The voltage at the base is

$$V_1 = V_{BE} + V_4 = 0.8 + 0.344 = 1.144 \text{ V}$$

Although the solution has not been difficult, it is certainly time-consuming. If some of the parameters of the circuit change, the solution must again be found. PSpice will allow these repetitive solutions to be found more easily. The input file is

```
BJT Biasing Circuit
VCC 2 0 12V
R1 2 1 40k
R2 1 0 5k
RC 2 3 1k
RE 4 0 100
Q1 3 1 4 QN
.MODEL QN NPN(BF=80)
.dc VCC 12V 12V 12V
.OP
.OPT nopage
.PRINT dc I(R1) I(R2) I(RC) I(RE)
.END
```

The chosen name for the BJT must begin with *Q*. The numbers *3, 1,* and *4* are the nodes of the *collector, base,* and *emitter,* respectively. *QN* is our chosen model name for an *npn* transistor. The .MODEL statement contains our chosen model name and the required NPN designation for the *built-in* model of an *npn* BJT. The *BF=80* gives a dc beta of 80. The results of the PSpice analysis are shown in Fig. P.9. The values given for the currents and voltages are close to those predicted.

In the chapter on bipolar junction transistors (Chapter 3), this circuit will be the foundation for a case study that will involve using the BJT as a common-emitter amplifier. Among other things, the voltage gain, current gain, input resistance, and output resistance will be found.

DC Circuit Analysis

Direct-current circuits are important not only in themselves, but also because many of the techniques used in dc analysis will carry over into ac circuit analysis. In fact, many of the electronic devices and circuits can be analyzed using these same methods.

A BEGINNING EXAMPLE

Figure 1.1 shows a series circuit with a dc voltage source and three resistors connected in series. The most important feature of this circuit is that all elements carry the same current. If the current through any element is known, the current through all other elements will of necessity be the same. Another important feature of this circuit is that the applied voltage (in this case 50 V) will divide among the resistors in direct proportion to their respective resistances. For example, the voltage drop across the 150-Ω resistor will be three times as great as the voltage drop across the 50-Ω resistor. Using the concept of voltage division, it is easy to find the voltage drops without knowing the circuit current. Thus the voltage drop across R_3 is

$$V_{R3} = V\left(\frac{R_3}{R_1 + R_2 + R_3}\right) = 50 \cdot \frac{150}{100 + 50 + 150} = 25 \text{ V}$$

Likewise, the voltage drop across R_2 is

$$V_{R2} = V\left(\frac{R_2}{R_1 + R_2 + R_3}\right) = 50 \cdot \frac{50}{100 + 50 + 150} = 8.333 \text{ V}$$

Fig. 1.1 Series circuit with
three resistors.

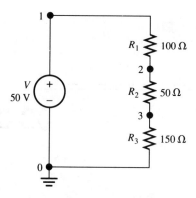

The current is easily found from either of these voltages:

$$I = \frac{V_{R3}}{R_3} = \frac{25}{150} = 0.1667 \text{ A}$$

or

$$I = \frac{V_{R2}}{R_2} = \frac{8.333}{50} = 0.1667 \text{ A}$$

USING SPICE TO INVESTIGATE THE CIRCUIT

An editor such as EDLIN (see Appendix C) is used to enter the statements for the
SPICE analysis. The file might be named PROB1.CIR and should contain the
following statements:

```
Spice Analysis of a Series Circuit
V 1 0 50V
R1 1 2 100
R2 2 3 50
R3 3 0 150
.OP
.END
```

There is a statement for each element of the circuit. Each line of the input file
is a statement. If you have need for a very long statement, begin the second line
with a plus (+) sign. The statements include the voltage source and the three
resistors. The beginning statement is necessary for PSpice and serves to identify
the circuit. The V statement is for the independent voltage. Nodes 1 and 0 are the
plus and minus nodes, respectively; $50V$ is for the value 50 volts. Note that there
is no space between 50 and V. The V may be omitted if desired. Each R statement
identifies a particular resistor and its associated nodes along with the resistance in
ohms. The .OP statement is a control statement which will give the maximum
amount of information, and the .END statement is needed as a signal that there
are no more statements.

When you have finished with the file, exit the editor, and use the PSpice program by typing

pspice prob1

This will call the PSpice program, which will analyze each of your statements and proceed with the analysis. If there are no errors, the output will be contained in a file PROB1.OUT. Bring this into your editor so that you can look at it on the screen and edit out some of the unnecessary lines. You might want to delete all lines containing a form feed (which wastes paper). Also delete the repetitive lines for the title (Spice Analysis of a Series Circuit), the lines that contain the "TEMPERATURE = 27.000 DEG C," and the lines containing all asterisks.

Obtain a printed copy of the output and look at the node voltages.

NODE VOLTAGE	NODE VOLTAGE	NODE VOLTAGE
(1) 50.0000	(2) 33.3330	(3) 25.0000

Node voltage 1 is the voltage V_{10}, the source voltage. Node voltage 2 is the voltage V_{20}, the voltage drop across both R_2 and R_3. Node voltage 3 is the voltage V_{30}, the voltage drop across R_3.

We predicted that V_{R3} (which is V_{30}) would be 25 V, so the PSpice analysis seems correct. How can you check for V_{R2}? This will be $V_2 - V_3$, which is actually $V_{20} - V_{30}$.

$$V_2 - V_3 = 33.333 - 25.000 = 8.333 \text{ V}$$

The PSpice analysis also gives the voltage-source current; the voltage source is called V, and the current is given as $-1.667E - 01$. The current has the correct numerical value, but what is the reason for the minus sign? SPICE shows source currents from plus to minus inside the source. Since the current is actually from minus to plus inside the source, it is given as a negative value. Or simply stated, when a source current is negative, it is from the positive terminal to the external circuit.

Note that the statement describing the voltage source was correctly given as

V 1 0 50V

As previously stated, this means that the positive terminal is at node 1, the negative terminal is at node 0, and the value of the voltage is 50 V (dc implied). Incidentally, independent voltages must begin with the letter V and are typically written as $V1$, $V2$, VIN, and so forth.

Notice that the total power dissipation is also given in the SPICE analysis as 8.33 watts. This is simply the product of V and I, $50 \cdot 0.1667 = 8.33$ W.

How much more information can you obtain for the circuit of Fig. 1.1? After you have thought of other things that might be found, look at another simple, common circuit.

Another Simple Circuit for Analysis

Consider the circuit of Fig. 1.2. This is a tee (T) circuit with a 50-V source and a load resistance $R_4 = 150\ \Omega$. This resistor is the load attached to the T portion of the circuit and might be changed to other values as needed. The load resistor is sometimes thought of as the output resistor.

Can you find the voltage across the load resistor and the current through it? To be specific, find V_3 and I (the current in the direction from node 3 to 0).

The input resistance R_{in} is found by adding R_2 and R_4 (sum of 200 Ω), putting this in parallel with R_3 (200 $\parallel$ 200 = 100 Ω), and adding R_1 (sum of 200 Ω). Thus $R_{in} = 200\ \Omega$.

The source (input) current is $V/R_{in} = 50/200 = 0.25$ A (out of the plus side of V).

The voltage drop across R_1 is $IR_1 = 0.25 \cdot 100 = 25$ V. The voltage drop across R_3 is $V - V_{R1} = 50 - 25 = 25$ V.

The voltage drop across R_4 is found by voltage division as

$$V_{R4} = \frac{V_{R3}R_4}{R_2 + R_4} = 25 \cdot \frac{150}{50 + 150} = 18.75\ \text{V}$$

The current I is found as $V_{R4}/R_4 = 18.75/150 = 0.125$ A.

In Fig. 1.2, the voltage across R_4 is called V_3, meaning precisely V_{30}. You may analyze this circuit by other methods, and you are encouraged to do so.

After you have obtained your pencil-and-paper results, it is time to look at what can be done with PSpice. Create a file called PROB2.CIR containing the following statements:

```
Spice Analysis of a Tee Circuit
V 1 0 50V
R1 1 2 100
R2 2 3 50
R3 2 0 200
R4 3 0 150
.OP
.OPTIONS NOPAGE
.TF V(3) V
.END
```

Fig. 1.2 Tee (T) circuit.

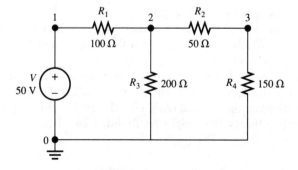

As usual, the file contains a title statement and is completed with the .END statement. Another statement that requires consideration is the .TF statement. This is a transfer function; it contains an output and input reference, respectively. The output is V(3), representing the voltage across R_4, and the input is V, which was the source voltage. It is your choice as to what you will call the output variable; it might have been the voltage across another resistor, for example. Simply stated, the transfer function will give the ratio V(3)/V. In this example it will be $18.75/50 = 0.375$.

The .OPTIONS statement with the option *NOPAGE* prevents unnecessary banners and page headings. You may want to include such a statement in all of your input files. It is preferable to edit out all unnecessary lines from the output file. In this case, it makes little difference whether the NOPAGE option is used or not. For this reason this statement will not be found in the ensuing examples.

Run the PSpice analysis and look at the file PROB2.OUT. Remember to strip the file of unwanted lines and form feeds; then obtain a printed copy for further study. Verify the voltage drop across R_3. This is V(2) in the PSpice output. Also verify the voltage drop across R_4. This is V(3) as shown in Fig. 1.2. The voltage-source current is given as $-2.5E-1$, for -0.25 A. Is this the expected value?

Now look for the bonus information obtained by using the .TF statement. The ratio V(3)/V is given as 0.375. Is this correct?

Also, we now have values shown for both input and output resistances. What is meant by each of these? The input resistance is the resistance as seen by the source V, which is referred to as the second parameter in the .TF statement. This parameter is always an input source name. Recall our calculation of $R_{in} = 200\ \Omega$. Your PSpice value should confirm this.

What is meant by output resistance? The output variable was specified in the .TF statement as V(3). We must look back into the circuit between nodes *3* and *0* with the source voltage V deactivated (shorted, not simply removed). Thus R_1 and R_3 are in parallel, and this combination is in series with R_2, with that combination in parallel with R_4. Verify that this will yield $R_{out} = 65.63\ \Omega$.

In many circuits, it is desirable to compare an output variable with an input source. Sometimes this is referred to as a gain. If both of these values are voltages, the result is called a voltage gain. In circuits containing only passive devices such as the resistors of Fig. 1.2, along with the independent source, this gain will be less than one. The value of the voltage gain is 0.375.

In summary, we have looked at simple resistive circuits for the purpose of seeing how they are analyzed by pencil-and-paper methods and how the results are verified using PSpice. Keep in mind that you will not want to use the PSpice tool as a substitute for understanding basic theory. If you do not understand how resistors in series and parallel may be combined to find equivalent resistances, do not assume that a computer analysis will take you through. Actually, you will find that in order to get the most out of a tool such as SPICE, you must know a great deal about circuit analysis.

At this point, you might ask, "Why bother with SPICE?" There are two reasons. After you apply SPICE to simple circuits and thoroughly understand what you are doing, you can use the same tools on circuits that are far more complicated than you care to solve by other methods. In addition, SPICE is so widely used that it is considered by many to be a necessary tool-of-the-trade.

Remember, independent-voltage statements begin with V and resistance statements begin with R. Use labels for voltage and resistance that will be easy to recognize in the circuit you are modeling. VS or VIN might be chosen for source voltages, and RS could represent the internal resistance of the source.

Next, we will look at some important circuit-analysis techniques and how SPICE can be used to verify important theorems.

BASIC CIRCUIT LAWS

In the study of electric circuits, it is well known that the algebraic sum of the voltages around a closed path is equal to zero. This is Kirchhoff's voltage law. Kirchhoff's current law is that the algebraic sum of the currents entering a junction is equal to zero. Figure 1.3 demonstrates the correctness of these two laws. Since the circuit contains three meshes and four nodes (in addition to the reference node), do not attempt to solve it by pencil-and-paper methods at this time, but set up the SPICE file instead. Do this yourself; then compare your results with the file given here.

```
Bridge Circuit for Use with Basic Circuit Laws
V 3 0 25V
R1 1 2 100
R2 1 0 75
R3 2 3 50
R4 4 0 60
R5 2 4 150
R6 1 4 200
.OP
.END
```

Fig. 1.3 Circuit with three meshes.

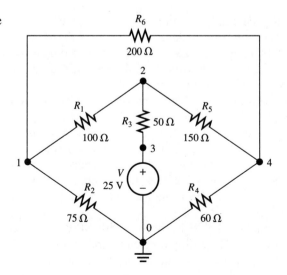

Run the PSpice analysis and obtain a printed copy for further study and work. We recommend that you leave room on your PSpice printed copy to show a sketch of the circuit, with all known values indicated. Show the location of the various nodes that go with the statements. It is easy to see the node labels if they are shown in a different color.

Find the sum of the voltages around the loop at the left. This involves

$$V_{12} + V_{23} + V_{30} + V_{01}$$

Remember that V_{12} is actually $V_1 - V_2$, and so forth. Check your values against these respective numbers:

$$-9.7039 - 8.632 + 25.000 - 6.6641 = 0$$

The zero sum is a verification of Kirchhoff's voltage law. Now write the symbolic equation for the loop on the right and check to see that the sum is zero. After you have done this, consider the following loop voltages:

$$V_{13} + V_{34} + V_{41} = 0$$

In the figure, note that V_{13} can be found as $V_1 - V_3$; you do not have to follow a wired path, going from node *1* to node *2* and then from node *2* to node *3*. In the laboratory, if you measured the voltage V_{13}, you would place the red lead at node *1* and the black lead at node *3*. The voltmeter should read -18.34 V. Check your calculations for the sum of the voltages which should be

$$-18.3359 + 19.9727 - 1.6368 = 0$$

Remember that the order of the subscripts determines whether a voltage will be given as a positive or negative value. That is, if V_{12} is positive (say, 6.5 V), then V_{21} is negative (-6.5 V). The importance of giving the proper sign to a voltage cannot be overstated.

Find the sum of the currents entering node *1*. Call these I_{21}, I_{01}, and I_{41}. Show these in symbolic form; then substitute the values. Thus

$$I_{21} = \frac{V_2 - V_1}{R_1} = 97.039 \text{ mA}$$

$$I_{01} = \frac{-V_1}{R_2} = -88.855 \text{ mA}$$

$$I_{41} = \frac{V_4 - V_1}{R_6} = -8.184 \text{ mA}$$

The sum of the currents equals zero, verifying Kirchhoff's current law. Incidentally, the value for I_{01} is rounded to five significant digits. Otherwise, the sum will be slightly different.

Currents are often shown with a single subscript rather than a double subscript. If a single subscript is to be used, you should show the direction of the currents in the diagram; otherwise, the solution is ambiguous! This is just as important as giving the proper sign to a voltage.

To check the current law again, find the sum of the currents at node *2*. Set up the symbolic form of the equations; then substitute the values.

CIRCUIT WITH TWO VOLTAGE SOURCES

Figure 1.4 shows a circuit with two voltage sources. Although it is not an involved circuit, finding the various currents and voltages will require significant effort. We assume that you will not attempt mesh or nodal analysis at this time. (More on these methods will follow.) Another technique, almost intuitive, is to consider the effects of each source taken one at a time. That is, consider the circuit *a* with source V_1 in place and source V_2 inactive (replaced by a short circuit); then consider the circuit *b* with source V_2 in place and source V_1 inactive.

Draw the original circuit; then draw the circuit *a* and also the circuit *b*. Find the voltage at node *2* for each of the circuits. After you have done this, compare your values with these values: $V_2(a) = 6.75$ V, $V_2(b) = 5.06$ V. By using the concept of *superposition*, the actual voltage at node *2* will be the sum of these two values; thus $V_2 = 11.81$ V.

You can find the source current from V_1. In symbols,

$$I_{12} = \frac{V_1 - V_2}{R_1}$$

$$I_{12} = \frac{20 - 11.81}{100} = 81.9 \text{ mA}$$

Superposition is useful in circuits containing resistors and more than one source; however, three or more sources might make the calculations very tedious.

Now use SPICE to verify your work. Your file should look something like this:

```
Circuit with Two Voltage Sources
V1 1 0 20V
V2 3 0 12V
R1 1 2 100
R2 2 3 80
R3 2 0 140
.OP
.TF V(2) V1
.END
```

Fig. 1.4 Circuit with two voltage sources.

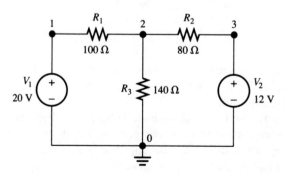

The PSpice results will show V(2) = 11.807 V, in close agreement with the calculations using superposition. The voltage-source current for V_1 is given by PSpice as $-8.193E-2$. Recall that this means that the current will be from the positive node of the source V_1.

What does the PSpice value for the input resistance mean? The resistance is referred to source V_1. It is the resistance seen by this source when the other source V_2 is inactive. This places the 80-Ω resistor and the 140-Ω resistor in parallel, and adds the 100-Ω resistor to that combination, giving $R_{in} = 150.9\ \Omega$.

Can you explain the output resistance? Recall that the .TF statement listed V(2) as the output variable. Visualize the output resistance as being seen between nodes 2 and 0 with all voltage sources inactive (shorted out). This gives R_1, R_2, and R_3 all in parallel. It is easily verified that the equivalent resistance of this combination is 33.7 Ω.

THEVENIN'S THEOREM AND APPLICATIONS

What is Thevenin's theorem, and why is it so important and useful? If you consider a nontrivial circuit and would like to work with a variety of load resistances for the circuit, Thevenin's theorem provides the ideal method.

Refer to Fig. 1.5(a), which contains a voltage source and several resistors, including a load resistor, R_L. Find the voltage across R_L and the current through it. If you find the equivalent resistance of the circuit, then the source current, and next the voltage drop across R_1, and so forth, you will eventually be able to find the voltage drop across R_L. However, if R_L changes in value, the entire solution must be reworked. The Thevenin theorem will help solve this problem.

Begin by removing the load resistance. Your method is to be independent of R_L, and this is important. Now find the voltage V_{30}. Simply stated, this is the voltage across the load terminals with the load resistance removed. This may be called V_{Th}.

Next, find the resistance as seen at the load terminals. This might be called R_{Th}.

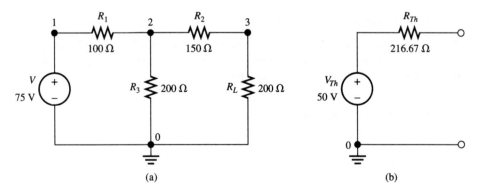

(a)   (b)

Fig. 1.5 (a) Circuit to illustrate Thevenin's theorem. (b) Thevenin voltage and series resistance.

Now replace the circuit with a practical voltage source consisting of V_{Th} in series with R_{Th}. Then the load resistance R_L may be put back in the circuit. As far as this resistance is concerned, the voltage drop across it, the current through it, and the power consumed by it will be the same as in the original circuit.

For Fig. 1.5, find V_{Th} and R_{Th}. Remove R_L; then use voltage division to find $V_{20} = 50$ V. To find R_{Th}, make V inactive by replacing it with a short circuit. Now looking at the network from terminals 3 and 0, you figure the resistance is $R_{Th} = 216.67$ Ω. The practical voltage source consists of V_{Th} in series with R_{Th}. Thus in Fig. 1.5(b) you have the new circuit. It can now be easily solved with any value of R_L in place. For example, when $R_L = 200$ Ω, use voltage division to find $V_{30} = 24$ V. On the other hand, when $R_L = 300$ Ω, $V_{30} = 29$ V.

SPICE and Thevenin's Theorem

Continuing with the circuit of Fig. 1.5, now use PSpice to verify your solution. Instead of simply removing R_L, replace the actual R_L with a very large load resistance, say, $R_L = 1$ teraohm (1E12). The circuit file will be

```
Thevenin Circuit for Spice
V 1 0 75V
R1 1 2 100
R2 2 3 150
R3 2 0 200
RL 3 0 1E12
.OP
.TF V(3) V
.END
```

After running the PSpice analysis, note that V(2) = 50.0000 V and V(3) = 50.0000 V. Be sure that you can explain this before going on. What is the value of V_{Th}?

The .TF statement gave the output resistance at V(3) as 216.7 Ω. This is the R_{Th}. Note that R_4 was many orders of magnitude larger than any of the other resistors in the circuit. Thus R_4 has a negligible loading effect on the circuit. You may want to use a smaller value for this resistance and compare the results.

Practical Application of Thevenin's Theorem

The previous example was easy enough to solve without calling on PSpice. If the problem gets more involved, like Fig. 1.6, PSpice can save a great deal of time. Set up the file for solving this problem; then check your results against those given here.

```
Thevenin Analysis of Bridged-Tee Circuit
V 1 0 80V
R1 2 1 20
R2 2 3 100
R3 3 0 200
R4 3 4 100
R5 2 4 400
R6 4 0 1E8
.OP
.TF V(4) V
.END
```

Fig. 1.6 Bridged-tee circuit and Thevenin equivalent.

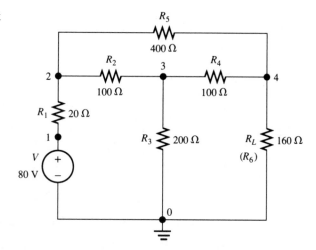

The output file is shown in Fig. 1.7. The voltage V(4) = 57.143 V is V_{Th}. The output resistance at V(4) = 128.6 Ω is R_{Th}. Note that both the open-circuit voltage and the resistance looking into the circuit at the load terminals with the load resistance removed are found by the simple technique of replacing the actual load resistance with a very large resistance value.

From the results of the analysis, draw the Thevenin practical voltage source, containing V_{Th} and R_{Th} in series. The original diagram showed a value R_L = 160 Ω. You have not needed this value until this point in the solution. Now, with R_L in place, you may find the values of load voltage and load current.

Was it worth the effort of using PSpice to find the Thevenin values for this problem? Try to find V_{Th} and R_{Th} by pencil-and-paper methods, and you will agree it was.

What do you think would happen if you omitted R_6 in the circuit file for the problem you just finished? You should try this, verifying that the results are the same. The reason that R_6 can be omitted is that node *4* will not be left hanging without a return path to ground.

Circuit for Thevenin Replacement

The circuit of Fig. 1.8 offers another opportunity to use Thevenin's theorem. In this circuit, several different values of R_L are to be used, and you desire to find the voltage and current associated with each one. Create your PSpice file for the circuit. You will see that R_L can be omitted without using a very large resistance in its place. Check your file against the one shown here.

```
Bridge Circuit for Thevenin
V 4 3 40V
R1 1 2 100
R2 2 0 150
R3 1 4 200
R4 4 0 200
R5 2 3 50
.OP
.TF V(1) V
.END
```

```
Thevenin Analysis of Bridged-Tee Circuit

****      CIRCUIT DESCRIPTION

V 1 0 80V
R1 2 1 20
R2 2 3 100
R3 3 0 200
R4 3 4 100
R5 2 4 400
R6 4 0 1E8
.OP
.OPT nopage
.TF V(4) V
.END

****      SMALL SIGNAL BIAS SOLUTION        TEMPERATURE =   27.000 DEG C

  NODE   VOLTAGE        NODE   VOLTAGE        NODE   VOLTAGE        NODE   VOLTAGE

(    1)   80.0000   (    2)   74.7250   (    3)   52.7470   (    4)   57.1430

    VOLTAGE SOURCE CURRENTS
    NAME           CURRENT

    V              -2.637E-01

    TOTAL POWER DISSIPATION   2.11E+01  WATTS

****      OPERATING POINT INFORMATION       TEMPERATURE =   27.000 DEG C

****      SMALL-SIGNAL CHARACTERISTICS

    V(4)/V =  7.143E-01

    INPUT RESISTANCE AT V =  3.033E+02

    OUTPUT RESISTANCE AT V(4) =  1.286E+02

          JOB CONCLUDED

          TOTAL JOB TIME          7.30
```

Fig. 1.7

Fig. 1.8 Circuit for Thevenin replacement.

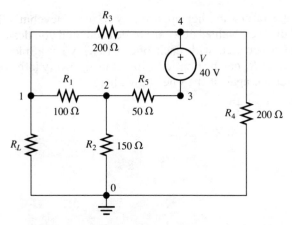

Fig. 1.9 Thevenin values for
Fig. 1.8.

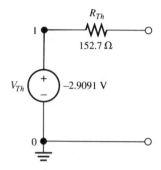

Run the analysis and then draw the Thevenin practical voltage source. Be sure to include the labels for the nodes involved. The results should agree with Fig. 1.9. The nodes are labeled *1* and *0*. Note that the open-circuit voltage at node *1* is negative with respect to node *0*. The PSpice results show V(1) = −2.9091 V. The .TF statement gave the output resistance at V(1) as 152.7 Ω. This is R_{Th}. Now you may choose a wide range of values for R_L and complete the circuit analysis.

Using Thevenin's theorem allows you to replace a complicated network with a practical voltage source. You recall that as far as R_L is concerned, it does not matter whether it is being supplied from the original circuit or from the Thevenin practical voltage source. The two circuits are not equivalent, however.

Return to the example of Fig. 1.5, our first Thevenin example. With the load removed, V_{Th} = 50 V and R_{Th} = 216.7 Ω. With R_L = 200 Ω, the current is 0.12 A. Since this is the current throughout the series circuit, the power furnished by the source voltage V_{Th} is 6 W. Since the load power is 2.88 W, the remaining 3.12 W is used by R_{Th}. But, in the original circuit, the source voltage is 75 V and the source current will be 0.33 A. This requires a source power of 24.8 W. Since the 200-Ω load requires only 2.88 W, the rest of the power goes to the three resistors in the tee.

The point of this is simply that the Thevenin practical voltage source is not equivalent to the original circuit. But the statement is still valid that as far as R_L is concerned the results are the same.

PRACTICAL CURRENT SOURCE VS. PRACTICAL VOLTAGE SOURCE

So far, you have worked with only a single type of source of electrical energy, the voltage source. In many situations, circuits behave as though they were supplied from a practical current source instead. Figure 1.10 shows a practical voltage source. It has an open-circuit voltage of 10 V. This is also called its ideal-voltage source value. To make it a practical voltage source, it must contain an internal resistance in series with it. This is given as R_i = 5 Ω. The circuit is completed with the addition of R_L = 15 Ω.

Solving for V_{20} = 7.5 V and I_L = 0.5 A, now attempt to find a practical current source that will take the place of the practical voltage source. This means that the load resistance should not know the difference as far as its voltage and current (and power) are concerned. It is easily verified that the ideal current

Fig. 1.10 Practical voltage source with load.

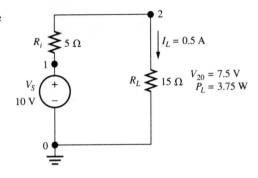

Fig. 1.11 Practical current source with same load as Fig. 1.10.

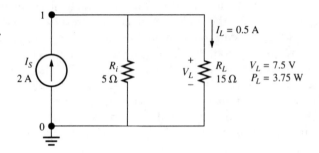

source of 2 A in parallel with $R_i = 5\ \Omega$ will do the job. To find I_s simply find the ratio V_s/R_i.

Figure 1.11 shows the practical current source. In each case the voltage across the load resistance and the current through this resistance are the same, with $V_{20} = 7.5$ V and $I_L = 0.5$ A. The power to the load resistance is $V_{20}I_L = 3.75$ W. Are the practical voltage source of Fig. 1.10 and the practical current source of Fig. 1.11 equivalent? To answer that, find the power furnished by each source. Verify that in the former case the power is 5 W, while in the latter case the power is 15 W. To explain the difference, compare the power in R_i in each case.

SPICE ANALYSIS OF CIRCUIT WITH CURRENT SOURCE

The solutions for circuits with current sources are more readily done with nodal analysis rather than loop analysis. The SPICE solutions are based on nodal analysis. Recall that each node in a circuit must be identified and that all circuit elements must be accounted for in terms of their respective nodes. With a voltage source, the first node listed in the SPICE statement is the positive node. With a current source, the first node listed is the one at the tail of the current arrow. A simple example is given in Fig. 1.12. Solve the circuit for currents and voltages.

Since the two resistive branches contain 200 Ω each, the 500-mA source current divides equally, giving $I_1 = I_2 = 250$ mA. Find $V_{10} = R_i I_1 = 200 \cdot 0.250 = 50$ V. Then find $V_{20} = R_L I_2 = 100 \cdot 0.250 = 25$ V.

Fig. 1.12 Simple circuit with current source.

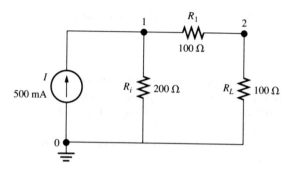

Now, we will look at the PSpice file that gives the circuit solution.

```
Simple Circuit with Current Source
I 0 1 500mA
RI 1 0 200
R1 1 2 100
RL 2 0 100
.OP
.TF V(2) I
.END
```

Note the use of *mA* as the symbol for *milliamperes*. Sometimes this is given instead as *M* or *MA*. Be careful! The symbol for *mega-* is *MEG*. The current source is shown with the nodes *0* and *1*, respectively, for tail-to-tip. The transfer function shows V(2) as the output variable and *I* as the input source. A statement such as this gives the transfer function as well as the input and output resistances.

Run the analysis and note these results: V(1) = 50 V, V(2) = 25 V. Then note that V(2)/I = 50. This transfer function is the ratio of output voltage to input current (ohms). It may not be of interest in this analysis. The input resistance is 100 Ω, as is easily verified. The output resistance is 75 Ω. This is the resistance as seen from nodes *2* and *0* with the current source inactive. This means that the current source is opened (or removed from the circuit). Verify that the value of the output resistance is 75 Ω.

Your output will show a heading for voltage-source currents, but none is named because there are no voltage sources. Next is a line that gives the total power dissipation as zero. This is easily misunderstood. It means that the product of source voltages and their respective currents is zero. The .OP statement gives powers for voltage sources only. Can you determine the total power dissipation? Use the sum of the I^2R values for the three resistors and verify that 25 W is the total power. A simpler method of finding the total power is to use the product of I, the source current, and V(1), the voltage across the source. Verify that this gives the same results of 25 W.

NORTON'S THEOREM

Norton's theorem is used to produce a practical current source with its accompanying shunt resistance as an alternative to the Thevenin practical voltage source with its series resistance. The relationship between the sources is

$$I_N = \frac{V_{Th}}{R_{Th}}$$

and the resistance value for both sources is the same. The technique for finding I_N is to replace the load resistance with a short circuit and find the current through this short-circuited branch. This current is I_N. In some circuits it is more convenient to find I_N, while in others it is more convenient to find V_{Th}. After one or the other is found, the conversion between the two is made by applying the equation given above.

Using Norton's Theorem

The circuit of Fig. 1.13, with the load resistor R_4 removed, is to be replaced by the Norton equivalent. In order to find the short-circuit current, a short is placed across node *3*, thereby eliminating this node. The input file becomes

```
Norton's Theorem; Find Isc
V 1 0 48V
R1 1 2 20k
R2 2 0 20k
R3 2 0 5k
.DC V 48V 48V 48V
.OP
.OPT nopage
.PRINT DC I(R3) V(1,2)
.END
```

Run the analysis and verify that the short-circuit current is the current through R_3 and that it is I(R3) = 1.6 mA.

Short-Circuit Current in Missing Element

Refer to the circuit of Fig. 1.6, where the load resistor R_L must be shorted in order to find the short-circuit current. The obvious problem is that there is no remaining element that will carry this current. In situations like this, R_L may be replaced by a very small resistor. The input file will be

```
Norton's Theorem with RL Replaced by Small R
V 1 0 80V
```

Fig. 1.13 Tee circuit for Norton analysis.

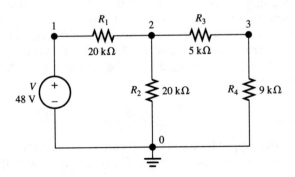

```
R1 2 1 20
R2 2 3 100
R3 3 0 200
R4 3 4 100
R5 2 4 400
RL 4 0 0.001
.DC V 80V 80V 80V
.OP
.PRINT DC I(RL)
.END
```

The PSpice analysis gives I(RL) = 0.444 A, which is the desired short-circuit current, I_N. In the input file shown above, it would not be helpful to include a .TF statement in an attempt to find the output resistance at V(4). The reason is that the output resistance includes the R_L value of 0.001 Ω which was used to find the short-circuit current.

We conclude that finding V_{Th} and R_{Th} using PSpice is easier, since one input file provides both values.

CIRCUIT WITH CURRENT AND VOLTAGE SOURCES

Circuits with both current and voltage sources can be solved by superposition. If the circuits are not complicated, this provides a simple and convenient method of solution. Figure 1.14 shows a circuit with a current source I and a voltage source V. Use superposition to find the voltage V_{10}. Check your results against this solution. With I active and V replaced with a short circuit, $V_{10}(a) = 5$ V; with V active and I replaced with an open circuit, $V_{10}(b) = 10$ V. Add these two values to obtain the actual voltage, $V_{10} = 15$ V.

Turn to PSpice for an analysis of the circuit. Your input file should look like this:

```
Simple Current and Voltage Sources
I 0 1 1A
V 2 0 20
R1 1 0 10
R2 1 2 10
.OP
.TF V(1) V
.END
```

Fig. 1.14 Simple current and voltage sources.

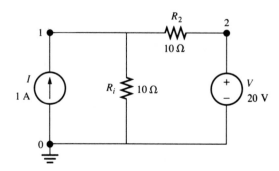

The output file shows V(1) = 15 V, in agreement with the superposition results. The .OP statement allowed you to find the voltage-source current of 0.5 A. Verify that this is the correct value. Remember that the *total power dissipation* given by PSpice is for the voltage source only; it is simply the product of V and the source current, which is 10 W. Use I^2R for each resistor and verify that $P_1 = 22.5$ W, $P_2 = 2.5$ W, giving $P_t = 25$ W. The .TF statement allows you to find input and output resistances. Check these values, remembering that the current source is replaced by an open circuit to find R_{in} and R_{out}.

In summary, when current sources are present, beware of how you use the total power dissipation. It represents only the power associated with the voltage sources. If there are two or more voltage sources, the power shown in PSpice is for all voltage sources.

MAXIMUM POWER TRANSFER

In a circuit where a load resistance can be made to vary or selected to fill a particular need, the question is sometimes asked, "What value of R_L will allow maximum power to be developed in the load resistor?" Figure 1.15 shows R_L as such a resistor. When the value of R_L is chosen to be equal to the resistance looking back into the network from the load terminals (nodes *3* and *0*), then the power to R_L will be the maximum.

The resistance of the network as seen at the load terminals with the load removed is simply the Thevenin resistance. For this circuit it is 30 Ω. Thus when $R_L = 30$ Ω, it will dissipate the maximum power. This means that with all other values in the circuit fixed, either a smaller or a larger value of R_L will take less power. We will demonstrate this with a SPICE analysis. The input file is

```
Maximum Power Transfer to R Load
V  1 0 12V
R1 1 2 20
R2 2 0 20
R3 2 3 20
RL 3 0 30
.OP
.END
```

The results show V(3) = 3 V, from which $P_L = V(3)^2/R_L = 0.3$ W. Now you see the real advantage of using SPICE when R_L is made either smaller or larger.

Fig. 1.15 Maximum power transfer to *R* load.

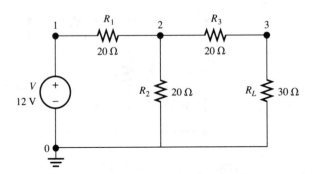

Go back to the input file and replace only the value of 30 Ω and quickly see the new results.

For example, with $R_L = 29$ Ω, V(3) = 2.9492 V, and $P_L = 0.2999$ W; on the other hand, with $R_L = 31$ Ω, V(3) = 3.0492 V, and again $P_L = 0.2999$ W. Try several other values for R_L and notice that the power to the load resistor will invariably be less than 0.3 W. Remember that the maximum power to R_L occurred when R_L was matched to the resistance seen looking back into the circuit with R_L removed. This is easily found for more complicated circuits by using the Thevenin technique.

The computer analysis allows more time for us to concentrate on understanding the principles that govern circuit behavior, spending less time on the mundane, repetitive calculations.

DEPENDENT SOURCES IN ELECTRIC CIRCUITS

A dependent (controlled) source might represent either a voltage source or a current source that is dependent on another voltage or current somewhere else in the circuit.

Voltage-Dependent Voltage Source

An example, Fig. 1.16 shows a circuit with a source voltage V, an independent source, along with another source voltage E, which is a dependent (or controlled) source. It is also labeled $2V_a$. In what way is E dependent? It is a function of the voltage drop across resistor R_1, which is shown as V_a. The factor 2 means that the voltage E will be twice the voltage drop V_a. The factor 2 is also called k.

The circuit voltages and currents may be found by conventional analysis. In the loop on the left,

$$V = R_1 I_{12} + E = R_1 I_{12} + 2V_a$$

where I_{12} is the current through R_1. Since $V_a = R_1 I_{12}$, the expression becomes

$$V = R_1 I_{12} + 2R_1 I_{12} = 3R_1 I_{12}$$
$$10 \text{ V} = (3)(250 \text{ Ω})I_{12}$$
$$I_{12} = 13.33 \text{ mA}$$
$$V_{12} = V_a = R_1 I_{12} = (250 \text{ Ω})(13.33 \text{ mA}) = 3.333 \text{ V}$$
$$E = 2V_a = 6.667 \text{ V}$$

Fig. 1.16 Voltage-controlled voltage source.

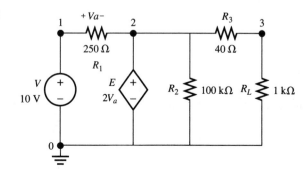

Since this is the voltage applied across R_2, the current through R_2 and the current through the R_3 and R_L branch are readily found:

$$I_{R_2} = \frac{6.667 \text{ V}}{100 \text{ k}\Omega} = 66.67 \text{ } \mu\text{A}$$

$$I_{R_3} = \frac{6.667 \text{ V}}{1.04 \text{ k}\Omega} = 4.41 \text{ mA}$$

The current through the source E is directed downward and is

$$I_E = 13.33 \text{ mA} - 66.67 \text{ } \mu\text{A} - 6.41 \text{ mA} = 6.85 \text{ mA}$$

How is a problem of this kind solved using PSpice? The input file for Fig. 1.16 is

```
Voltage-Controlled Voltage Source
V 1 0 10V
E 2 0 1 2 2
R1 1 2 250
R2 2 0 100k
R3 2 3 40
RL 3 0 1k
.OP
.TF V(3) V
.END
```

The new statement in this input file is a description of the dependent source E. The nodes *2* and *0* are its *plus* and *minus* nodes, respectively. The nodes *1* and *2* are the *plus* and *minus* nodes of the voltage upon which E is dependent. The output file shows V(2) = 6.6667 V and V(3) = 6.4103 V, as predicted. The voltage-source current for the source V is given as 13.33 mA, also as predicted. Note that the source current for E is given as 6.856 mA. It is shown as positive which means from plus to minus inside E.

The PSpice analysis shows the input resistance as 750 W. This is simply the ratio of V to I_{12}. In finding the output resistance, each voltage source is considered as a short. This puts a short between nodes *2* and *0*, giving $R_3 \parallel R_L$ for a resistance of 38.46 Ω.

Figure 1.17 shows a modification of the circuit. The input file for this circuit is

```
Another Voltage-Controlled Voltage Source
V 1 0 10V
E 3 0 2 0 2
R1 1 2 250
R2 2 0 100k
R3 3 4 40
RL 4 0 1k
.OP
.TF V(4) V
.END
```

In this simplified model for a voltage amplifier, the results are easily predicted by circuit analysis. The current in the loop on the left is

$$I_{12} = \frac{V}{R_1 + R_2} = \frac{10 \text{ V}}{250 \text{ } \Omega + 100 \text{ k}\Omega} = 99.75 \text{ } \mu\text{A}$$

Fig. 1.17 Revised circuit with voltage-controlled voltage source.

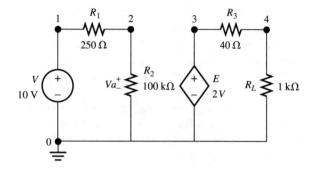

The voltage drop across R_2 is

$$V_2 = V_a = I_{12}R_2 = (99.75 \ \mu A)(100 \ k\Omega) = 9.975 \ V$$

and

$$E = 2V_a = (2)(9.975 \ V) = 19.95 \ V$$

From the PSpice analysis, we find $V(2) = 9.9751$ V, $V(3) = 19.95$ V, and $V(4) = 19.183$ V. The source current through V is 99.75 μA, and the source current through E is -19.18 mA. The negative sign indicates that the current from plus to minus inside E is negative, or alternately, that the positive current is upward through E.

Current-Dependent Voltage Source

Sometimes a dependent voltage source is dependent on a *current* somewhere else in the circuit as shown in Fig. 1.18. The dependent source is shown to have a value $0.5I$, where I is the current through the resistor R_1. The current is in the direction from node *1* toward node *2*. The dependent voltage source has its positive terminal at node *3*, which means that it tends to furnish current clockwise in the loop on the right. It is important to note these features, since they play a role in assigning polarities and directions in the solution.

The circuit is easily analyzed. In the loop on the left, the 15-V source will give a current $I = V/(R_1 + R_2) = 15/(10 + 5) = 1$ A. The dependent voltage source

Fig. 1.18 Circuit with current-dependent voltage source.

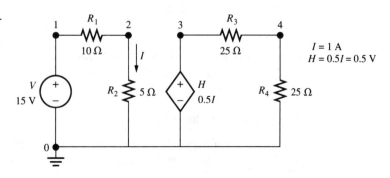

is marked $0.5I$. This will be equal to $0.5 \cdot 1 = 0.5$ V. Thus $V_{30} = 0.5$ V. How is this dimensionally a voltage? The value $k = 0.5$ is not dimensionless; it is in ohms. In general, the notation kI will be used for this CDVS (current-dependent voltage source). The k factor must be in ohms for the product of k and I to give volts.

The current in the loop on the right is found to be $I_L = 0.5/(25 + 25) = 10$ mA. The voltage $V_{40} = R_L I_L = 0.25$ V.

Using PSpice, solve the problem as an introduction to other, more elaborate circuits. The input file will contain a statement for the current-dependent voltage source. Here is what the file should look like:

```
Circuit with Current-Dependent Voltage Source
V 1 0 15V
H 3 0 V -0.5
R1 1 2 10
R2 2 0 5
R3 3 4 25
R4 4 0 25
.OP
.TF V(4) V
.END
```

Look carefully at the entry for the CDVS shown on the line beginning with the symbol H. On the same line the *3* and *0* represent the plus and minus terminals of the dependent voltage source. This is followed by the symbol V, which represents the independent voltage through which the controlling current I is carried. The last entry on this line is -0.5. The 0.5 is the factor k, but the minus sign requires explanation. In SPICE, a voltage-source current such as the current through V would be positive if it were directed from plus to minus through V. Since the reference current I is oppositely directed, the H statement requires the minus sign for the factor k. Study this feature, as it can easily be confusing. Had the minus sign been omitted, voltages V(3) and V(4) would have had the wrong sign in your PSpice results.

When you are sure you are ready to continue, run the analysis and look at the results. Note V(3) = 0.5 V and V(4) = 0.25 V. The voltage-source current of V is -1.0 A. Recall that this means that the current is actually out of the plus side of V, and that I in Fig. 1.18 is a positive value. Thus V(3) is positive.

At the expense of being repetitive, also recall that the value shown in the output file for total power is simply VI and is not really total power. Under the section in the output file called *Current-Controlled Voltage Sources*, there are two lines of entries. The first gives a V-source entry of 0.5 V. This is the voltage value of the dependent-source voltage. More puzzling is the next entry, which gives an I-source value of -10 mA. Can you figure out what this means? It means that the current through the CDVS is 10 mA from the minus to the plus terminal inside the source. This is in agreement with what we already know about how SPICE shows source currents.

Current-Dependent Current Source

Another type of dependent source, one that is frequently encountered in electronics, is the current-dependent current source. Figure 1.19 shows the basic circuit.

Fig. 1.19 Current-controlled current source.

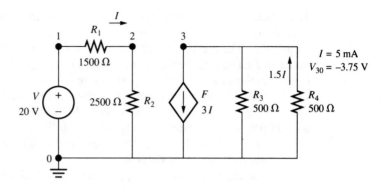

The dependent source has a value of $3I$, where I is the current through the resistor R_1. As in the previous example, the current is in the direction from node 1 toward node 2, clockwise in the loop on the left. The factor $3I$ is generally given as kI, where k is the multiplier for the referenced current that appears somewhere else in the circuit. Simply stated, if $I = 2$ A in the loop on the left, then the current through F will be $(3)(2) = 6$ A in the direction of the arrow shown in F. In this example, we can easily solve for $I = 20/(1500 + 2500) = 5$ mA, as the current in the loop on the left. The current in F is thus $(5\text{ mA})(3) = 15$ mA (downward). This current will divide evenly between R_3 and R_4, giving 7.5 mA through each resistor. The currents go from node 0 toward node 3, putting node 3 at a negative potential. The voltage $V(3) = (-500)(7.5\text{ mA}) = -3.75$ V.

Many texts fail to show the two forms of Ohm's law that are illustrated in Fig. 1.20. Note that in one case $v = RI$, while in the other case $v = -RI$. The difference is, of course, in the reference direction of the current.

As a prelude to more elaborate circuits with dependent sources, how will Fig. 1.19 look in a PSpice file? Here it is:

```
Current-Controlled Current Source
V 1 0 20V
F 3 0 V -3
R1 1 2 1500
R2 2 0 2500
R3 3 0 500
R4 3 0 500
.OP
.TF V(3) V
.END
```

Your output should show $V(2) = 12.5$ V and $V(3) = -3.75$ V. Under the heading of *Current-Controlled Current Sources*, the I-source called F has a value of 15 mA. Since this is three times the current in the loop on the left, it is correct.

Fig. 1.20 Two forms of Ohm's law.

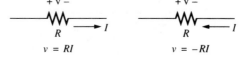

The positive value means that the current is in the direction from node *3* toward node *0* inside *F*. It is necessary to show the proper values in the input-file statement for *F*. On the line describing *F*, the first two values are for the tail and tip nodes of *F*. The next value, *V*, refers to the voltage source that carries the current *I*. This current *I* is related to the dependent-source current by the relationship *kI*. The factor *k* is a multiplier for *I*; it is dimensionless. The *k* in this example has a value of −3 because of how the voltage *V* relates to the current *I* as explained in several previous examples. Study this carefully, as it is where mistakes are often made. If you fully understand the simpler examples, you will be able to handle the more involved ones with confidence.

In the input file the value given for R_{in} is 4 kΩ, which is obviously correct. Also of interest is the value for $R_{\text{out}} = 250$ Ω. In the portion of the circuit on the right, looking in on the far right, we see R_3 and R_4 in parallel. The current source is made inactive by opening it (or removing it).

Another Current-Dependent Current Source Situation

A slightly different situation involving the CDCS often appears in electric circuit analysis. It is where the controlling current is in a branch containing no independent voltage source, that is, no *V*-type element. Figure 1.21(a) shows a typical example. Note that *I* is the current through R_3. It is this current that appears with its multiplier *k* through the current-dependent current source *F*. Recall that the

Fig. 1.21 (a) Another CDCS example. (b) The circuit modified to include zero-volt source.

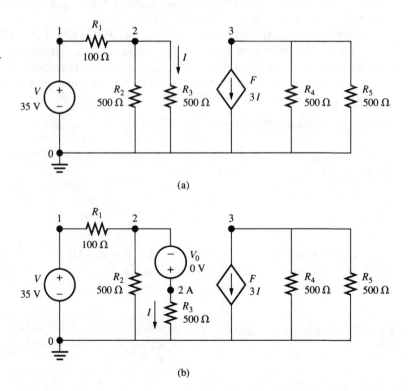

(a)

(b)

line in the input file that describes F must show a *V-type* value. How do you handle this? Simply insert a zero-valued voltage source in the branch where I appears, as shown in the detail of Fig. 1.21(b). Call this source *V0* to remind you that it is a zero-valued source. Look at the PSpice input file for this circuit.

```
Another CDCS Example
V  1  0  35V
V0 2A 2 0V
F  3  0  V0 -3
R1 1  2  100
R2 2  0  500
R3 2A 0  500
R4 3  0  500
R5 3  0  500
.OP
.TF V(3)  V
.END
```

Notice the line describing F. It refers to the independent voltage *V0* since this branch of the circuit contains the current I, the controlling current. Compare this statement with the F statement of the previous example to see the difference. Also note the addition of a line describing *V0*. The two nodes are *2A* and *2*. Note carefully that the plus node (always the first node) tends to furnish current in the direction shown for I in the figure. This follows the same convention used in the previous example, where I was also shown out of the plus source node.

We have not previously used an alphabetic character to designate a node, but this is all right. In fact, combinations of letters and numbers may be used. Thus nodes might be called *a1, b12, 1c,* and so forth. Also, the fact that this is a zero-valued source means that its presence does not alter the behavior of the circuit. The line describing *R3* has been changed to show the presence of the new node *2A*.

Run the simulation and look at the results. The values are V(2) = 25 V, V(3) = −37.5 V, and as expected V(2A) = 25 V also. You should be able to easily verify that I = 50 mA, giving $3I$ = 75 mA; this is the current in F. This current splits equally between R_4 and R_5, giving a current of 37.5 mA upward (node *0* toward node *3*) in each. Thus V(3) = −37.5 V.

As an additional exercise, see what happens when two lines of the input file are changed as follows:

```
V0 2 2A 0V
F  3 0 V0 3
```

This is simply an alternate way of describing the relationship between the CDCS and its controlling current. Try to visualize that they are equally valid, and use whichever one you feel more comfortable using.

Voltage-Dependent Current Source

A dependent current source that is a function of a voltage somewhere else in the circuit is described by a G entry for SPICE. Figure 1.22 shows an example. This circuit is easily analyzed by pencil-and-paper methods. The voltage v_2 is found by

Fig. 1.22 Voltage-controlled current source.

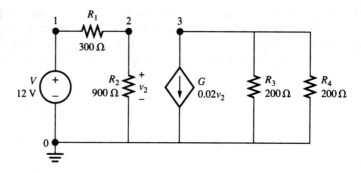

voltage division to be 9 V. The current through the dependent source is thus (0.02)(9) = 180 mA. The *k* value of 0.02 is dimensionally conductance or ohms^{-1}. This current divides equally between R_3 and R_4, giving 90 mA through each resistor. This produces $V_{30} = (-0.09)(200) = -18$ V. The PSpice input file is

```
Voltage-Controlled Current Source
V 1 0 12V
G 3 0 2 0 0.02
R1 1 2 300
R2 2 0 900
R3 3 0 200
R4 3 0 200
.OP
.TF V(3) V
.END
```

The line showing *G* gives the first two nodes as *3* and *0*. These are for the tail and tip, respectively, of the dependent current arrow. The next two nodes are *2* and *0*, for the plus and minus nodes of the controlling voltage, v_2. Run the PSpice analysis and verify V(2) = 9 V, V(3) = −18 V, R_{in} = 1200 Ω, and R_{out} = 100 Ω.

The ratio V(3)/V = −1.5 is the gain. Later you will see how this type of analysis can be used for active circuits containing transistors and various integrated circuits.

Another Current-Dependent Voltage Source

Recall that a voltage source that is controlled by a current elsewhere in the circuit is called a current-controlled, or current-dependent, voltage source. Figure 1.23

Fig. 1.23 Current-controlled voltage source.

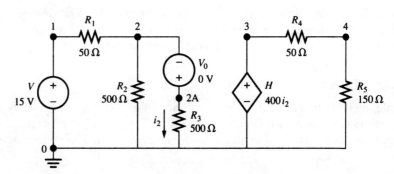

gives a type of example different from the one previously considered. The controlling current in this example is i_2, the current down through R_3 (toward node 0). The CCVS is marked H. The value of k is 400. Solving for the source current from V, you get $I_s = 15/(50 + 250) = 50$ mA. This current divides equally between R_2 and R_3, giving $i_2 = 25$ mA as the controlling current. The value of H is then $(400)\cdot(25$ mA$) = 10$ V. The current in the loop at the right of the diagram is $10/(200) = 50$ mA. The PSpice input file is

```
Current-Controlled Voltage Source
V  1  0  15V
V0  2A  2  0V
H  3  0  V0  -400
R1  1  2  50
R2  2  0  500
R3  2A  0  500
R4  3  4  50
R5  4  0  150
.OP
.TF  V(4)  V
.END
```

Run the analysis and confirm that V(2) = 12.5 V, V(3) = 10 V, and V(4) = 7.5 V. The PSpice statement for H gives 3 and 0 as the plus and minus nodes of the controlled voltage. It also shows $V0$ as a named voltage source in the path of the controlling current i_2. The polarity of $V0$ is correct to give i_2 as a positive value in the direction shown in Fig. 1.23. The last value in the H statement is -400 representing k. The minus sign is required, as in earlier examples, in keeping with how SPICE treats currents through voltage sources.

Note that the output file gives the current through V as -50 mA, which means 50 mA out of the positive terminal of V, and the current through $V0$ as -25 mA, which means 25 mA out of the positive terminal of $V0$. Under the heading of *Current-Controlled Voltage Sources* you find a voltage value of 10 V; this is shown as *V*-source. The value of -50 mA, which is shown as *I*-source, is the current through H. Once more, the minus sign means that the direction of positive current through H is from the minus to the plus terminal.

In summary, we have looked at each of the four types of controlled sources, E (for VCVS), F (for CCCS), G (for VCCS), and H (for CCVS). Controlled sources play an important part in the analysis of most circuits with active devices such as transistors. It is important to be able to analyze simple circuits with such devices, thus paving the way for an understanding of more elaborate circuits. The basic ideas are best treated in the dc circuits which we are dealing with in this chapter.

POLYNOMIAL DEPENDENT SOURCES

This topic is usually not included in an introductory course. If you are not interested in polynomial sources at this time, skip over this material, moving ahead to the section on mesh analysis.

The possibility of using a nonlinear dependent source might sometimes arise. You can see what this means and how it may be applied to an actual example. Figure 1.24 shows a voltage source V supplying two equal resistors $R_1 = R_2 =$

Fig. 1.24 Circuit with polynomial-controlled source.

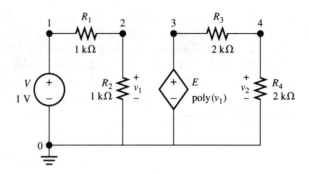

$1\ k\Omega$. The VCVS is E, and in this case there is no simple k factor relating E to another voltage in the circuit. Suppose that E is to be based on V_{20} in a nonlinear fashion, given by the polynomial

$$f(x) = 3 + 2x + x^2$$

Let E be related to V_{20} by the polynomial expression. This would mean, for example, that if $V_{20} = -1$ V, $E = 2$ V, or if $V_{20} = 2$ V, $E = 11$ V. Verify this in the equation before going on. Now consider the input file.

```
Circuit with Controlled Source
V  1  0  1V
E  3  0  POLY(1)  2,0  3  2  1
R1  1  2  1k
R2  2  0  1k
R3  3  4  2k
R4  4  0  2k
.DC  V  -4  4  1 ;  this is a dc "sweep" of the source voltage V
.PRINT  DC  V(2)  V(3)  V(4)
.END
```

Before running this analysis, look carefully at the statement describing E. The nodes *3* and *0* are the plus and minus nodes of E, as we would expect. The *POLY(1)* means that we will use a polynomial to describe the functional relationship between E and some other voltage. The *(1)* means that only one pair of nodes is referenced as a controlling voltage. The *2,0* identifies the controlling *plus* and *minus* nodes for the voltage v_1.

Now for the polynomial itself, the *3 2 1* values are for the *a, b,* and *c* values of the general case

$$f(x) = a + bx + cx^2$$

giving us the desired polynomial shown above the input file. Note that if either of the values *a* or *b* is zero, a zero must be shown. Otherwise the statement would not completely describe the degree of the polynomial. This means that a 3rd-degree polynomial would require 4 values (for *a, b, c,* and *d*).

Since the SPICE manuals are not very clear on this subject, you may want to spend some extra time looking over this example. Remember this when the need for the nonlinear source arises.

The .DC statement gives a range of voltages for V from -4 V to 4 V. This is called a *sweep* of voltage, and this statement will override the V statement that shows $V = 1$ V. The last value in the .DC statement shows the increment of V, which in the example is 1 V. Other uses of the .DC statement in this chapter will provide further clarification.

Run the analysis and verify our predictions for the relationship between $V(3)$ and $V(2)$.

Dependent Source as a Function of Two Other Voltages

Letting a dependent source depend on more than one other source can be done using a POLY form in the dependent statement. For example, in Fig. 1.24 make E a function of both v_1 and v_2. This requires part of the statement to be POLY(2) 2,0 4,0. The commas are optional and are added for clarity. The rest of the statement must contain coefficients. Determining what these coefficients represent will require some thought and planning. When two controlling voltages are involved, the terms represent k_0, $k_1 v_1$, $k_2 v_2$, $k_3 v_1^2$, $k_4 v_1 v_2$, and $k_5 v_2^2$. The pattern may look complicated, but with a little study it becomes clear. The k factors are multipliers for each possible voltage or combination of voltages. The voltages are listed in order of degree beginning with the first voltage, called v_1. In our example, v_1 will represent V_{20}, and v_2 will represent V_{40}.

Now consider the entire statement involving the two controlling voltages:

```
E 3 0 POLY(2) 2,0 4,0 0 2 3
```

The last three values *0 2 3* represent k_0, k_1, and k_2. This stands for $0k_0 + 2v_1 + 3v_2$. Thus the dependent voltage E will be the sum of twice the voltage drop across R_2 and three times the voltage drop across R_4. Note that the commas are optional. They are often added for clarity.

The input file is

```
Polynomial Form for Two Inputs
V 1 0 1V
E 3 0 POLY(2) 2,0 4,0 0 2 3
R1 1 2 1k
R2 2 0 1k
R3 3 4 2k
R4 4 0 2k
.DC V -4 4 1
.PRINT DC V(2) V(3) V(4)
```

Run the analysis and verify that E is given by the expression $2V_{20} + 3V_{40}$.

You might need to use this multidependent tool when there are circuits where the addition, subtraction, or multiplication of several voltages or currents is to take place. Using the proper choice of E, F, G, and H for the dependent source along with the POLY feature will allow the circuit to be simulated.

MESH ANALYSIS AND PSPICE

Introductory circuits courses contain a treatment of how to write loop and mesh equations to solve for currents in circuits. The standard form to solve for three mesh currents would be

$$R_{11}I_1 + R_{12}I_2 + R_{13}I_3 = V_1$$
$$R_{21}I_1 + R_{22}I_2 + R_{23}I_3 = V_2$$
$$R_{31}I_1 + R_{32}I_2 + R_{33}I_3 = V_3$$

where R_{11} is the self-resistance of mesh 1, R_{12} is the mutual resistance between meshes 1 and 2, R_{13} is the mutual resistance between meshes 1 and 3, and V_1 is the net source voltage tending to furnish the current in the direction of I_1. Similar statements can be made about the other equations, where R_{22} and R_{33} represent self-resistances and all other R terms represent mutual resistances. A pencil-and-paper solution for three simultaneous equations is tedious and error prone. When there are more than three equations, the job becomes a much more difficult chore. Numerous versions of computer programs are available for solving these equations.

As an exercise, you may wish to write the mesh equations for Fig. 1.25 and solve for I_1, I_2, and I_3, using a computer program or a calculator designed for this purpose.

Can you solve such equations using SPICE? In a word, no. However, you can use a few tricks to get the job done as the next example will show. Figure 1.25 shows the circuit with its three meshes. If you set up three equations following the standard format, you could solve for I_1, I_2, and I_3. You may wish to do this as an exercise before looking at the input file for PSpice, which is given here.

```
Mesh Analysis with PSpice
V1 1 0 50V
V2 4 0 30V
R1 1 2 100
R2 2 0 200
R3 2 3 400
R4 3 0 200
R5 3 4 100
.OP
.DC V1 50 50 10
.PRINT DC I(R1) I(R3) I(R5)
.END
```

Two interesting statements are shown in the input file. The first is the .DC statement. This is a dc-sweep statement. It allows you to sweep through a set of

Fig. 1.25 Mesh analysis with PSpice.

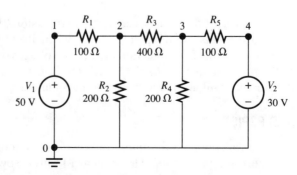

source voltages V_1. The next two values, *50* and *50*, are for the start and stop voltage values of the sweep, and the last value *10* is an increment. If you had wanted to sweep V_1 in 10-V increments, beginning at zero, the statement would have been

```
.DC V1 0 50 10
```

Since you are not actually interested in a sweep, you force a sweep of a single voltage. The concept of the sweep is needed for the next statement. The .PRINT statement includes the term DC, which is the sweep parameter previously chosen. In this type of .PRINT statement, you can obtain such things as currents I(R1) and voltages V(2,3) as needed. If you omit the .DC sweep, the print statement of this form will be invalid.

Run the PSpice analysis and look at the results. Values obtained include I(R1) = 0.1833 A, I(R3) = 25 mA, and I(R5) = −83.33 mA. These are the mesh currents that would be obtained from the standard mesh analysis when you solve three simultaneous equations. The PSpice solution is actually more like a nodal analysis, but you have forced more calculations to be made, giving the desired mesh currents (actually, branch currents also).

DC SWEEP

Since the mesh problem introduced the concept of a dc sweep, look at an example where the sweep is used in the normal fashion, with a range of input voltages. The familiar tee circuit of Fig. 1.26 will be used. You do not need a preliminary analysis, so look at the input file for the PSpice analysis.

```
Spice Sweep Analysis of Tee Circuit
V 1 0 50V
R1 1 2 100
R2 2 3 50
R3 2 0 200
R4 3 0 150
.OP
.TF V(3) V
.DC V 0 50 10
.PRINT DC V(2,3) I(R3)
.END
```

Fig. 1.26 Tee circuit and a voltage sweep.

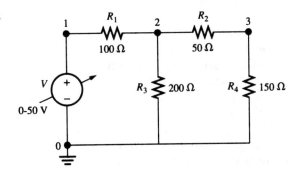

The dc sweep will begin with $V = 0$ and continue through $V = 50$ in 10-V increments. Two output tables will be produced, the first showing the relation between V and V(2,3) and the second showing the relation between V and I(R3) as directed by the print statement. Run the PSpice analysis and look at the results. Notice how the tables are shown. How might the tables be used? If you are looking for the V required to give a current I(R3) of 50 mA, the table shows that $V = 20$ V is required. You could easily ratio this result yourself, but if there are many *ifs* to look at, PSpice can do the work for you.

A PREVIEW OF THE .PROBE STATEMENT

In order to use Probe, you must have a graphics card such as monochrome, Hercules, CGA, EGA, or VGA for use with your monitor. Also you should follow the directions given in Appendix C for setting up the file PROBE.DEV.

PSpice includes a powerful graphing capability that may be invoked by including a .PROBE statement in your input file. Up until now, you have not needed this feature. But in a dc sweep, where a range of voltages V is being used, you can get a preview of using Probe. Simply revise the input file so that it looks like this:

```
SPICE Sweep Analysis of Tee Circuit
V 1 0 50V
R1 1 2 100
R2 2 3 50
R3 2 0 200
R4 3 0 150
.OP
.TF V(3) V
.DC V 0 50 10
.PROBE
.END
```

Now perform the PSpice analysis and see what happens. As soon as the calculations are done, you will be taken to what is sometimes called the *on-screen oscilloscope* feature. This is the Probe screen.

Using this feature is easy if you have a copy of your input file for reference. This is helpful in identifying nodes, voltages, currents, and so forth. For this example, when the Probe screen appears you will see V identified along the horizontal axis. The reason is that you chose to sweep V in the dc sweep statement.

The legend across the bottom of the screen identifies your options. Choose the option Add Trace by pressing the *a* key or [*Rtn*]. Now you can simply type the variable of interest. Type I(R3) and press [*Rtn*]. This will give a plot of the current through resistor R_3. See if you can add a plot of the current I(R1). When both plots appear on the screen, verify that when $V = 50$, I(R1) = 250 mA and I(R3) = 125 mA.

Follow the directions to obtain a printed copy of your V, I graph. From the main menu this is chosen by pressing *h* for Hard_copy, then *l* for 1_page_long. In earlier versions of PSpice, you choose options by pressing numeric keys.

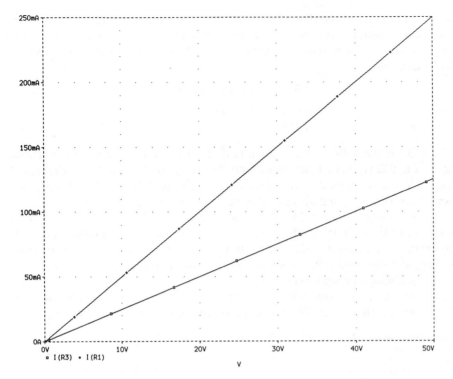

Fig. 1.27 SPICE Sweep Analysis of Tee Circuit.

The traces are shown in Fig. 1.27. This graph is not very interesting because it is linear; however, you can see that more information is available than you had asked for previously in the .PRINT statement. You may recall that the statement, "All voltages and currents are available," appeared on the screen at the beginning of the Probe display. As you continue to look at a variety of circuits, the Probe feature will prove to be invaluable.

As another interesting example of using Probe, go back to Fig. 1.24 and the example where we introduced the dependent polynomial voltage source. Replace the .PRINT statement with a .PROBE statement and look at the graphical display of the relationships among the voltages V(2), V(3), and V(4). Work with the graphs until you feel comfortable using Probe. Verify the numerical results compared to using the .PRINT statement.

Many more features are available in Probe. These features will be explored in the examples that follow throughout the text.

NODAL ANALYSIS AND PSPICE

Introductory circuits courses cover the topic of nodal analysis using standard equations. These equations are more easily written if all practical voltage sources are replaced by practical current sources. This has the disadvantage of physically changing the circuit, but the advantage is that fewer nodes appear and this allows for fewer equations. After you find the node voltages, if necessary you can convert the sources back to their original forms. The standard form for node voltages looks like this:

$$G_{11}V_1 + G_{12}V_2 + G_{13}V_3 = I_1$$
$$G_{21}V_1 + G_{22}V_2 + G_{23}V_3 = I_2$$
$$G_{31}V_1 + G_{32}V_2 + G_{33}V_3 = I_3$$

where G_{11} is the self-conductance at node 1, G_{12} is mutual conductance between nodes 1 and 2, G_{13} is mutual conductance between nodes 1 and 3, and the term I_1 is the net source current directed toward node 1. In nodal analysis, all self-conductances are positive, and all mutual conductances are negative.

The circuit of Fig. 1.28 will be used for the nodal example. As an exercise, see if you can write the nodal equations and solve them with either a computer program or a calculator designed for the solution of simultaneous equations. Writing the standard equations and understanding them is important, but solving such equations over and over is not fun.

The PSpice solution for the circuit of Fig. 1.28 is simple, involving nothing that is new at this point. The input file should look something like this:

```
Nodal Analysis of Circuit with Several Current Sources
I1 0 1 20mA
I2 0 2 10mA
I3 0 3 15mA
R1 1 0 500
R2 1 2 500
R3 2 0 400
R4 2 3 500
R5 3 0 300
.OP
.END
```

The input file contains enough information to allow for all of the node voltages to be found. Run the analysis and verify the voltages: V(1) = 7.694 V, V(2) = 5.3947 V, V(3) = 4.8355 V.

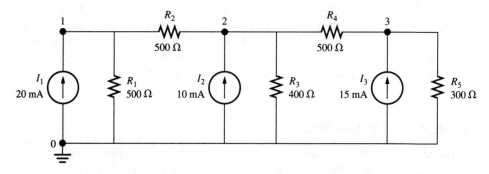

Fig. 1.28   Nodal analysis with several current sources.

Fig. 1.29 Current sources converted to voltage sources.

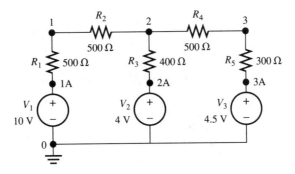

It is instructive to convert the current sources to voltage sources in Fig. 1.28 and create an input file for this resulting circuit. Figure 1.29 shows the revised circuit. You will notice that nodes *1, 2,* and *3* have been preserved for easy reference. There are three additional nodes, and your input file must be modified accordingly.

```
Nodal Analysis with Current Sources Converted to Voltage Sources
V1 1A 0 10V
V2 2A 0 4V
V3 3A 0 4.5V
R1 1A 1 500
R2 1 2 500
R3 2 2A 400
R4 2 3 500
R5 3 3A 300
.OP
.END
```

When you run this analysis, verify that V(1) = 7.6974 V, V(2) = 5.3947 V, and V(3) = 4.8355 V, as before. Three additional node voltages are given as V(1A) = 10 V, V(2A) = 4 V, and V(3A) = 4.5 V as given in the input file for the source voltages. As a bonus, you obtain the three source currents. For example, the current in V_1 is given as −4.605 mA. This means that positive current in the amount of 4.605 mA is out of the plus side of V_1. Verify that the other source currents are correct, and state their true directions as positive numbers. Since all sources were voltage sources instead of current sources, the total power is correctly given as 27.1 mW.

A NONPLANAR CIRCUIT

When a circuit is nonplanar, it cannot be drawn in two dimensions without a crossing of the lines connecting the nodes. Such a circuit is shown in Fig. 1.30, which contains a voltage source and eight resistors for a total of nine elements. Using circuit-analysis techniques, mesh analysis may be applied to planar circuits, but nodal or loop analysis must be used on nonplanar ones. PSpice is a more convenient tool for handling circuits with large numbers of elements. For the

Fig. 1.30 A nonplanar circuit.

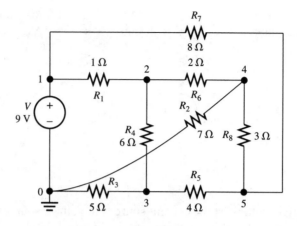

circuit illustrated, the input file is

```
Nonplanar Circuit Containing Nine Elements
V  1  0  9V
R1 1  2  1
R2 4  0  7
R3 3  0  5
R4 2  3  6
R5 5  3  4
R6 2  4  2
R7 1  5  8
R8 4  5  3
.DC V 9V 9V 9V
.PRINT DC V(1,2) V(2,3) V(2,4)
.PRINT DC V(4,5) V(5,3) V(1,5)
.OP
.OPT nopage
.END
```

This input file produces the voltage drops across the individual elements, which is desirable if branch currents are needed. Alternately, the print statements could call for the currents through the resistors, such as I(R1). Note that without the .OP entry, the node voltages would not be obtained.

Verify that V(1,2) = 1.367 V, V(2,4) = 1.685 V, and V(1,5) = 3.031 V. Also verify numerically that V(1) = V(1,2) + V(2,3) + V(3).

SUMMARY OF PSPICE STATEMENTS USED IN THIS CHAPTER

The designation [. . .] means an optional item; <. . .> is a required item.

E[*name*] **<+*node*>** **<−*node*>** **<[+*controlling node*><−*controlling node*]>** **<*gain*>**

For example,

```
E  2  3  1  0  5
```

means a voltage-controlled voltage source between nodes *2* and *3*, dependent on the voltage between nodes *1* and *0*. The voltage *E* will be five times the voltage V_{10}.

F[*name*] *<+node>* *<−node>* *<controlling V device name>* *<gain>*

For example,

    F 4 2 VA 50

means a current-controlled current source between terminals *4* and *2* with the current through the source from node *4* to node *2*. The current is dependent on the current through source VA, having a gain factor of 50. If necessary, VA may be set at zero volts.

G[*name*] *<+node>* *<−node>* *<+controlling node>* *<−controlling node>* *<transconductance>*

For example,

    G 5 1 2 0 0.05

means a voltage-controlled current source between nodes *5* and *1*, which is dependent on the voltage between nodes *2* and *0*. The g_m value is 50 mS.

H[*name*] *<+node>* *<−node>* *<controlling V device name>* *<transresistance>*

For example,

    H 6 4 VB 20

means a current-controlled voltage source between nodes *6* and *4*, which is dependent on the current through the voltage source VB. If necessary, VB may be set at zero volts. The transfer-resistance value is 20.

I[*name*] *<+node>* *<−node>* **[DC]** *<value>*

For example,

    I 0 1 DC 2A

means a direct current from an ideal current source directed from node *0* to node *1* internally. The value of the current is 2 A.

R[*name*] *<+node>* *<−node>* *<value>*

For example,

    R1 1 2 100

means a resistor between nodes *1* and *2* with a value of 100 Ω. Since a resistor is bilateral, either node may be called the +node (or the −node). If it is desired to find the current through the resistor, the node designation is significant. In this example if the current is called I(R1) and it is actually moving from node *1* to node *2*, it will be shown as a positive number in PSpice.

V[*name*] <+*node*> <−*node*> [DC] <*value*>

For example,

```
V 1 0 DC 50V
```

means a source voltage with the positive terminal at node *1* and the negative terminal at node *0*. It is an ideal dc voltage source of 50 V. [DC] appears in brackets, meaning that this portion of the statement is optional. The V following the 50 is also optional but should be included for clarity. There are other forms of the *V* statement found in later chapters. Some involve ac voltages, and some include a transient specification.

DOT COMMANDS USED IN THIS CHAPTER

.DC[LIN] [OCT] [DEC] <*sweep variable*> <*start*> <*end*> <*increment*>

For example,

```
.DC LIN VS 0V 10V 0.1V
```

means that the source voltage VS will assume values from 0 V to 10 V, in 0.1-V increments. The sweep in voltages is linear.

.END

This statement must come at the end of the input file. It tells PSpice that there are no more statements in the file.

.OP

This statement is used to produce detailed bias-point information.

.OPTIONS

For example,

```
.OPTIONS NOPAGE
```

means the option has been selected that will suppress paging and printing of a banner for each major section of output. Other options are

> ACCT to produce accounting information
>
> LIST giving a summary of circuit devices
>
> NODE producing a node table
>
> NOECHO to suppress listing of the input file
>
> NOMOD suppressing listing of model parameters
>
> OPTS to show which options are chosen
>
> WIDTH to set the number of columns for output

.PRINT DC *<output variable>*

For example,

```
.PRINT DC V(5) I(RL)
```

means that the output file will contain a listing that includes V(5), the voltage at node *5*, and I(RL), the current through resistor RL.

.PROBE

means that the post-processor Probe is to be used in the analysis. All voltages and currents are available for use in Probe. It is up to you to select which item to plot, assuming that a range of values has been included in the analysis. The range could be one of input voltages or, in the case of ac, a frequency range.

In Probe, arithmetic expressions of output variables may be used. The simple $+$, $-$, $/$, $*$ (add, subtract, divide, and multiply) operators are frequently used. Available functions include ABS(x), SGN(x), DB(x), EXP(x), LOG(x), LOG10(x), PWR(x, y), SQRT(x), SIN(x), COS(x), TAN(x), ARCTAN(x), d(x), s(x), AVG(x), and RMS(x).

.TF *<output variable> <input source>*

For example,

```
.TF V(5) VS
```

means that the transfer function V(5)/VS is produced. This is described as a small-signal transfer function. It also produces the input and output resistances.

The **POLY** form of source

For example,

```
E1 5 2 POLY(1) 3 1 1 2 3
```

means that the VCVS E1 is dependent on the voltage between nodes *3* and *1* in a nonlinear form as described by a polynomial. The (1) means that there is a single controlling voltage V_{31}. The next three terms are for k_0, k_1, and k_2 in the formula

$$k_0 + k_1 v_1 + k_2 v_1^2$$

If there are more k values, it means that the polynomial is of higher degree.

When a line in an input file begins with an asterisk (*), it is a comment line rather than a statement to be processed. A comment may also be placed at the end of a statement if the comment is preceded by a semicolon (;).

In describing PSpice statements, the notation [. . .] means an optional item; the notation <. . .> means a required item. When an asterisk is placed at the end of a PSpice statement, it means that the item may be repeated.

PROBLEMS

1.1   For the circuit of Fig. 1.31, find the current I. Your PSpice input file should include a method for finding the current directly. Verify the results by finding the current as V_{12}/R_1 and V_{23}/R_2.

Fig. 1.31

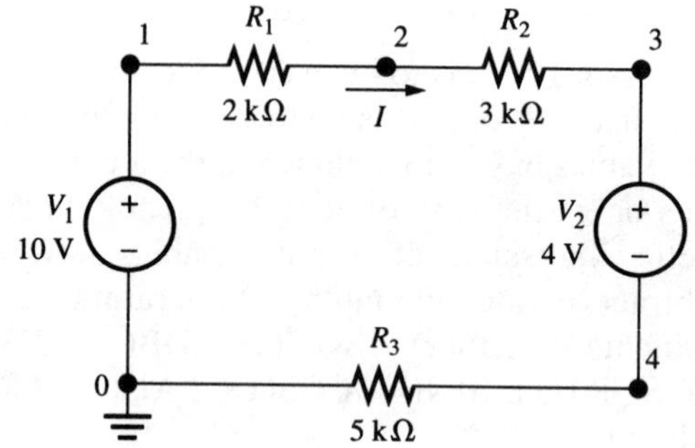

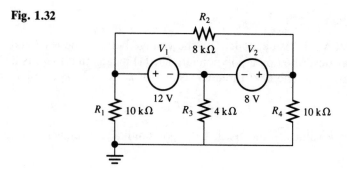

1.2   For the circuit of Fig. 1.32, find the power furnished by each of the voltage sources V_1 and V_2. Your input file should provide for finding the current through each voltage source. Verify the results by finding the power absorbed by all of the resistors. Find each resistor power by using either I^2R or V^2/R.

Fig. 1.32

1.3   Find V_{ab} (the Thevenin voltage) and R_{ab} (the Thevenin resistance) for the circuit of Fig. 1.33. Your input file should follow the method suggested in the text for finding these values.

Fig. 1.33

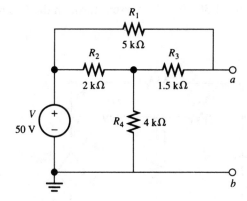

1.4 For the ladder network shown in Fig. 1.34, find R_{in}, the resistance seen by the source. Include the necessary statements in your PSpice input file to find the input resistance directly.

Fig. 1.34

$R = 1.5\,k\Omega$

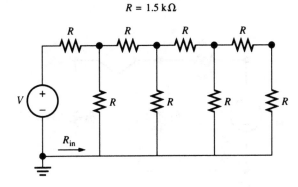

1.5 For the circuit of Fig. 1.35, find the voltage V_{12}. This may be found simply as $V(1) - V(2)$. Verify the results by including provisions in your input file to find the current through R_2 directly.

Fig. 1.35

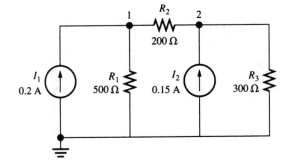

1.6   For the circuit of Fig. 1.36, find the current through the 6-Ω resistor and the voltage v.

Fig. 1.36

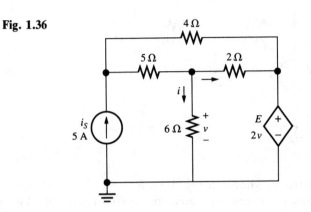

1.7   For the circuit of Fig. 1.37, find the voltage v_{23} and the current i.

Fig. 1.37

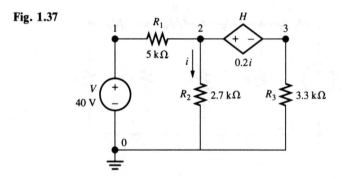

1.8   For the circuit of Fig. 1.38, find the voltage across resistor R_4.

Fig. 1.38

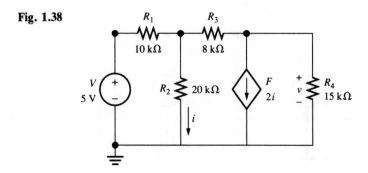

1.9 For the circuit of Fig. 1.39, find the voltage v_{ab}.

Fig. 1.39

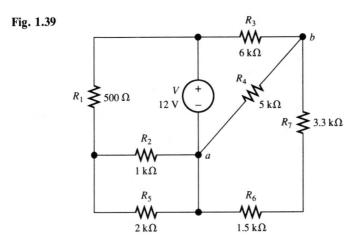

1.10 For the circuit of Fig. 1.40, find V_{ab} and R_{ab}, the Thevenin equivalents.

Fig. 1.40

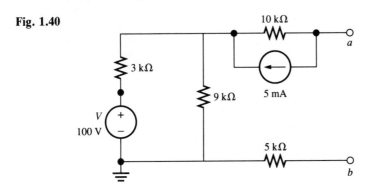

1.11 For the circuit of Fig. 1.41, find the Thevenin equivalent circuit at terminals ab.

Fig. 1.41

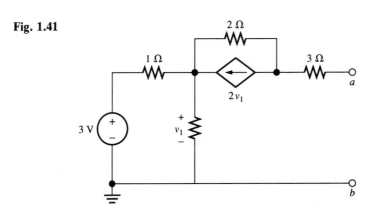

2

AC Circuit Analysis (for Sinusoidal Steady-State Conditions)

SPICE gives dc node voltages without any special requirements, since the dc voltages are part of the normal calculations for operating-points that are necessary for transistor biasing and the like. If you want an ac analysis, you must specifically ask for it. An introductory example will show how this is done.

SERIES AC CIRCUIT WITH *R* AND *L*

The series circuit of Fig. 2.1 shows an ac source voltage of 1 V in series with a resistance and an inductance. The *R* and *L* combination might be a coil, for example. Values are $R = 1.5\ \Omega$, $L = 5.3$ mH, and $f = 60$ Hz. Find the current in the circuit and the impedance of the coil. The input file might be

```
AC Circuit with R and L in Series (Coil)
V 1 0 AC 1V
R 1 2 1.5
L 2 0 5.3mH
.AC LIN 1 60Hz 60Hz
.PRINT AC I(R) IR(R) II(R) IP(R)
.END
```

The .AC statement gives a LIN (linear) sweep for 1 frequency only, with the beginning and ending frequencies of 60 Hz. The .PRINT statement with the sweep parameter AC is required to print the variables chosen. This is what they are:

I(R) is the magnitude of the current through *R*.

IR(R) is the real component of the current through *R*.

II(R) is the imaginary component of the current through *R*.

IP(R) is the phase angle in degrees of the current through *R*.

69

Fig. 2.1   Series R and L.

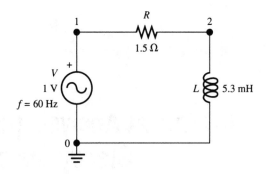

Fig. 2.2   Phasor diagram for series R and L.

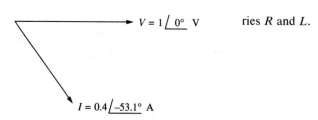

Run the PSpice analysis. The node voltages are shown as zero. This means that there are no biasing, dc values. Voltage-source currents and power dissipation also refer to biasing. These values are also zero. The interesting part of the result shows FREQ = 60 Hz, I(R) = 0.4002 A for the magnitude of the ac current, IR(R) = 0.2403 A for the real part of this current, II(R) = −0.3201 A as the imaginary part of the current, and IP(R) = −53.1° as the angle of the current.

In ac problems of this type, it is helpful to draw phasor diagrams. Figure 2.2 shows such a diagram. The voltage is taken as the reference at zero degrees. The current is shown at its angle of −53.1°. You may find the impedance of the coil as

$$Z = \frac{V}{I} = \frac{1\underline{/0^\circ}}{0.4\underline{/-53.1^\circ}} = 2.5\underline{/53.1^\circ}\,\Omega$$

Since the voltage was given as a unit value, this is the same as the reciprocal of the I phasor.

SERIES AC CIRCUIT WITH *R* AND *C*

If a capacitor is used in place of an inductor, the circuit is as shown in Fig. 2.3. The values are $R = 5\ \Omega$, $C = 100\ \mu\text{F}$, and $f = 318$ Hz. The input file then might become

```
AC Circuit with Resistance and Capacitance in Series
V 1 0 AC 1V
R 1 2 5
C 2 0 100uF
.AC LIN 1 318Hz 318Hz
.PRINT AC I(R) IP(R) V(2) VP(2)
.END
```

Fig. 2.3 Series R and C.

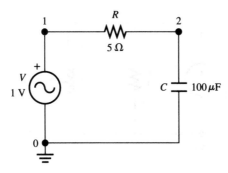

Fig. 2.4 Phasor diagram for series R and C.

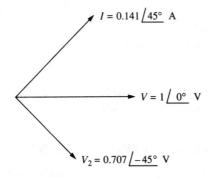

In the .PRINT statement, V(2) and VP(2) are for the magnitude and phase of the capacitor voltage. Run the PSpice analysis; then plot the phasor diagram of current and voltages. Your results should agree with Fig. 2.4.

PARALLEL BRANCHES IN AC CIRCUIT

The next example, shown in Fig. 2.5, is for R and L in parallel, using a current source. Values chosen are $I = 100\underline{/0°}$ mA, $R = 8\frac{1}{3}$ Ω, and L = 6.36 mH. For this circuit, find the voltage across the parallel branches, the currents through each branch, and the admittance of the RL combination. The input file should be like this:

```
AC Circuit with Parallel Branches R and L
I 0 1 AC 100mA
R 1 0 8.33333
L 1 0 6.36mH
.AC LIN 1 500Hz 500Hz
.PRINT AC V(1) VP(1) I(R) IP(R) I(L) IP(L)
.END
```

The voltage is found as V(1) and VP(1) for the magnitude and phase angle, and the current through each branch is found in the usual way. Run the analysis, verifying that V(1) = 0.7691 and VP(1) = 22.64°. The admittance of the RL combination is $Y = I/V(1) = (0.100\underline{/0°})/(0.7691\underline{/22.6°}) = (0.13\underline{/-22.6°})$ S.

Fig. 2.5 Parallel R and L.

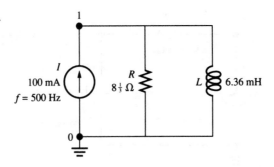

Fig. 2.6 Phasor diagram for parallel R and L.

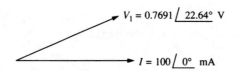

Construct the phasor diagram for this circuit, showing all currents and the voltage V_1. Check your results as shown in Fig. 2.6.

PARALLEL BRANCHES WITH CAPACITIVE BRANCH

Figure 2.7 shows another parallel circuit. Values are $I = 100\underline{/0°}$ mA, $R = 8\frac{1}{3}\ \Omega$, $C = 14.14\ \mu\text{F}$, and $f = 500$ Hz.

Before running the PSpice analysis, calculate the admittance of the RC combination. This is given as $Y = G + jB$, where $G = 1/R$ and $B = 2\pi f C$. The input file is

```
AC Circuit with Parallel R and C
I 0 1 AC 100mA
R 1 0 8.333333
C 1 0 14.14uF
.AC LIN 1 500Hz 500Hz
.PRINT AC V(1) VP(1) I(R) IP(R) I(C) IP(C)
.END
```

Run the analysis; then construct a complete phasor diagram for the circuit. Compare your results with Fig. 2.8. The results show $V = 0.7815\underline{/-20.3°}$ V. Verify your predicted value of Y. Use the formula $Y = I/V$.

Fig. 2.7 Parallel R and C.

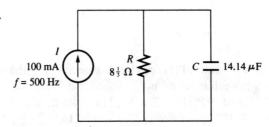

Fig. 2.8 Phasor diagram for parallel R and C.

$I = 100 \underline{/\ 0°}\ $ mA

$V = 0.7815 \underline{/\ -20.3°}\ $ V

MAXIMUM POWER TRANSFER IN AC CIRCUIT

Recall that in a dc circuit, maximum power is delivered to a load resistance when $R_L = R_s$. In an ac circuit, the source impedance is often complex, or the Thevenin equivalent of everything excluding the load is most likely complex, containing resistance and reactance. The maximum power theorem tells you that for maximum power to be delivered in the case of a source impedance, which contains resistive and reactive components, the load should be its conjugate. For example, if $Z_s = (600 + j150)\ \Omega$, for maximum power to be delivered to the load, it should be adjusted to $Z_L = (600 - j150)\ \Omega$. For this circuit, refer to Fig. 2.9. Of course, the X values must be converted to inductance and capacitance values. If we choose $f = 1$ kHz, then $L = 23.873$ mH, and $C = 1.061\ \mu$F. The input file is

```
Maximum Power Transfer in AC Circuits
V 1 0 AC 12V
RS 1 2 600
L 2 3 23.873mH
RL 3 4 600
C 4 0 1.06uF
.AC LIN 1 1000Hz 1000Hz
.PRINT AC I(RL) V(3) VP(3)
.END
```

Run the analysis and verify that the current in the circuit is 10 mA at an angle of (almost) zero degrees. Also, observe that V(3) = 6.18 V at an angle of $-14°$. The power delivered to the load is easily found as $P = |I|^2 R = 60$ mW.

RESONANCE IN SERIES *RLC* CIRCUIT

Series resonance occurs in a circuit with R, L, and C when the input impedance is purely resistive. Thus the inductive reactance and the capacitive reactance can-

Fig. 2.9 Maximum power to load impedance.

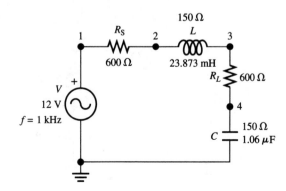

cel, and the current is a maximum. As a result, the phase angle of the circuit is zero. That is, the current and the applied voltage are in phase.

The resonant frequency is easily found from

$$f_o = \frac{1}{2\pi\sqrt{LC}}$$

Figure 2.10 shows such a circuit. The values are $R = 50\ \Omega$, $L = 20$ mH, and $C = 150$ nF. For this example, this gives $f_0 = 2905.8$ Hz. The applied voltage is chosen to be $1\underline{/0°}$ V. Work with the following input file:

```
Series Resonance with RLC
V 1 0 AC 1V
R 1 2 50
L 2 3 20mH
C 3 0 150nF
.AC LIN 99 100Hz 5000Hz
.PROBE
.END
```

The .AC statement calls for a linear sweep beginning at 100 Hz and ending at 5000 Hz in 99 steps, giving 50-Hz increments.

Run the PSpice analysis and then look for the Probe screen display. There are many interesting variables that can be displayed. As an exercise, do the following:

1. Plot IP(R) over a linear frequency range extending from 2 kHz to 4 kHz. The vertical axis should show phase angles from $-100°$ to $100°$. Feel free to experiment with the on-screen plot features. These are menu driven and are easily learned by trial-and-error. If you get a graph that you do not want, simply remove it and try something else. Your graph should pass through zero degrees at a frequency near 2.9 kHz. Obtain a printed copy of this and the next two parts.

2. Plot 1/I(R) over the same frequency range (2 kHz to 4 kHz). Force the Y-axis values to the range of 0 to 300. Since $V = 1$, the plot will represent the magnitude of Z. Check these (approximate) values: at 2 kHz, $Z = 284\ \Omega$; at 2.9 kHz, $Z = 50\ \Omega$; and at 4 kHz, $Z = 243\ \Omega$.

Fig. 2.10 Series resonance with *RLC*.

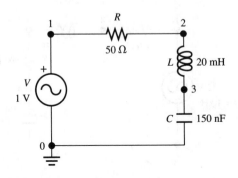

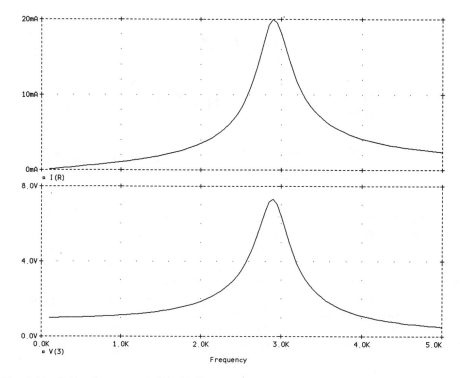

Fig. 2.11 Series Resonance with RLC.

3. Plot two graphs on the screen together, one giving V(3) and the other giving I(R) (use plot control to add the second graph). Use a frequency range extending from 0 to 5 kHz. Your results should show a maximum for V(3) of about 7.3 V and a maximum for I(R) of 20 mA. Check these values for accuracy using pencil-and-paper methods. See Fig. 2.11 for comparison.

FREQUENCY SWEEP FOR SERIES-PARALLEL AC CIRCUIT

Figure 2.12 shows another ac network. Values are $V = 100\underline{/0°}$ V, $R_1 = 10\ \Omega$, $R_2 = 10\ \Omega$, $L = 100$ mH, and $C = 10\ \mu$F. Assume that the resonant frequency is unknown and that a preliminary investigation is required. The input file might be

```
Series-Parallel AC Circuit
V 1 0 AC 100V
R1 1 2 10
R2 2 3 10
L 3 0 100mH
C 2 0 10uF
.AC LIN 100 50Hz 1000Hz
.PRINT AC I(R1) IP(R1)
.END
```

Fig. 2.12   Series-parallel ac circuit.

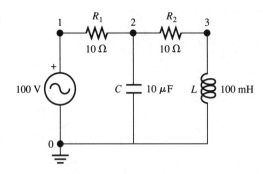

The .AC statement is a guess for the resonant frequency somewhere in the range between 50 Hz and 1000 Hz. If this is not close to the actual range, you can choose a new set of values. The .PRINT statement calls for the circuit current magnitude and phase. This will allow you to look for unity power factor, since when this condition occurs, the angle on the current will be zero.

Run the analysis and look at the output file. Do not bother to print the file, since you will want to change the range of frequencies before you are through. You should confirm that between $f = 155$ Hz and $f = 165$ Hz, resonance occurs. Now revise the input file by modifying the ac sweep statement to become

```
.AC LIN 101 100 200
```

Now look at all the whole-number frequencies between 100 Hz and 200 Hz. Run the analysis and see that resonance comes between $f = 158$ Hz and $f = 159$ Hz. Also the current close to resonance is approximately 98 mA.

Now you are in a position to look at a powerful feature of the PSpice analysis. Does resonance occur at the series-resonance-formula predicted value of $1/(2\pi \sqrt{LC})$? Use your calculator to compute this formula value. It should be $f = 159.155$ Hz. This does not quite agree with our prediction that f_0 lies between 158 Hz and $f = 159$ Hz. Is the difference merely round-off error?

Revise the input file to give the ac sweep as

.AC LIN 51 155Hz 160Hz

This will give an increment of frequency of 0.1 Hz in the analysis. Run the study again and find where the sign of the current angle IP(R1) changes. The results should show that this is between a frequency of 158.3 Hz and 158.4 Hz. From your study of resonance, you should confirm that the series-resonance formula is incorrect for this series-parallel circuit. Note that the minimum current comes not at the resonant frequency but at $f = 159.2$ Hz, where the phase angle of the current is about 5.6°.

For an interesting exercise, replace the .PRINT statement with a .PROBE statement in this analysis. You will be able to show in graphical form what you have seen with the numerical results. The graphs have the advantage of allowing you to look at many variables without having to revise the input file.

EFFECT OF CHANGES IN COIL RESISTANCE

In Fig. 2.12, one of the parallel branches contains $R = 10\,\Omega$ and $L = 100$ mH. This might be a coil with a small resistance. The question might be asked, "What effect does the resistance of the coil have on the behavior of the circuit?" Modify the input file by making $R_2 = 50\,\Omega$ and confirm that $f_o = 138$ Hz. Then make $R_2 = 80\,\Omega$ and see that $f_o = 95$ Hz. Would you have guessed that the change in resonant frequency would have been that great?

Note how easily PSpice lets you vary the circuit parameters and carry out a new set of calculations. If you want numerical accuracy, the .PRINT statement provides it, and if you want to see the actual plot of variables, the .PROBE statement is invaluable.

A PARALLEL-RESONANT CIRCUIT

The equations for analyzing a parallel resonant circuit are considerably more tedious than those for the series-resonant circuit. You may refer to an introductory text for a full treatment of these equations. However, the analysis using PSpice will allow you to find the resonant frequency and the input impedance at resonance (or near resonance) with ease. In this example you will also look at the cursor available in Probe.

The circuit shown in Fig. 2.13 contains one branch with a coil and another branch with a capacitor. Values are $R_L = 10\,\Omega$, $L = 2.04$ mH, $R_C = 5\,\Omega$, and $C = 0.65\,\mu$F. Note that a small sampling resistor $R = 1\,\Omega$ has been added in series with the voltage source, making V a 1-V practical source. A preliminary estimate indicates that the resonant frequency is between 4 kHz and 5 kHz. Consider the input file as shown here.

```
Parallel Resonant Circuit
V 1 0 AC 1V
RL 1A 2 10
RC 1A 3 5
R 1 1A 1
L 2 0 2.04mH
C 3 0 0.65uF
.AC LIN 1001 4000Hz 5000Hz
.PROBE
.END
```

Fig. 2.13 Parallel resonant circuit.

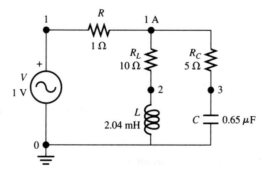

The input file shows a frequency sweep extending from 4 kHz to 5 kHz in 1-Hz increments. Run the analysis; then plot IP(R). Since the *X*-axis shows a range extending from 1 kHz to 10 kHz, the trace appears over a small portion of the graph. Change the *X*-axis to show the desired range, 4 kHz–5 kHz, by choosing *X_axis, Set_range;* then type *4k 5k [Rtn]*. Then select *Exit*. The desired location of the resonant frequency is now easier to find. By inspection, f_o is near 4.3 kHz.

In earlier versions of SPICE (and PSpice), line-printer type plots were used to obtain crude plots along with coordinate pairs. The coordinate pairs in this case would be for frequency and phase angle. In later versions of PSpice, there is a greatly improved method for finding coordinate pairs from the graph.

Using the Probe Cursor

Select *Cursor;* then notice the cross hairs that appear on the left of the screen along with a window at the bottom right of the screen. In the beginning, both *C1* and *C2* refer to the original cursor locations. As the cursor is moved with the arrow keys, the values of *C2* will remain the same while the values of *C1* change to indicate the new cross-hair location. In this example, the beginning values for *C1* and *C2* are 4.0000 (kHz) and -27.015 (degrees). This means that at $f = 4$ kHz, the

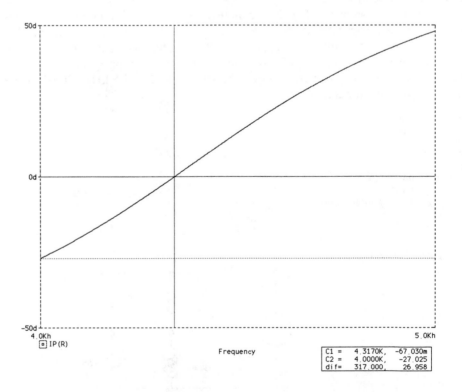

Fig. 2.14 Parallel Resonant Circuit.

phase angle of the input current is $-27.025°$ relative to the input voltage with an implied angle of zero.

With the right-arrow key, move the hairline until you find the point where the phase shift is (almost) zero degrees. This will give a numerical value for the resonant frequency that is more accurate than you could see from trying to interpolate the graph visually. You should confirm $C1 = 4.3170$ kHz in the first column and -67.030 (millidegrees) in the second column. This is as close to zero as you can get without running the analysis over a smaller range of frequencies using a smaller increment of frequency. This would hardly be worth the added effort.

In earlier versions of PSpice, a hard copy of the graph would not show the cursor box or the cross hairs. In release 5.1 this desirable feature has been added. While you are still in the cursor mode, select *Hard_copy, 1_page_long* to produce the desired plot. Refer to Fig. 2.14 for this plot.

It is also of interest to find the input impedance (or admittance) of the circuit at resonance. Recall that the value should be purely resistive (or conductive) at resonance. With this in mind, plot I(R). Since $V = 1$ V, the circuit current can also be thought of as representing the input admittance. Explain why. From this plot, again use the cursor mode. Move the hairline until you find the predicted $f_o = 4.317$ kHz. What is the value of I at this frequency? Confirm that $I = 4.683$ mA at the resonant frequency. Also confirm that $Z_0 = R_0 = 213.5\ \Omega$.

FINDING THE INPUT IMPEDANCE OF AN AC CIRCUIT

The PSpice analysis can give you magnitude, real parts, and imaginary parts of voltage and current, but it cannot give Z except as a ratio of voltage to current. If a network is driven by an independent current source of your choice, you can make it $I = 1$ A at zero degrees; then since $Z = V/I$, the impedance has the same numerical value as the voltage. On the other hand, you might drive the network by an independent voltage source of $V = 1$ V at zero degrees; then since $Y = I/V$, the admittance has the same numerical value as the current. These techniques might not be suitable for all situations, so we would like to develop a more general method for finding input impedance.

Let $V = a + jb = x\underline{/\alpha}$ and $I = c + jd = y\underline{/\beta}$. It is easily shown that

$$Z = \frac{V}{I} = \frac{ac + db}{y^2} + \frac{j(bc - ad)}{y^2}$$

Consider the circuit suggested by Fig. 2.15. The box on the right contains the unknown network for which you desire to find the input impedance Z. The resistor can be a small resistor that you will use to sample the current. This will allow you to find Z from V_1/I, where both voltage and current are phasors. Following the method given in the previous paragraph, in PSpice notation we need for the real part of Z

$$(VR(1)*IR(R)+VI(1)*II(R))/(I(R)*I(R))$$

Fig. 2.15 Box contains unknown impedance network.

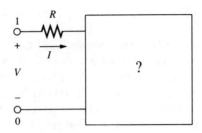

and for the imaginary part of Z

$$(VI(1)*IR(R)-VR(1)*II(R))/(I(R)*I(R))$$

These equations may be used by running a PSpice analysis and calling on Probe to
construct a plot. An example will help to clarify the technique.

Input-Impedance Example

The circuit of Fig. 2.16 contains a voltage source $V = 10$ V, at $f = 1$ kHz, a
resistance $R = 4\ \Omega$, and an inductance $L = 0.47746$ mH. Now rather than simply
finding X from L, we will attempt to show the method involved in finding Z using
PSpice. For this simple circuit it is hardly worthwhile, but for more complicated
circuits, it will prove an asset. The input file will be

```
Input Impedance to ac Circuit Using V/I
V 1 0 AC 10V
R 1 2 4
L 2 0 0.47746mH
.AC LIN 3 500Hz 1500Hz
.PROBE
.END
```

Note that the .AC statement calls for three frequencies rather than one. This
will give a usable graph rather than just a point. The actual frequency of 1 kHz is
included as the center frequency. Run the analysis and in Probe plot

$$VR(1)*IR(R)/I(R)/I(R)+VI(1)*II(R)/I(R)/I(R)$$

to obtain the real part of the input impedance. Then plot

$$VI(1)*IR(R)/I(R)/I(R)-VR(1)*II(R)/I(R)/I(R)$$

Fig. 2.16 Circuit with *unknown*
impedance.

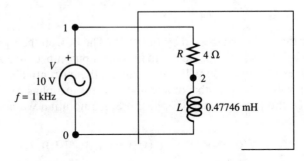

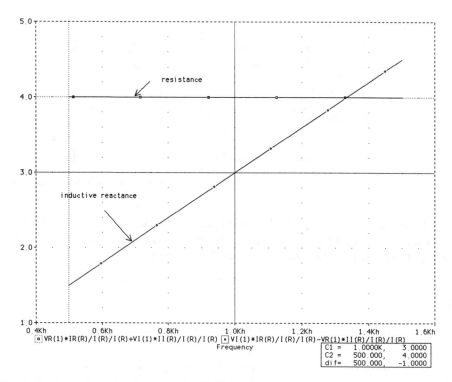

Fig. 2.17 Input Impedance to ac Circuit Using V/I.

to obtain the imaginary part of the input impedance. (These equations look a little different from those given previously. The only difference is in the way the terms are grouped.) Since your main interest is finding the input impedance at $f = 1$ kHz, use the cursor mode and verify that $Z_{in} = R + jX = (4 + j3)$ Ω. Of course, in this one-branch circuit R is fixed at 4 Ω, and X increases in linear fashion with frequency, being inductive reactance. See Fig. 2.17 for this plot.

Input Impedance Using a Small, Current-Sensing Resistor

If you have a more complicated circuit, you may want to include a small resistor in series with an arbitrary voltage source to find the input impedance. This technique is used in the next example. Figure 2.18 shows a circuit containing R and C in series, but the circuit could be of any degree of complexity and you would use the same technique. The input file is

```
Input Impedance Using a Small Current-Sensing Resistor
V 1 0 AC 1V
R 1 2 0.001
RL 2 3 100
C 3 0 1.9894uF
.AC LIN 501 500Hz 1500Hz
.PROBE
.END
```

Fig. 2.18 Circuit with current-sensing resistor.

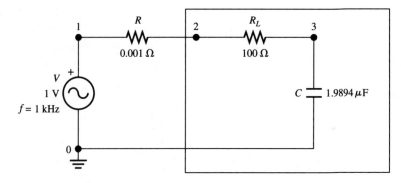

Using an input voltage of 1 V at zero degrees for the reference will allow you to use an abbreviated form of the equations for the real and imaginary portions of the input impedance. Run the analysis and in Probe plot

$$(IR(R))/(I(R)*I(R))$$

and

$$-(II(R))/(I(R)*I(R))$$

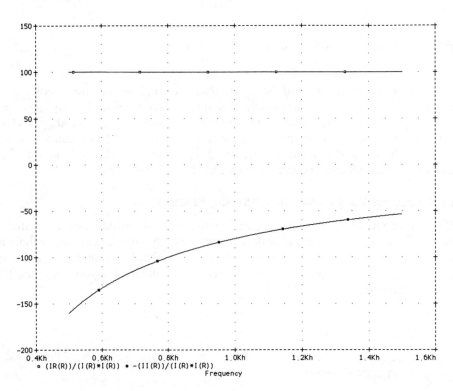

Fig. 2.19 Input impedance to ac network.

giving the real and imaginary parts of Z_{in}, respectively. Do you understand why the abbreviated equations work in this case? Keep in mind that this is possible only because $V = 1\underline{/0^\circ}$ is the input voltage. Use the cursor mode and verify that when $f = 1$ kHz, $R = 100$ Ω and $X = -80$ Ω. See Fig. 2.19 for this plot. (*Note:* Before leaving this plot, refer to the section "Figure 2.19 Revisited" at the end of this chapter.)

In this case you might prefer to plot the imaginary part of Z_{in} by removing the minus sign from the second of the two equations. This will allow the details of both the R and X plot to show up better on the graph. Try this and decide which plot you like better. Of course, you must remember to change the sign of your answer for X if you do this. Figure 2.20 shows the alternate plot.

Input Impedance of a Two-Branch Network

The examples used so far in finding input impedance have been needed for introducing the method. The real power of this feature becomes more obvious in circuits such as the one shown in Fig. 2.21. It has two branches for impedances, the first branch being capacitive and the second branch being inductive. The source voltage is $V = 12$ V, with an internal resistance $R_s = 50$ Ω and a frequency

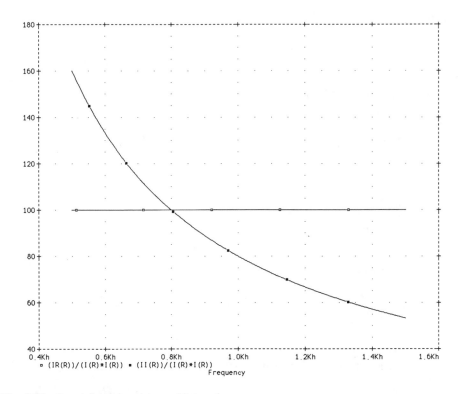

Fig. 2.20 Input Impedance to ac Network.

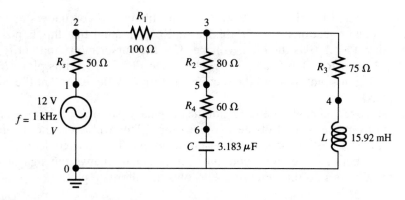

Fig. 2.21 Input impedance of two-branch network.

$f = 1$ kHz. Using the values of the components shown in the diagram, the input file becomes

```
Input Impedance of Two-Branch Network
V  1 0 AC 12V
RS 1 2 50
R1 2 3 100
R2 3 5 80
R3 3 4 75
R4 5 6 60
```

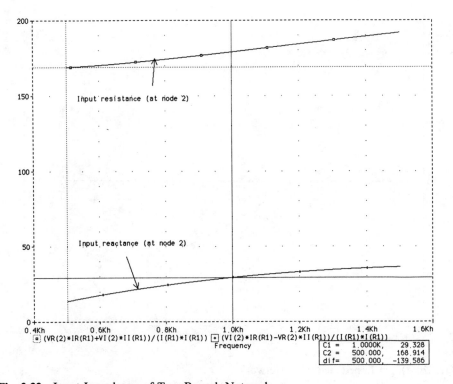

Fig. 2.22 Input Impedance of Two-Branch Network.

```
L  4  0  15.92mH
C  6  0  3.183uF
.AC LIN 501 500Hz 1500Hz
.PROBE
.END
```

Run the analysis and use the long form of the equations to obtain the real and imaginary parts of Z_{in}. That is,

$$(VR(2)*IR(R1)+VI(2)*II(R1))/(I(R1)*I(R1))$$

and

$$(VI(2)*IR(R1)-VR(2)*II(R1))/(I(R1)*I(R1))$$

Use the cursor mode to find Z_{in} at $f = 1$ kHz. You should verify that $Z_{in} = (178.9 + j29.33)$ Ω. You may prefer to show complex values simply as an ordered pair; in this case $Z_{in} = (178.9, 29.33)$ Ω is the result. Compare your answers to Fig. 2.22.

A PHASE-SHIFT NETWORK

A simple phase-shifting circuit using only capacitors and resistors is shown in Fig. 2.23. This bridged-T circuit uses the following values: $C_1 = C_2 = 10$ nF, $R_1 = 200$ Ω, $R_2 = 250$ Ω, $R_L = 100$ Ω, and $R = 1$ Ω (sampling resistor). The PSpice analysis is to investigate the phase-shifting property of this network. Determine at what frequency the network will produce the maximum phase shift of current relative to input voltage and what the phase shift will be. Trial-and-error may be used to find the proper range of frequencies for this analysis. This is the input file:

```
Phase-Shift Network
V  1  0  AC 1V
R  1  1A 1
R1  1A 3 200
R2  2  0 250
RL  3  0 100
C1  1A 2  10nF
C2  2  3 10nF
.AC LIN 501 5kHz 500kHz
.PROBE
.END
```

Fig. 2.23 Phase-shift network.

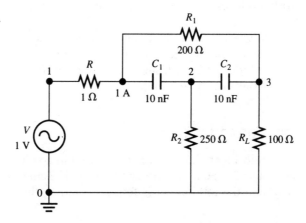

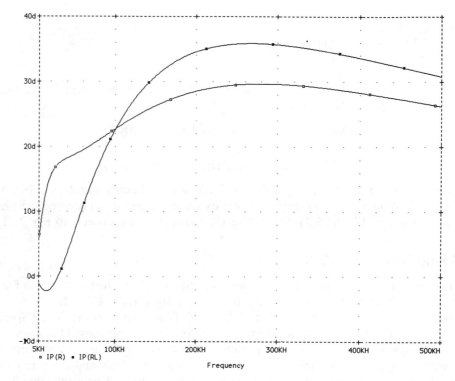

Fig. 2.24 Phase-Shift Network.

Run the PSpice analysis; then plot IP(R) using a linear frequency range of 5 kHz to 500 kHz. You can readily see that the maximum phase shift is slightly less than 30° and that this occurs at a frequency near 300 kHz. For a more accurate answer, use the cursor mode and verify that the maximum phase shift is 29.67° at $f = 284.2$ kHz.

While still in the plot mode, obtain a graph of IP(RL). This will indicate the phase shift of the bridged-T network itself. Find the frequency at which the network produces zero phase shift. Notice that this occurs at a frequency of less than 50 kHz.

Revise the input file to allow for an ac sweep beginning at 5 kHz and ending at 50 kHz. Now run the analysis and determine the frequency for zero phase shift, with a plot of IP(RL). Using the cursor, verify that $f = 29.32$ kHz for zero phase shift. Your plot should look like Fig. 2.24.

LOCUS OF ADMITTANCES

A graphical technique that is often used in ac circuit analysis is based on finding the locus of impedances or admittances. If elements are in series, impedances are used, and the total effect is found by adding the various loci. If elements are in

Fig. 2.25 Circuit for locus of admittances.

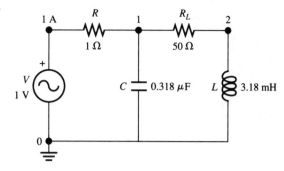

parallel, admittances are used since the total effect is again found by adding the admittances of the various branches.

The circuit of Fig. 2.25 contains two parallel branches with values $C = 0.318\ \mu F$, $R_L = 50\ \Omega$, and $L = 3.18$ mH. The input file is created using an ac sweep from 5 Hz to 10 kHz and is

```
Locus of Admittances
V 1A 0 AC 1V
R 1A 1 1
RL 1 2 50
L 2 0 3.18mH
C 1 0 0.318uF
.AC LIN 201 5Hz 10kHz
.PROBE
.END
```

Run the analysis and plot IP(R) to find the resonant frequency. Using the cursor, verify that $f_o = 3.334$ kHz.

Now change the X-axis, letting it represent the real component of admittance. Since $V = 1$ V, $Y = I/V = I/1$. Thus Y is numerically the same as the current, and you can plot I and Y interchangeably. In the Y plane, G is shown on the horizontal axis and B on the vertical axis.

Change the X-axis to represent IR(R) with a linear range. It will automatically go to 20 mA. Plot II(R), which will auto-range to 16 mA. Since I and Y are numerically the same, the axes can be thought of as the Y-plane values for G (X-axis) and B (Y-axis). The graph contains important information. The point at the top left is for $f = 10$ kHz. Moving down the graph takes you to lower frequencies where the values of B and G may be obtained. Move the cursor to the point where $B = 0$ and confirm that $G = 5$ mS. This indicates that the impedance of the network is 200 Ω at the resonant frequency of 3.334 kHz.

Obtain a printed copy of the Y-plane graph. Mark the X-axis G (in mS) and the Y-axis B (in mS) for further study. Your graph should look like Fig. 2.26. You know where two frequencies are located on this graph, but how can you identify others? A simple method involves going back to the input file and changing the top frequency in the ac sweep statement. For starters, let the top frequency be 6 kHz and run the analysis again. When you see the Y-plane values, carefully note where the graph begins (6 kHz), and transfer this to your previous graph using B and G

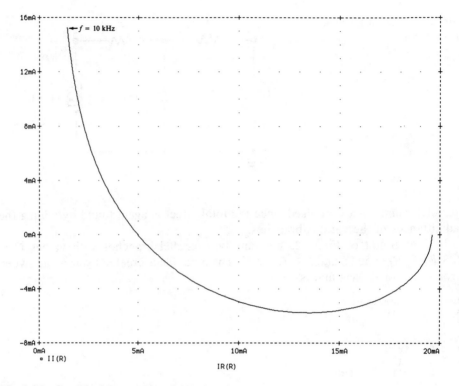

Fig. 2.26 Graphical Treatment of Admittance.

values. Try this for several other top frequencies, marking each new top value of f on the original graph.

Admittance Locus for Series *RLC*

The admittance locus for a series *RLC* circuit is interesting. Can you predict what it will look like for a sweep of frequencies? It will be a circle (if *B* and *G* are shown to the same scale). The circuit shown in Fig. 2.27 uses the values $R = 50\ \Omega$, $L =$

Fig. 2.27 Series circuit for admittance locus.

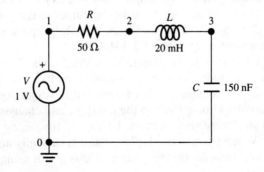

20 mH, and C = 150 nF. The input file is

```
Admittance Locus for Series RLC Circuit
V 1 0 AC 1V
R 1 2 50
L 2 3 20mH
C 3 0 150nF
.AC LIN 501 100Hz 10kHz
.PROBE
.END
```

The sweep of frequencies takes you well beyond the resonant frequency of 2906 Hz. Run the analysis and plot IR(R) on the X-axis vs. II(R) on the Y-axis. Watch carefully as the curve is drawn on the screen. Note that the curve begins on the left of the screen and moves in a clockwise direction. The first portion of the curve is drawn quickly, but the last part, completing the circle, appears to be drawn slowly. This is due to the relative locations of low and high frequencies on the graph. From the 9 o'clock position moving clockwise to the 3 o'clock position, there is a range of frequencies from 0 Hz to the resonant frequency of 2906 Hz. The entire lower half of the circle extends from the resonant frequency to the upper frequency limit.

Obtain a printed copy of the graph for further study. Note that you can make the graph look more like a true circle by careful selection of the X- and Y-ranges,

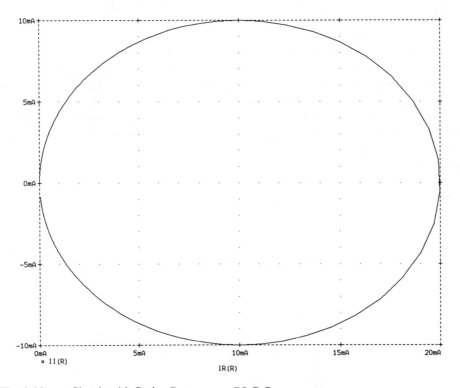

Fig. 2.28   ac Circuit with Series Resonance, RLC Components.

although this may mean losing some portion of the circle. Figure 2.28 shows this graph.

MULTIPLE SOURCES IN AC NETWORKS

When there is more than a single source in an ac network, you must specify the relative phase angles of the sources. In each statement describing a voltage source in the example of Fig. 2.29, note that the value of the voltage is shown with a magnitude followed by an angle. Thus V_2, which is 10 V at an angle of $-90°$, is shown as 10V -90. For this example, your assignment is to find the current through each of the elements C, L, and R and to find the voltage V(2). The input file is

```
AC Network with More Than One Source
V1 1 0 AC 20V 0
V2 0 4 AC 10V -90
V3 3 0 AC 40V 45
R 2 4 3
L 2 3 7.96mH
C 1 2 663uF
.AC LIN 1 60Hz 60Hz
.PRINT AC I(C) IR(C) II(C)
.PRINT AC I(L) IR(L) II(L)
.PRINT AC I(R) IR(R) II(R)
.PRINT AC V(2) VR(2) VI(2)
.END
```

Run the analysis and look at the results in the output file. What are the directions of the currents that have been found? You must look at the element statements to be sure of this. For example, the statement describing the capacitor shows the nodes as *1* and *2*, respectively. This means that on the circuit diagram you should show the current in the direction from node *1* toward node *2*. Failure to do this would leave you with an incomplete (or ambiguous) solution. Looking at the statements for *R* and *L*, assign current-direction arrows to these elements to complete the solution.

In a later section, three-phase networks will be considered. The source voltages will be specified in much the same way as in the present example.

Fig. 2.29 An ac network with more than one source.

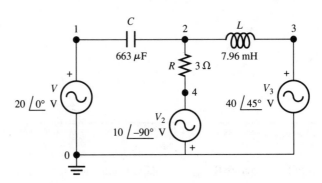

TRANSFORMERS

When using transformers in SPICE, you need a descriptive entry to show the self-inductance of each of the two windings, primary and secondary, and the coefficient of coupling k. Figure 2.30 is an example showing a voltage source of 20 V at a frequency of 1 kHz. Values for the transformer are known to be $R_1 = 20 \ \Omega$, $L_1 = 25$ mH, $R_2 = 20 \ \Omega$, $L_2 = 25$ mH, and $M = 20$ mH (mutual inductance). Find current in the primary, current in the secondary, power to the secondary, and power to the load impedance.

The coefficient of coupling may be found as

$$k = \frac{M}{\sqrt{L_1 L_2}}$$

For the example, this will be 20/25 = 0.8. With this, you are ready to create the input file.

```
Circuit with Mutual Inductance
V 1 0 AC 20V
R1 1 2 20
R2 3 4 20
L1 2 0 25mH
L2 3 0 25mH
RL 4 5 40
CL 5 0 5.3uF
K L1 L2 0.8
.AC LIN 1 1kHz 1kHz
.PRINT AC I(R1) IR(R1) II(R1)
.PRINT AC I(R2) IR(R2) II(R2)
.END
```

Run the analysis and obtain a printed copy of the output file. Verify that the primary current is (0.1767, −0.1441) A and that the secondary current is (0.1979, −0.04904) A. Note that these answers are simply given as ordered pairs for the real and imaginary components. The results do not give the desired powers directly, since the .PRINT statements cannot contain expressions such as $I*I*R$. Use the calculator to verify that the power to the secondary is 2.49 W and that the power to the load impedance is 1.66 W.

Fig. 2.30 Circuit with mutual inductance.

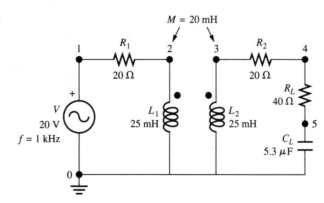

The dot convention is important in some transformer situations. In this example the dots are assumed to be at node *2* for L_1 and node *3* for L_2. The dots are thus assumed to be at the first-named node for each L.

FREQUENCY RESPONSE OF TUNED CIRCUIT

Tuned circuits are used in various electronic circuits such as those found in radio and television sets. Capacitors are placed in shunt with the transformer windings to create resonant conditions. At and near the resonant frequency, the power delivered to the secondary is large, but away from the resonant frequency, little power is delivered. Figure 2.31 shows a typical circuit fed from an ac current source: $I = 19.6$ mA, $R_1 = R_2 = 1\,\Omega$, $L_1 = L_2 = 25$ mH, $C_1 = C_2 = 1.013\,\mu\text{F}$, $R_L = 5\,\text{k}\Omega$, and $k = 0.05$. The LC combination produces a resonant frequency $f_o = 1$ kHz.

You are interested in investigating the behavior of the circuit at a range of frequencies near the resonant frequency. This is done by using the input file shown here.

```
Frequency Response of Tuned Circuit with Mutual Inductance
I 0 1 AC 19.6mA
C1 1 0 1.013uF
C2 3 0 1.013uF
R1 1 2 1
R2 3 4 1
L1 2 0 25mH
L2 4 0 25mH
K L1 L2 0.05
RL 3 0 5k
.AC LIN 401 800Hz 1200Hz
.PROBE
.END
```

The frequency sweep will extend from 800 Hz to 1200 Hz. Run the analysis and plot V(3) over the desired range of frequencies. Use a linear *X*-axis. Examine the shape of the voltage across the load. Notice that it rises on either side of the

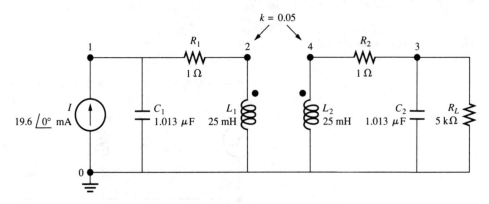

Fig. 2.31   Tuned circuit with mutual inductance.

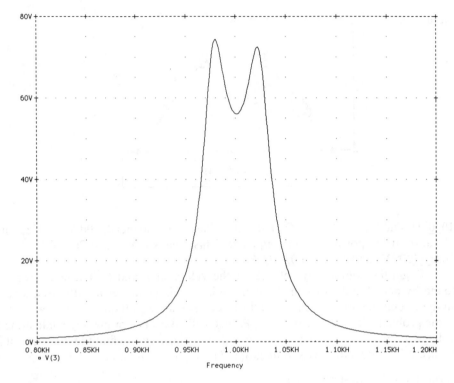

Fig. 2.32 Frequency Response of Tuned Circuit with Mutual Inductance.

resonant frequency, dipping at f_o. Obtain a printed copy of this graph for further study. Figure 2.32 shows the graph.

The value of $k = 0.05$, which was specified for this example, produced overcoupling. This gives a usable band of frequencies that are passed with little attenuation. For frequencies where the voltage level falls below 0.7 of its peak value, we have moved outside the usable bandwidth. Can you determine the bandwidth for this tuned circuit?

In the input file, change the value of k to 0.03. Run the analysis and note that the peak is higher and the bandwidth is smaller. The peak voltage to the load will be achieved at a value of k giving critical coupling. For this circuit this occurs with $k = 0.0155$. What is the peak value of V(3) for this value of coupling?

THREE-PHASE AC CIRCUITS

Three-phase ac circuits may be treated as though they were single-phase circuits if the load is equally balanced among the phases. When the load is unbalanced, the solution becomes more tedious. This example will show the method of solution for the unbalanced case.

Figure 2.33 shows an unbalanced three-phase load. In the statement of the problem the delta-connected impedances are given as $Z_{ab} = 25\underline{/40°}$ Ω, $Z_{bc} =$

Fig. 2.33 Unbalanced three-phase load.

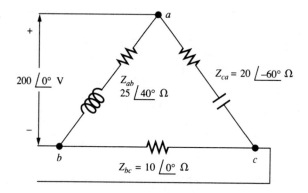

$10/0°$ Ω, and $Z_{ca} = 20/-60°$ Ω. The line voltages are balanced, 200 V, 60 Hz, with V_{ab} at 0°, using positive phase sequence. This means that $V_{ab} = 200/0°$ V, $V_{bc} = 200/-120°$ V, and $V_{ca} = 200/120°$ V.

Begin the solution by solving for the values of L and C. These are readily found by pencil-and-paper methods using known impedances and the frequency. In the circuit diagram, include small source resistances although none was stated in the problem. If these are omitted, PSpice will give an error message indicating a voltage loop. The line resistances are added to allow you to find the line currents. Figure 2.34 shows the revised diagram. The input file then becomes

```
Three-Phase Unbalanced Load
VAB 12 2 AC 200V 0
VBC 20 0 AC 200V -120
VCA 10 1 AC 200V 120
RS1 12 1 0.01
RS2 20 2 0.01
RS3 10 0 0.01
RA 1 3 0.01
```

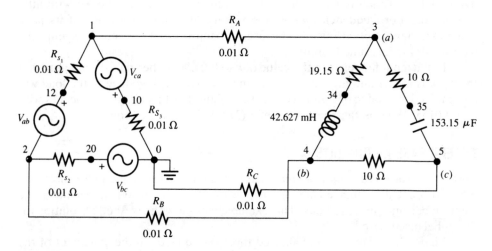

Fig. 2.34 Unbalanced three-phase load revised.

```
RB 2 4 0.01
RC 0 5 0.01
RAB 3 34 19.15
LAB 34 4 42.627mH
RBC 4 5 10
RCA 3 34 10
CCA 35 5 153.15uF
.AC LIN 1 60Hz 60Hz
.PRINT AC I(RA)  IR(RA)  II(RA)
.PRINT AC I(RB)  IR(RB)  II(RB)
.PRINT AC I(RC)  IR(RC)  II(RC)
.END
```

Run the PSpice analysis and verify that I(RA) = (16.09, −5.136) A, I(RB) = (−16.09, −12.15) A, and I(RC) = (−0.003, 17.28) A. Show the reference directions for each of the currents on your circuit diagram, for the solution is incomplete without these. As a check, add the line currents to see that they total zero. Expect slight round-off errors in the results.

POWER-FACTOR IMPROVEMENT

An induction motor requires a larger-than-necessary current when used without a capacitor. For example, if a 5-hp induction motor takes 53 A at 117 V, operating single phase with an efficiency of 78.5%, let us first use the calculator in preparation for an analysis. The power input to the motor is

$$P_{in} = \eta/P_{out} = \frac{(5)(746)}{0.785} = 4.75 \text{ kW}$$

The volt-ampere product will be

$$S = VI = (117)(53) = 6.2 \text{ kVA}$$

With P and S known, Q is found as the missing side of the volt-ampere triangle:

$$S = P + jQ$$

giving

$$Q = 3.985 \text{ kVA}$$

The resistance of the motor is

$$R = \frac{V^2}{P} = \frac{(117)^2}{4750} = 2.88 \text{ } \Omega$$

and the reactance of the motor is

$$X = \frac{V^2}{Q} = \frac{(117)^2}{3985} = 3.44 \text{ } \Omega$$

At a frequency $f = 60$ Hz, this represents an inductance of

$$L = \frac{X}{\omega} = 9.12 \text{ mH}$$

Since the problem has been analyzed with pencil-and-paper techniques, it might appear that there is little left to do with PSpice. However, in order to see the effects of placing various capacitors across the line, the computer analysis will be helpful. Refer to Fig. 2.35, which shows the resistance R and the inductance L of the motor, along with a pair of sensing resistors, R_A and R_B. Their roles will become apparent when a capacitor is added.

An input file is needed that will show the total current and the branch currents in relation to the applied voltage. This file is

```
Single-Phase Motor, 5 hp
V 1 0 ac 117V
RA 1 2 0.001
RB 2 3 0.001
R 3 0 2.88
L 3 0 9.12mH
.ac LIN 1 60Hz 60Hz
.PRINT ac i(RA) ip(RA) i(RB) ip(RB)
.PRINT ac i(R) ip(R)
.PRINT ac I(L) ip(L)
.END
```

The PSpice output file will show the following currents:

$$I(RA) = 5.296E+01, \ IP(RA) = -3.992E+01$$
$$I(RB) = 5.296E+01, \ IP(RB) = -3.992E+01$$
$$I(R) = 4.060E+01, \ IP(R) = 3.331E-02$$
$$I(L) = 3.401E+01, \ IP(L) = -8.997E+01$$

The current I(RA) is the line current, in close agreement with the stated current of 53 A, at an angle near $-40°$. The power factor (pf) is found as the cosine of the angle of line current with respect to line voltage, so

$$pf = \cos(-40°) = 0.76$$

The two branch currents, through the motor resistance and inductance, have a phasor sum equal to the line current. It is now easy to show the effect of placing a capacitor across the line, which will be from node 2 to node 0. Add a statement to the input file

```
C 2 0 380uF
```

Fig. 2.35 Circuit for power-factor improvement.

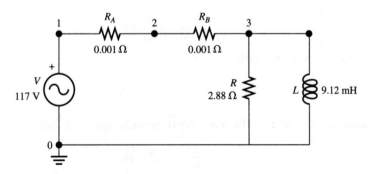

and modify one of the print statements to include the current through the capacitor. Now run the analysis again. The output file will show the following:

$$I(RA) = 4.411E+01, \ IP(RA) = -2.299E+01$$
$$I(RB) = 5.296E+01, \ IP(RB) = -3.993E+01$$
$$I(C) = 1.676E+01, \ IP(C) = 9.001E+01$$
$$I(R) = 4.060E+01, \ IP(R) = -2.510E-01$$
$$I(L) = 3.401E+01, \ IP(L) = -8.997E+01$$

Observe that the line current I(RA) has been reduced to 44.11 A at a lagging angle near 23°, clearly showing the effects of power-factor improvement. The power factor is now

$$pf = \cos(-23°) = 0.92$$

The capacitor draws a current of 16.76 A at an angle of 90°, bringing about the change in line current. Note that the current through the sensing resistor R_B is the same as the line current before the capacitor was added, as we would expect.

THREE-PHASE POWER-FACTOR IMPROVEMENT

A three-phase motor is represented by the components shown on the right side of Fig. 2.36, with a delta connection assumed. R_1 and L_1 are the per-phase resistance and inductance, respectively, of the motor. The other phases have the same values of components. Line-voltage dropping resistors are included in each of the three-phase lines. An input file is prepared which will show the various voltages and currents. It is shown along with the output in Fig. 2.37.

The order of the subscripts in each statement must be given careful attention. For each passive element the subscripts are in keeping with the current directions shown in Fig. 2.36. A phasor diagram of currents and voltages is shown

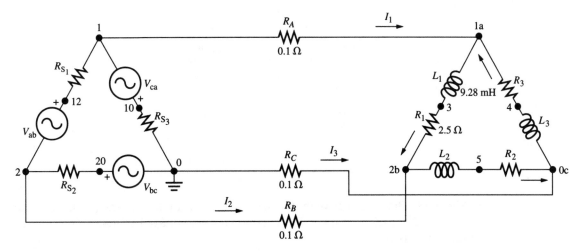

Fig. 2.36 A delta-connected, three-phase motor.

Fig. 2.37

```
Circuit for Power-Factor Correction

VAB 12 2 AC 240V 0
VBC 20 0 AC 240V -120
VCA 10 1 AC 240V 120
RS1 12 1 0.01
RS2 20 2 0.01
RS3 10 0 0.01
RA 1 1a 0.1
RB 2 2b 0.1
RC 0 0c 0.1
R1 3 2b 2.5
R2 5 0c 2.5
R3 4 1a 2.5
L1 1a 3 9.28mH
L2 2b 5 9.28mH
L3 0c 4 9.28mH
.AC LIN 1 60Hz 60Hz
.PRINT AC I(RA) IP(RA)
.PRINT AC I(RB) IP(RB)
.PRINT AC I(RC) IP(RC)
.PRINT AC V(1a,2b) VP(1a,2b)
.PRINT AC V(2b,0c) VP(2b,0c)
.PRINT AC V(0c,1a) VP(0c,1a)
.PRINT AC I(R1) IP(R1)
.PRINT AC I(R2) IP(R2)
.PRINT AC I(R3) IP(R3)
.OPT nopage
.END

    FREQ         I(RA)        IP(RA)
    6.000E+01    9.264E+01    -8.123E+01

    FREQ         I(RB)        IP(RB)
    6.000E+01    9.264E+01    1.588E+02

    FREQ         I(RC)        IP(RC)
    6.000E+01    9.264E+01    3.877E+01

    FREQ         V(1a,2b)     VP(1a,2b)
    6.000E+01    2.300E+02    3.222E+00

    FREQ         V(2b,0c)     VP(2b,0c)
    6.000E+01    2.300E+02    -1.168E+02

    FREQ         V(0c,1a)     VP(0c,1a)
    6.000E+01    2.300E+02    1.232E+02

    FREQ         I(R1)        IP(R1)
    6.000E+01    5.348E+01    -5.123E+01

    FREQ         I(R2)        IP(R2)
    6.000E+01    5.348E+01    -1.712E+02

    FREQ         I(R3)        IP(R3)
    6.000E+01    5.348E+01    6.877E+01
```

in Fig. 2.38. The angle between the phase voltage V(1a,2b) and the phase current I(R1) is 3.22 + 51.23 = 54.45°. The current lags the voltage by 54.45°. The cosine of this angle is the power factor

$$pf = \cos(-54.45°) = 0.581$$

Fig. 2.38 Phasor diagram of current and voltage.

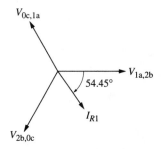

It is our desire to improve the power factor by adding a bank of capacitors as shown in Fig. 2.39. The input file is modified to show the presence of the capacitors. When the PSpice analysis is run, the results will be as shown in Fig. 2.40.

Observe that each of the line currents is 75.51 A compared to 92.64 A before the addition of the capacitor bank. The reduction in current is accompanied by an improvement in the power factor. The power-factor angle will be found as the angle between phase voltage and phase current, as was the case before the addition of the capacitors. The phase voltage will be taken as $V(1a,2b) = 230\underline{/2.26°}$ V. The phase current is found (indirectly) from the current $I(RA) = 75.52\underline{/-72.2°}$ A. Since this is a line current, the corresponding phase current has a magnitude

$$I_f = \frac{75.52}{\sqrt{3}} = 43.6 \text{ A}$$

at an angle of $-42.2°$. This angle is obtained by adding 30° to the angle on the line current. Both the magnitude and the angle values are based on the presence of a balanced load. The power factor angle is $2.26° + 42.2° = 44.46°$, and the power factor is

$$pf = \cos(-44.46°) = 0.71$$

An alternate approach to finding the phase current, which would apply to unbalanced loads as well, is to add the current in one phase of the load to the corresponding current in a capacitor. Thus by adding $I(R1)$ and $I(C1)$, we obtain

$I(R1) + I(C1) = 53.53\underline{/-52.19°} < \text{A} + 13.02\underline{/92.226°} \text{ A} = 43.6\underline{/-42.18°} \text{ A}$

Fig. 2.39 Motor circuit with capacitor bank added.

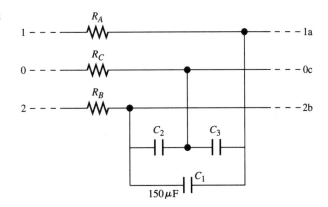

```
Circuit for Power-Factor Correction with Capacitor Bank Added
VAB 12 2 AC 240V 0
VBC 20 0 AC 240V -120
VCA 10 1 AC 240V 120
RS1 12 1 0.01
RS2 20 2 0.01
RS3 10 0 0.01
RA 1 1a 0.1
RB 2 2b 0.1
RC 0 0c 0.1
R1 3 2b 2.5
R2 5 0c 2.5
R3 4 1a 2.5
L1 1a 3 9.28mH
L2 2b 5 9.28mH
L3 0c 4 9.28mH
C1 1a 2b 150uF
C2 2b 0c 150uF
C3 0c 1a 150uF
.AC LIN 1 60Hz 60Hz
.PRINT AC I(RA) IP(RA) I(C1) IP(C1)
.PRINT AC I(RB) IP(RB) I(C2) IP(C2)
.PRINT AC I(RC) IP(RC) I(C3) IP(C3)
.PRINT AC V(1a,2b) VP(1a,2b)
.PRINT AC V(2b,0c) VP(2b,0c)
.PRINT AC V(0c,1a) VP(0c,1a)
.PRINT AC I(R1) IP(R1)
.PRINT AC I(R2) IP(R2)
.PRINT AC I(R3) IP(R3)
.OPT nopage
.END

   FREQ          I(RA)        IP(RA)       I(C1)       IP(C1)
   6.000E+01     7.552E+01   -7.220E+01    1.302E+01    9.226E+01

   FREQ          I(RB)        IP(RB)       I(C2)       IP(C2)
   6.000E+01     7.552E+01    1.678E+02    1.302E+01   -2.774E+01

   FREQ          I(RC)        IP(RC)       I(C3)       IP(C3)
   6.000E+01     7.552E+01    4.780E+01    1.302E+01   -1.477E+02

   FREQ          V(1a,2b)     VP(1a,2b)
   6.000E+01     2.302E+02    2.260E+00

   FREQ          V(2b,0c)     VP(2b,0c)
   6.000E+01     2.302E+02   -1.177E+02

   FREQ          V(0c,1a)     VP(0c,1a)
   6.000E+01     2.302E+02    1.223E+02

   FREQ          I(R1)        IP(R1)
   6.000E+01     5.353E+01   -5.219E+01

   FREQ          I(R2)        IP(R2)
   6.000E+01     5.353E+01   -1.722E+02

   FREQ          I(R3)        IP(R3)
   6.000E+01     5.353E+01    6.781E+01
```

Fig. 2.40

in agreement with the previous calculation. Before the capacitors were added, the power factor was 0.58.

If it is necessary, the PSpice analysis can readily be performed with other values of capacitance for comparison.

A THREE-PHASE RECTIFIER

A three-phase Y-connected rectifier is shown in Fig. 2.41. Each phase voltage has a peak value of 10 V at 60 Hz. The diodes allow current to reach the load without falling to a zero value at any time. The input file is

```
Three-Phase Rectifier
v1 1 0 sin(0 10V 60Hz 0 0 0)
v2 2 0 sin(0 10V 60Hz 0 0 −120)
v3 3 0 sin(0 10V 60Hz 0 0 120)
DA 1 4 D1
DB 2 4 D1
DC 3 4 D1
.MODEL D1 D
.TRAN 0.1ms 33.33ms
.PROBE
.END
```

Run the analysis and verify the results shown in Fig. 2.42. Also display the load current and verify that the current varies between a minimum value of 43.528 mA and a maximum value of 92.278 mA.

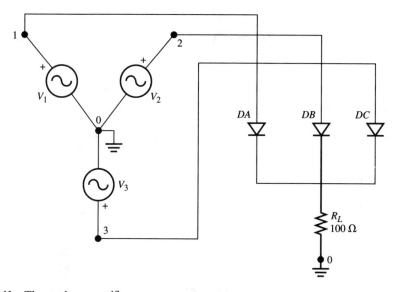

Fig. 2.41   Three-phase rectifier.

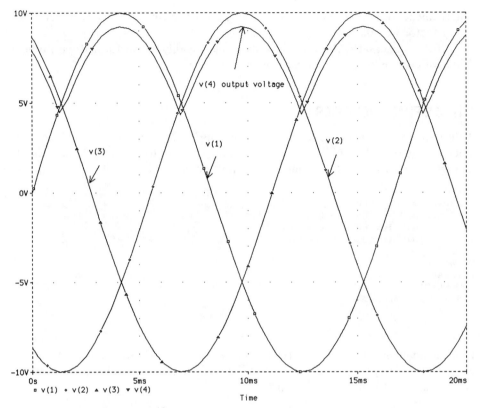

Fig. 2.42 Three-Phase Rectifier.

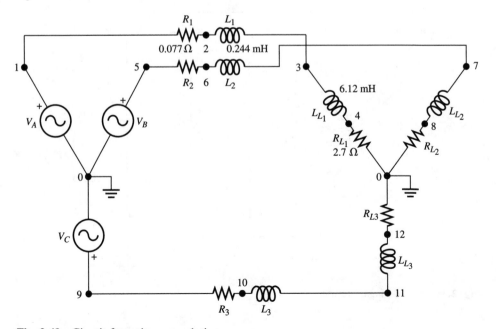

Fig. 2.43 Circuit for voltage regulation.

VOLTAGE REGULATION IN A THREE-PHASE SYSTEM

A power feeder line must be designed to allow for no greater voltage drop between the source and load than an allowed amount. Often the allowed amount must provide a voltage regulation of 5% or less. The circuit of Fig. 2.43 will be used to illustrate voltage-regulation requirements. Observe that each line contains both resistance and inductance. The Y-connected load represents a 440-V, three-phase, 60-Hz motor. The source voltage is given as 460 V, from which the phase voltage is obtained:

$$V_A = \frac{460}{\sqrt{3}} = 265.58 \text{ V}$$

The input file should require no further explanation. It is shown in Fig. 2.44 with the results of the analysis. The voltage regulation is

$$\text{Voltage regulation} = \frac{V_{NL} - V_{FL}}{V_{FL}} = \frac{265.58 - 257}{257} = 3.34\%$$

```
Voltage Regulation for Three-Phase Load

  ****        CIRCUIT DESCRIPTION

VA 1 0 AC 265.58V 0
VB 5 0 AC 265.58V -120
VC 9 0 AC 265.58V 120
R1 1 2 0.077
R2 5 6 0.077
R3 9 10 0.077
L1 2 3 0.244mH
L2 6 7 0.244mH
L3 10 11 0.244mH
RL1 4 0 2.7
RL2 8 0 2.7
RL3 12 0 2.7
LL1 3 4 6.12mH
LL2 7 8 6.12mH
LL3 11 12 6.12mH
.AC LIN 1 60Hz 60Hz
.PRINT AC I(R1) IP(R1) I(R2) IP(R2)
.PRINT AC I(R3) IP(R3)
.PRINT AC V(3) VP(3) V(7) VP(7)
.PRINT AC V(11) VP(11)
.OPT nopage
.END

  ****      AC ANALYSIS                    TEMPERATURE =   27.000 DEG C

   FREQ        I(R1)      IP(R1)      I(R2)       IP(R2)
   6.000E+01   7.237E+01  -4.083E+01  7.237E+01  -1.608E+02

   FREQ        I(R3)      IP(R3)
   6.000E+01   7.237E+01   7.917E+01

   FREQ        V(3)       VP(3)       V(7)        VP(7)
   6.000E+01   2.570E+02  -3.108E-01  2.570E+02  -1.203E+02

   FREQ        V(11)      VP(11)
   6.000E+01   2.570E+02   1.197E+02
```

Fig. 2.44

Fig. 2.45   A two-phase circuit.

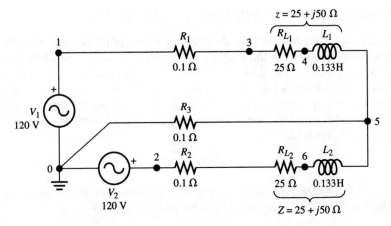

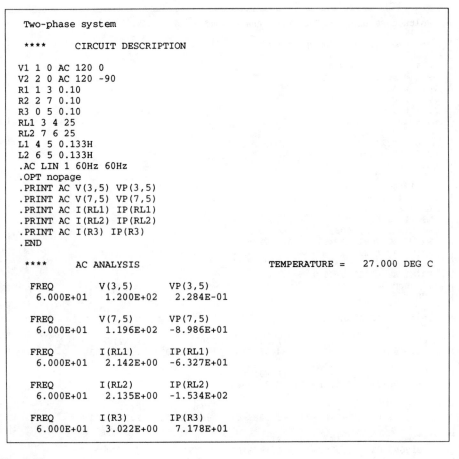

Fig. 2.46

A TWO-PHASE SYSTEM

A two-phase electrical system is little more than a curiosity, but its analysis can be readily undertaken using PSpice. The diagram in Fig. 2.45 shows such a system, where the load is shown in terms of equal impedances of $Z = (25 + j50)$ Ω for each phase. At a frequency of 60 Hz, the reactance of 50 Ω becomes an inductance $L = 0.133$ H. The preparation of the input file is similar to that of other examples previously shown. It is included in Fig. 2.46, which shows the output voltages and currents. The phase voltages at the load are almost the same value (120 V) and are about 90° apart. The line currents I(RL1) and I(RL2) are almost the same value (2.15 A) and are also about 90° apart. Note that I(RL1) is at an angle of $-63.27°$, which is the same as the angle on the impedance of the load. The neutral current I(R3) is greater than the two line currents.

$$Z = R + jX_L = 25 + j50 = 55.9\underline{/63.4°}\ \Omega$$

```
Two-phase system

****        CIRCUIT DESCRIPTION

V1 1 0 AC 120 0
V2 2 0 AC 120 -90
R1 1 3 10
R2 2 7 10
R3 0 5 10
RL1 3 4 25
RL2 7 6 25
L1 4 5 0.133H
L2 6 5 0.133H
.AC LIN 1 60Hz 60Hz
.OPT nopage
.PRINT AC V(3,5) VP(3,5)
.PRINT AC V(7,5) VP(7,5)
.PRINT AC I(RL1) IP(RL1)
.PRINT AC I(RL2) IP(RL2)
.PRINT AC I(R3) IP(R3)
.END

****      AC ANALYSIS                          TEMPERATURE =    27.000 DEG C

    FREQ        V(3,5)        VP(3,5)
    6.000E+01    1.110E+02    1.926E+01

    FREQ        V(7,5)        VP(7,5)
    6.000E+01    8.909E+01    -8.220E+01

    FREQ        I(RL1)        IP(RL1)
    6.000E+01    1.981E+00    -4.424E+01

    FREQ        I(RL2)        IP(RL2)
    6.000E+01    1.590E+00    -1.457E+02

    FREQ        I(R3)         IP(R3)
    6.000E+01    2.280E+00    9.265E+01
```

Fig. 2.47

Draw a phasor diagram showing the phase voltages at the load and each of the three line currents.

It is interesting to see what happens when the resistance of each line increases. Let us use 10-Ω values for R1, R2, and R3 and run the analysis again. This output file is shown in Fig. 2.47. Note that V(3,5) = 111$\underline{/19.3°}$ and V(3,7) = 89.1$\underline{/-82.2°}$. The voltages are now unbalanced and are 105.5° apart. The line currents are also unbalanced, and again the current in the neutral (ground) is larger than either of the other two currents.

FIGURE 2.19 REVISITED

In preparing for the second edition, the author ran the plot of Fig. 2.19 using version 5.1 of PSpice and Probe. The results were not the same as those shown in the text! The trace of reactance, which should have given negative values, gave positive values instead. This means that the plot was not properly responding to the minus sign for the second trace. After several trials-and-error, it was discovered that the term I(R), appearing in the denominator of each expression, could be changed to IM(R) and the results would be as expected. This version of the traces is shown in Fig. 2.19(a).

To shed more light on the problem, return to the example of Fig. 2.16, which shows resistance and inductive reactance in series. A plot of the circuit current can readily be obtained in Probe by using I(R) for the trace. Refer to Fig. 2.19(b)

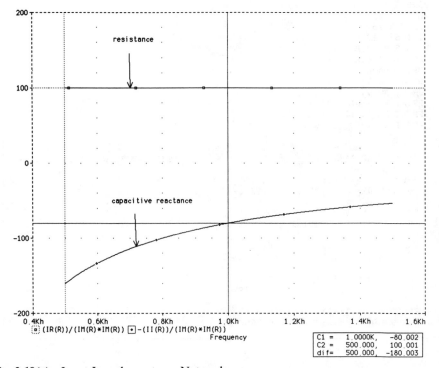

Fig. 2.19(a)   Input Impedance to ac Network.

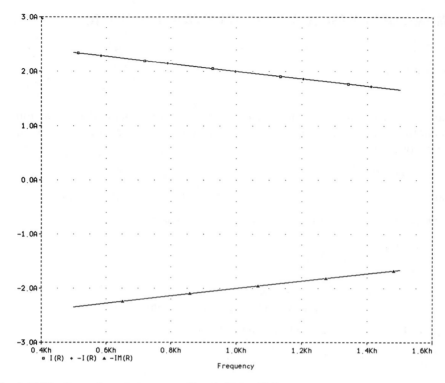

Fig. 2.19(b)   Input Impedance to ac Circuit Using V/I.

and note that at $f = 1$ kHz, the current has a magnitude of 2.0 A. Here the applied voltage was 10 V, and the impedance had a magnitude of $(4 + j3)$ Ω $= 5$ Ω. But when a trace of $-\mathrm{I(R)}$ is called for, the unexpected results are that the two expressions yield the same trace! Probe simply ignores the minus sign.

Now if a trace for $-\mathrm{IM(R)}$ is obtained, the minus sign is not ignored, and the trace is negative as expected. We conclude that the Probe version 5.1 program has changed in this regard compared to Probe version 4.1, which was used in the first edition. According to the official MicroSim *Analysis–Reference Manual*, p. 87, for ac analysis the suffix *none* gives a magnitude, and the suffix *M* also gives a magnitude. If there is a difference between the two, it is not mentioned. A complete list of Probe *Add_trace* commands from the *Analysis–Reference Manual* follows:

Suffix	Meaning
(none)	Magnitude
M	Magnitude
DB	Magnitude in decibels
P	Phase
G	Group delay ($-d$PHASE/dFREQUENCY)
R	Real part
I	Imaginary part

SUMMARY OF NEW PSPICE STATEMENTS USED IN THIS CHAPTER

C[*name*] <+*node*> <−*node*> <*value*>

For example,

```
C 4 5 0.5μF
```

means that a capacitor of value 0.5 μF is located between nodes *4* and *5*. Another form of the C statement adds *IC = value* at the end of the line for an initial voltage value. For example,

```
c 4 5 0.5μF IC=3V
```

means that the capacitor has an initial voltage of 3 V, with node *4* positive.

I[*name*] <+*node*> <−*node*> AC <*mag*> [<*phase*>]

For example,

```
IS 1 2 AC 0.35 45
```

means an ac source of 350 mA between nodes *1* and *2* at a phase angle of 45°. Remember that currents and voltages default to dc unless otherwise shown.

K[*name*] L[*name*] L[*name*] <*coupling value*>

For example,

```
K L1 L2 0.1
```

means a coupled circuit, perhaps a transformer, with two coupled inductors L1 and L2. The coefficient of coupling is $k = 0.1$. Another form of this statement that relates to a coupled circuit with an iron core will be introduced later.

L[*name*] <+*node*> <−*node*> <*value*>

For example,

```
L1 3 0 25mH
```

means an inductor of 25 mH between nodes *3* and *0*. To show an initial current, use *IC = value* at the end of the line.

V[*name*] <+*node*> <−*node*> AC <*mag*> [*phase*]

For example,

```
V2 4 1 AC 110 120
```

means an ac source of 110 V between nodes *4* and *1* at a phase angle of 120°.

DOT COMMANDS USED IN THIS CHAPTER

.AC [LIN] [OCT] [DEC] <*points*> <*f start*> <*f end*>

For example,

```
.AC DEC 20 1kHz 1MEG
```

means that PSpice will run an analysis with frequency as a variable. The frequency range is from 1 kHz to 1 MHz using 20 points per decade. If LIN is chosen, the *points* value represents the total number of points in the frequency range.

.PRINT <[DC] [AC] [NOISE] [TRAN]> *<output variable list>*

For example,

```
.PRINT AC V(2) V(5,4) VP(5,4) I(R1) IP(R1)
```

means to print to the output file the ac values shown. V(2) will give the magnitude of V_2; V(5,4) will give the magnitude of V_{54}; I(R1) will give the magnitude of the current through resistor R_1; and IP (R1) will give the phase angle of the current through R_1.

Note that one of the items (and only one) in the list DC, AC, NOISE, and TRAN must be chosen.

.PROBE

This statement was introduced in Chapter 1 but is being repeated here in more detail. When this statement is included in the input file and you run PSpice with the command *PSPICE fname*, after the calculations in PSpice have been completed and the output file created, you will be taken directly into the post-processor Probe.

The screen contains the successive messages, Checking data file, Loading data file, and Setting up data structures. Then a rectangular box will appear on the screen. In the box is the message, All voltages and currents are available. The *X*-axis is marked in terms of the variable that you have specified when creating the input file. For example, if you have chosen to run a frequency analysis using something like

```
.AC DEC 20 1kHz 1MEG
```

the *X*-axis will show this range of frequencies.

At the bottom of the screen you will see the following:

```
Exit Add_trace X_axis Y_axis Plot_control Display_control Macros Hard_copy
```

This list may vary depending on which version of PSpice you are using.

Now you will choose *Add_trace* by pressing the *a* key. Then decide which parameter you want to plot. For example, if you want to look at the voltage at node 5, type *V(5)* to produce the trace. Then you can modify the *X*-axis by selecting another range of frequencies or by changing from a log plot to a linear plot. You might decide to expand the *Y*-axis to allow for a better trace of a portion of the graph. You may elect to use the cursor mode by first choosing *X_axis*.

Perhaps the best way to learn more about the Probe post-processor is to experiment. Regardless of what you do in Probe, you will not modify the results of your analysis. You can always undo or modify a plot that does not suit your purposes.

When you have obtained a plot worth keeping, choose *Hard_copy* and get a printed version of what you see on the screen. Usually you will choose the page length of one page. This copy will also display the date and time of the run, the title line (which was your first line of the input file), and the temperature.

Refer to the description of the .PROBE statement in Chapter 1 for details of using arithmetic operators and functions in your plots.

As you work through the many examples in the text, you will be given specific guidelines for producing the desired traces in Probe. As you work the problems at the end of the chapters, it will often be your choice as to precisely what the trace will produce.

PROBLEMS

2.1 Find the equivalent impedance of the circuit shown in Fig. 2.48 as seen by the source. Since the inductive and capacitive values are given in ohms, use the frequency $f = 5$ kHz to solve for L and C values needed in the input file. Check your results using conventional circuit-solving methods.

Fig. 2.48

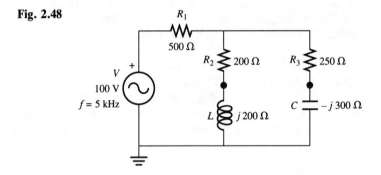

2.2 The circuit shown in Fig. 2.49 has a low Q. Find the resonant frequency by using a sweep of frequencies in the range from 3 kHz to 6 kHz. Verify that $f_o = 3.56$ kHz. Find the current at resonance and the minimum current. At what frequency does minimum current occur?

Fig. 2.49

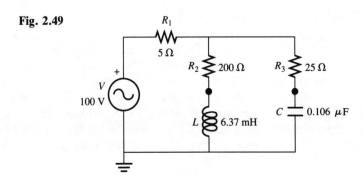

2.3   Rework Problem 2.2 when $R_2 = 20 \ \Omega$.

2.4   This problem investigates the voltage variations across R, L, and C in the vicinity of resonance. With the elements shown in Fig. 2.50, $f_o = 159.15$ Hz. Prepare an input file to obtain plots of V_R, V_L, and V_C for a frequency range from 10 Hz to 300 Hz. Demonstrate that V_R maximum is at f_o, while V_{Lmax} is below f_o, and V_{Cmax} is above f_o.

Fig. 2.50

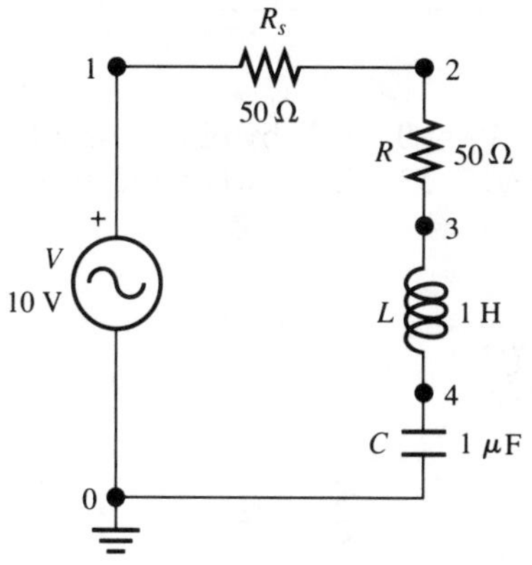

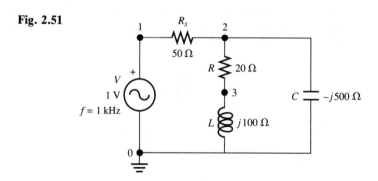

2.5   For the circuit shown in Fig. 2.51 find the impedance as seen by the source at $f = 1$ kHz.

Fig. 2.51

2.6   The circuit of Fig. 2.52 is to be used to obtain a plot of locus of admittances for a typical two-parallel-branch circuit. It is similar to one used in an example in this chapter. Run a Probe analysis and plot IP(R) to determine the resonant frequency. Then produce an admittance plot and find the value of G and B at resonance.

Fig. 2.52

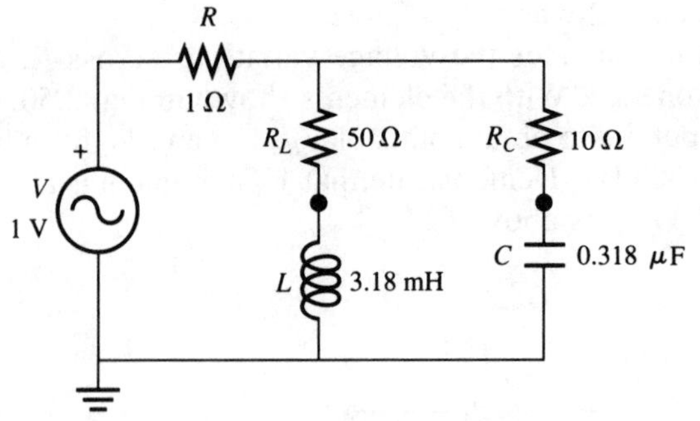

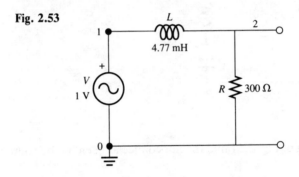

2.7 Find the frequency at which $V_2 = 0.707$ V by obtaining a Bode plot of V_2/V_1 for the circuit shown in Fig. 2.53. Show what the phase shift is at this frequency.

Fig. 2.53

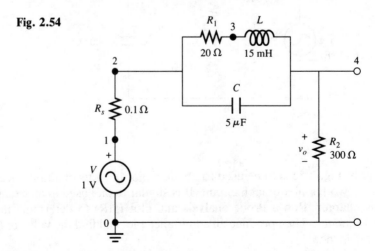

2.8 Find the frequency at which the output voltage is minimum, and find the value of the output voltage at this frequency (magnitude and phase). Find the stop band of frequencies, representing the frequency range where the output is down by 3 dB or more. Refer to Fig. 2.54.

Fig. 2.54

2.9   This is an example of a double-resonant filter. It has a pass-band frequency at 150 kHz. For the circuit of Fig. 2.55 obtain plots that show the details of the output voltage magnitude and phase in the region of interest.

Fig. 2.55

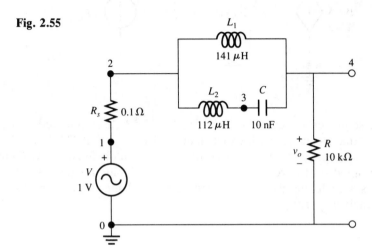

2.10 For the circuit of Fig. 2.56 find i_1 and i_2. Also find V_{40}. Hint: Since reactance values cannot be used directly in SPICE, assume that $\omega = 1000$ rad/s and solve for L and C.

Fig. 2.56

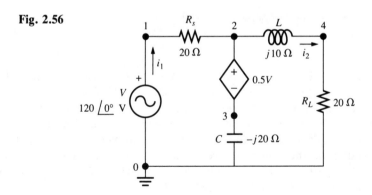

2.11 To check the answers found in Problem 2.10, find V_{20}; then use the voltage across L to find the current i_2. Compare this with the results previously obtained.

2.12 For the circuit shown in Fig. 2.57 find i and V_2. By converting the current sources to voltage sources, use conventional pencil-and-paper methods to verify your results.

Fig. 2.57

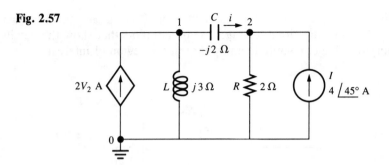

2.13 A phase-sequence indicator is shown in Fig. 2.58. It is assumed that $f = 60$ Hz and that R_1 and R_2 are identical lamps. Given $V_{12} = 100\underline{/0°}$ V and $V_{23} = 100\underline{/-120°}$ V, verify that the phase sequence (which obviously is ABC) may be determined by the relative brightness of the lamps R_1 and R_2. The sequence is said to be given as capacitor, bright lamp, dim lamp. From a PSpice analysis verify this.

Fig. 2.58

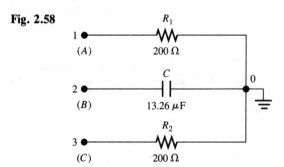

2.14 A three-phase, 60-Hz, unbalanced load is Y-connected and is fed from balanced supply voltages $V_{ab} = 208\underline{/0°}$ V, $V_{bc} = 208\underline{/-120°}$ V, and $V_{ca} = 208\underline{/120°}$ V. The phase impedances are $Z_{a0} = 8\underline{/30°}$ Ω, $Z_{b0} = 4\underline{/-50°}$ Ω, and $Z_{c0} = 6\underline{/20°}$ Ω. Find the three line currents and the neutral current. Hint: Convert the impedances to R and X values; then convert each X to either L or C depending on the sign of the reactance. Verify that phase a has $R = 6.928$ Ω and $L = 10.61$ mH, phase b has $R = 2.571$ Ω and $C = 865.7$ μF, and phase c has $R = 5.638$ Ω and $L = 5.433$ mH.

3

Transistor Circuits

SPICE has built-in models for bipolar-junction transistors and field-effect transistors. These models are more complicated than the models used in introductory electronics courses. It is customary to study biasing circuits and amplifier circuits separately. This is done in order to give the student a more complete understanding of the dc and ac analysis of BJTs and FETs.

The analysis of transistor circuits will be more meaningful if the built-in models are not used in the beginning. Therefore a simplified model for the dc forward-biased transistor will be used when it is appropriate.

THE BIPOLAR-JUNCTION TRANSISTOR (BJT)

Bipolar-junction transistors will be the first topic of study in this chapter. As an example, Fig. 3.1 shows a typical transistor biasing circuit. The transistor is type *npn* silicon, with $h_{FE} = 80$ and $V_{BE} = 0.7$ V (assumed for the active region). No other information is known about the transistor. Circuit values are $R_1 = 40$ kΩ, $R_2 = 5$ kΩ, $R_C = 1$ kΩ, $R_E = 100$ Ω, and $V_{CC} = 12$ V.

A Model Suitable for Bias Calculations

In order to use a SPICE analysis, we propose that you devise a suitable model for the bipolar-junction transistor. This model will allow you to find the quiescent values of voltages and currents in the bias circuit. Figure 3.2 shows the model along with the other components that you will need for the analysis. The transistor contains a current-dependent current source F to handle h_{FE} and an independent voltage source VA to represent the active-region voltage V_{BE}.

Fig. 3.1 Typical transistor biasing circuit.

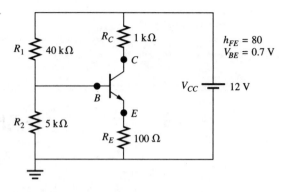

The input file is

```
Transistor Bias Circuit
VCC 4 0 12
VA 1 2 0.7
F 3 2 VA 80
R1 4 1 40k
R2 1 0 5k
RC 4 3 1k
RE 2 0 100
.OP
.END
```

Run the analysis; then verify that V_{CE} = 7.55 V. Draw current-direction arrows on the circuit diagram; then compute the collector current

$$I_C = \frac{V_4 - V_3}{R_C}$$

It should be 4.039 mA. What is the base current? Verify from the voltage-source currents that I_B = 50.59 μA. Compute I_B from the value of h_{FE} and compare it with this answer. Compute the emitter current $I_E = V_2/R_E$. It should be 4.089 mA. Verify that $I_E = I_B + I_C$.

Fig. 3.2 Bias model for bipolar *npn* transistor.

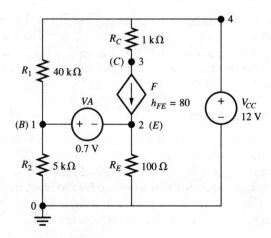

If you are interested in getting the currents directly from the PSpice analysis, you can use the dc sweep method as shown in this modified input file:

```
VCC 4 0 12
VA 1 2 0.7
F 3 2 VA 80
R1 4 1 40k
R2 1 0 5k
RC 4 3 1k
RE 2 0 100
.DC VCC 12 12 12
.OP
.PRINT DC I(RC) I(RE)
.END
```

Verify that this gives I(RC) = 4.039 mA and I(RE) = 4.089 mA as previously calculated. Note that in the first analysis you were able to easily calculate the currents. It may not be worth the extra effort to use the dc sweep technique; however, the choice is available.

Saturation Considerations

A precautionary note is needed for the situation involving biasing conditions that produce saturation. From your study of transistors you will recall that the value of h_{FE} in the active region of operation is not the same as h_{FE} in saturation. This means that if saturation occurs, the predicted value for I_C using the active region h_{FE} will be too large. You should suspect that this condition exists when the calculated V_{CE} drops below a few tenths of a volt. Several problems at the end of the chapter deal with the question of active-region vs. saturation biasing.

In summary, we have introduced the bias model for the *npn* Si transistor. This model can be used with various bias configurations and multistage amplifiers. Can you modify the model for (a) *pnp* Si transistors and (b) *pnp* Ge transistors?

Biasing Example for a Germanium Transistor

For another example of transistor biasing, see Fig. 3.3, which shows a *pnp* germanium transistor, with $h_{FE} = 60$ and $V_{BE} = -0.2$ V. Component values are $R_F =$

Fig. 3.3 Bias circuit using *pnp* germanium transistor.

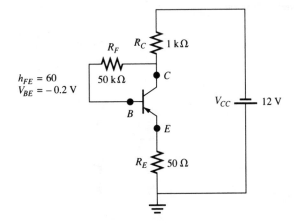

50 kΩ, $R_E = 50$ Ω, $R_C = 1$ kΩ, and $V_{CC} = -12$ V. Draw the SPICE model for the transistor and include the resistor values to complete the circuit. Figure 3.4 shows the results. Compare the treatment of the current-dependent current source shown here with the one of the previous example. Since this is a *pnp* transistor, the current arrow is away from the internal junction. Now decide how much information you would like to obtain from the PSpice analysis. Your input file might be

```
Transistor Bias Circuit for PNP Ge
VCC 0 4 12
VA 1 2 0.2
F 1 3 VA 60
RF 2 3 50k
RE 1 0 50
RC 3 4 1k
.DC VCC 12 12 12
.PRINT DC I(RC) I(RE) I(RF)
.OP
.END
```

Run the analysis; then draw current-direction arrows in their proper directions for the *pnp* transistor. Verify that $I_E = 6.311$ mA and $I_B = 103.5$ μA. Why are some of the resistor currents shown positive while others are shown negative? This has to do with the order of the subscripts in the *R* statements. For example, the statement

```
RE 1 0 50
```

produced a negative current I(RE). This is because the current is actually from node *0* toward node *1* in R_E. A careful comparison of the current-reference directions shown in the diagram should be made with those implied in the SPICE statements. Note that the current in R_C is actually the emitter current, rather than the collector current. Do you understand why? The collector current is shown in the PSpice output under *current-controlled current sources* as 6.208 mA. Add the base and collector currents, and compare the sum with the emitter current.

Fig. 3.4 Bias model for germanium *pnp* transistor.

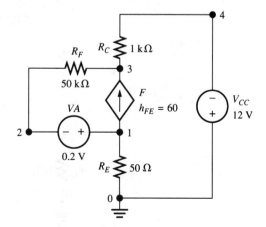

SMALL-SIGNAL *h*-PARAMETER MODEL OF THE BIPOLAR TRANSISTOR

An accurate model for the bipolar transistor, widely used in small-signal analysis, is the *h*-parameter model, shown in Fig. 3.5. This model, with appropriate values, is used for common-emitter, common-base, or common-collector configurations. Our task is to produce a version of this model for use with SPICE. The model will contain a current-dependent current source for use with h_f and a voltage-dependent voltage source for use with h_o. Figure 3.6 shows the model with *RI* for h_i, *E* to specify h_r, *RO* as $1/h_o$, and *F* to specify h_f.

Common-Emitter Transistor Analysis Using *h*-Parameter Model

Figure 3.7 shows a typical circuit for analysis. Even if the circuit is more complicated than this figure, you can often reduce it to this form by using various theorems and reduction techniques. Values given are $V_s = 1$ mV, $R_s = 1$ kΩ, $R_i = 1.1$ kΩ (h_{ie}), $h_r = 2.5 \times 10^{-4}$ (to be used with *E*), $h_f = 50$ (to be used with *F*), $R_o = 40$ k$\Omega = 1/h_o$, and $R_L = 10$ kΩ. $V0 = 0$ V is needed to give an independent source for the *F* statement.

Although we are interested in small-signal response, we will not use an ac analysis. The reason is simple and should be completely understood at this time. As long as you are dealing with small signals (ac steady state) and there are no reactive elements in the circuit, you can get more information from the PSpice analysis by letting the dc analysis represent either ac effective or peak values. The PSpice program cannot tell the difference! Be sure that you understand that the results will be small-signal results and have nothing to do with the dc biasing. Of

Fig. 3.5 The *h*-parameter model for the transistor.

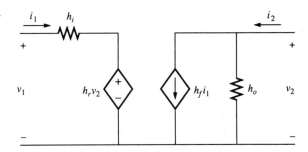

Fig. 3.6 The PSpice version of the *h*-parameter model.

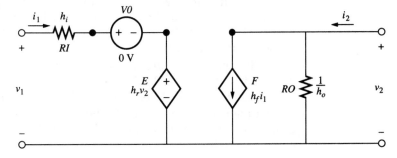

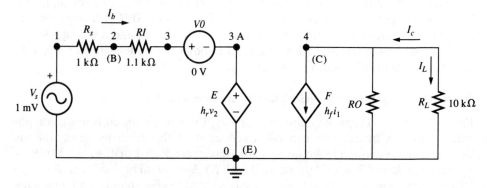

Fig. 3.7 *CE* *h*-parameter model with source and load.

course, we assume that the Q point has been established properly for active-region operation. Here is the input file:

```
Small-Signal Analysis of Transistor Circuit Using h Parameters
VS 1 0 1mV
VO 3 3A 0
E 3A 0 4 0 2.5E-4
F 4 0 VO 50
RS 1 2 1k
RI 2 3 1.1k
RO 4 0 40k
RL 4 0 10k
.OP
.TF V(4) VS
.END
```

Run the analysis and get a printed copy of the results for further study. Verify that $I_b = 0.5$ μA, $I_c = 20$ μA (from V(4)/R_L), the overall voltage gain is -200 (V(4)/VS), $R_i = 2$ kΩ, and $R_o = 8.4$ kΩ.

Since R_i includes R_s, what is the input resistance looking into the base of the transistor? It is $R_i - R_s = 1$ kΩ. Also, since R_o includes R_L, what is the output resistance looking into the collector (not including R_L)? Find this by using conductances. $1/R_o = 1.1905 \times 10^{-4}$; subtract $1/R_L = 1 \times 10^{-4}$ from this, giving $1/R_o' = 0.1905 \times 10^{-4}$. Thus $R_o' = 52.5$ kΩ.

The voltage gain from base to collector is V(4)/V(2) $= -400$. The current gain is $A_I = I_L/I_b = -20$ $\mu A/0.5$ $\mu A = -40$.

In summary, the PSpice analysis has saved you some calculations; but without an understanding of the current-reference directions and voltage polarities, your solution will be incomplete. You should have an understanding of *h*-parame-

Fig. 3.8 *CC* transistor amplifier.

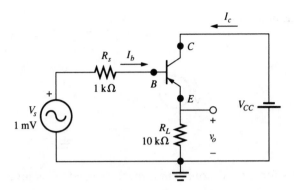

ter theory to go along with the PSpice model, which we have developed. Remember, the choice of *h*-parameter values depends on the configuration.

Some authors use models other than those based on the *h* parameters. These other models are often simpler and less accurate. However, you should have little trouble taking the other models through the analyses for comparison to the results given in this and other examples to follow. Problem 3.14 deals with an alternate simplified model and will serve as an introduction to this topic.

It is relatively easy to develop the models for the common-base and common-collector configurations.

Common-Collector Transistor Analysis Using *h*-Parameter Model

Another widely used circuit is the common-collector configuration, shown in Fig. 3.8. Again the circuit may be more elaborate than this, but it can often be reduced to this by use of Thevenin's theorem and other circuit simplification techniques. The input signal is fed through R_s to the base of the transistor, and the output is taken at the emitter. Figure 3.9 shows the circuit with the *h*-parameter model for the transistor. The circuit is almost identical to that of Fig. 3.7, but the *h* parame-

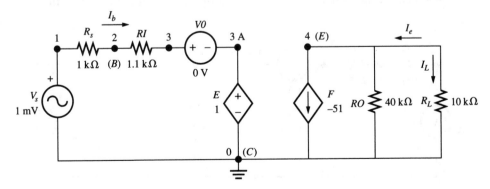

Fig. 3.9 *CC h*-parameter model with source and load.

Fig. 3.10   *CB* transistor amplifier.

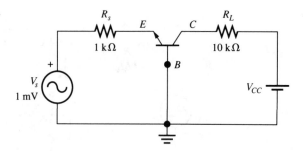

ters for the *CC* connection must be used. This gives the following input file:

```
Common-Collector Circuit Analysis with h Parameters
VS 1 0 1mV
VO 3 3A 0
E 3A 0 4 0 1
F 4 0 VO -51
RS 1 2 1k
RI 2 3 1.1k
RO 4 0 40k
RL 4 0 10k
.OP
.TF V(4) VS
.END
```

Run the PSpice analysis, and verify that $V(4)/VS = 0.9949$, $I_L = 9.949\text{E}{-8}$, $I_b = 2.438\text{E}{-9}$, $A_I = I_L/I_b = 40.8$, $R'_o = 40.97\ \Omega$ (including R_L), and $R'_i = 410\ \text{k}\Omega$ (including R_s). Determine the input resistance at the base and the output resistance excluding R_L. These should be $R_i = 409.1\ \text{k}\Omega$ and $R_o = 41.14\ \Omega$. Show the currents in the circuit diagram, and label the locations of the various input and output impedances. Note that the voltage gain is almost one with no phase reversal. The current gain also shows no phase reversal.

Common-Base Transistor Analysis Using *h*-Parameter Model

The common-base circuit shown in Fig. 3.10 uses the same values for external components as in the previous examples. Figure 3.11 shows the circuit with the *h*

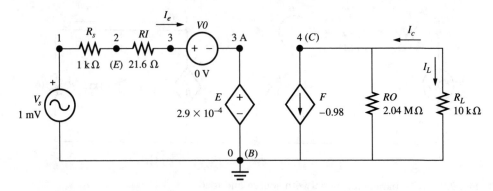

Fig. 3.11   *CB h*-parameter model with source and load.

parameters added. Using typical *h* parameters for the *CB* configuration, the input file becomes

```
Common-Base Circuit Analysis with h Parameters
VS 1 0 1mV
VO 3 3A 0
E 3A 0 4 0 2.9E-4
F 4 0 VO -0.98
```

```
                                                                        **

    Common-Base Circuit Analysis with h parameters

    ****        CIRCUIT DESCRIPTION

  VS 1 0 1mV
  VO 3 3A 0
  E 3A 0 4 0 2.9E-4
  F 4 0 VO -0.98
  RS 1 2 1k
  RI 2 3 21.6
  RO 4 0 2.04MEG
  RL 4 0 10k
  .OP
  .TF V(4) VS
  .END

   NODE   VOLTAGE     NODE   VOLTAGE     NODE   VOLTAGE     NODE   VOLTAGE

  (    1)     .0010  (    2) 23.85E-06  (    3) 2.761E-06  (    4)     .0095
  (   3A) 2.761E-06

      VOLTAGE SOURCE CURRENTS
      NAME           CURRENT

      VS           -9.762E-07
      VO            9.762E-07

      TOTAL POWER DISSIPATION   9.76E-10   WATTS

  **** VOLTAGE-CONTROLLED VOLTAGE SOURCES

  NAME         E
  V-SOURCE     2.761E-06
  I-SOURCE     9.762E-07

  **** CURRENT-CONTROLLED CURRENT SOURCES

  NAME         F
  I-SOURCE     -9.566E-07

    ****        SMALL-SIGNAL CHARACTERISTICS

        V(4)/VS =  9.520E+00

        INPUT RESISTANCE AT VS =   1.024E+03

        OUTPUT RESISTANCE AT V(4) =   9.924E+03
```

Fig. 3.12

Fig. 3.13 *CE* amplifier with bridging resistor.

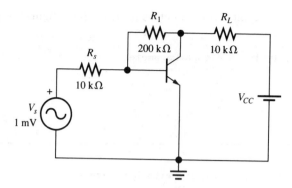

```
RS 1 2 1k
RI 2 3 21.6
RO 4 0 2.04MEG
RL 4 0 10k
.OP
.TF V(4) VS
.END
```

Run the analysis and verify that $A_V = 9.52$, $I_L = 0.95\ \mu A$, $I_e = 0.976\ \mu A$, $R_i' = 1024\ \Omega$, $R_o' = 9.924\ k\Omega$. Solve for R_i at the emitter and R_o without including R_L. These should be $R_i = 24\ \Omega$ and $R_o = 1.3\ M\Omega$. Show that the voltage gain from emitter to collector is $A_V = 406$.

Figure 3.12 shows the output from PSpice for this example. Unnecessary information was deleted from the file before it was printed.

In summary, we have presented the three basic transistor configurations, *CE*, *CC*, and *CB*. The *h*-parameter model for each is used in the PSpice solution. We have used typical *h* parameters for each configuration; sometimes these must be estimated, but data sheets should be used as available.

Other Configurations

When the transistor circuits do not simplify to the basic models of Figs. 3.7, 3.9, and 3.11, you must take care to maintain the positioning of the elements among the nodes. For example, in Fig. 3.13, a resistor is connected between the collector and the base in a *CE* circuit.

USING A CIRCUIT INVOLVING MILLER'S THEOREM

In your study of electronic circuits, you should learn that the bridging resistor R_1 interferes with an easy application of the gain equations. The resistor is often replaced with two other resistors using Miller's theorem. If you are familiar with Miller's theorem, use it to solve this problem with pencil-and-paper technique to see what is involved before continuing. Using SPICE, it is unnecessary to apply Miller's theorem. Keeping the bridging resistor in the circuit, the *h*-parameter model is shown in Fig. 3.14. The SPICE treatment is almost identical to the

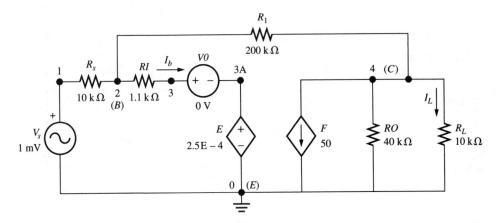

Fig. 3.14 *CE h-parameter model with bridging resistor.*

standard *CE* analysis. The input file becomes

```
Common-Emitter Circuit with Bridging Resistor
VS 1 0 1mV
VO 3 3A 0
E 3A 0 4 0 2.5E-4
F 4 0 VO 50
RS 1 2 10k
RI 2 3 1.1k
RO 4 0 40k
RL 4 0 10k
R1 2 4 200k
.OP
.TF V(4) VS
.END
```

Run the analysis and obtain the output file for comparison with the previous *CE* results. Verify that $A_V = $ V(4)/VS $= -12.7$ (voltage gain from source to load), $I_L = -1.27$ μA, and $I_b = 33.02$ nA, giving a current gain $A_I = I_L/I_b = -38.46$. From the overall input resistance $R'_i = 10.34$ kΩ, calculate the input resistance at the base, and from the overall output resistance $R'_o = 2.834$ kΩ, calculate the output resistance with R_L removed. These should be $R_i = 340$ Ω and $R_o = 3.95$ kΩ. If you worked this problem using Miller's theorem, you can appreciate how much work was saved using PSpice to obtain the results. It is difficult to go through the Miller method without making mistakes.

Compare the results of this analysis with those of the basic common-emitter amplifier (without the bridging resistor). Notice the effect of the bridging resistor on the gains and the input and output resistances. Figure 3.15 shows the output file.

The Dual of Miller's Theorem

Another circuit configuration is often analyzed using the dual of Miller's theorem. In Fig. 3.16, the emitter resistor R_e is replaced with two other resistors (one in

```
Common-Emitter Circuit with Bridging Resistor

****        CIRCUIT DESCRIPTION

VS 1 0 1mV
V0 3 3A 0
E 3A 0 4 0 2.5E-4
F 4 0 V0 50
RS 1 2 10k
RI 2 3 1.1k
RO 4 0 40k
RL 4 0 10k
R1 2 4 200k
.OP
.TF V(4) VS
.END

  NODE    VOLTAGE      NODE    VOLTAGE      NODE    VOLTAGE      NODE    VOLTAGE

(    1)     .0010   (    2) 33.15E-06   (    3)-3.175E-06   (    4)    -.0127
(   3A)-3.175E-06

    VOLTAGE SOURCE CURRENTS
    NAME          CURRENT

    VS          -9.669E-08
    V0           3.302E-08

    TOTAL POWER DISSIPATION   9.67E-11   WATTS

**** VOLTAGE-CONTROLLED VOLTAGE SOURCES

NAME          E
V-SOURCE    -3.175E-06
I-SOURCE     3.302E-08

**** CURRENT-CONTROLLED CURRENT SOURCES

NAME          F
I-SOURCE     1.651E-06

    ****      SMALL-SIGNAL CHARACTERISTICS

       V(4)/VS = -1.270E+01

       INPUT RESISTANCE AT VS =   1.034E+04

       OUTPUT RESISTANCE AT V(4) =   2.834E+03
```

Fig. 3.15

series with the base, the other in series with the collector). If you are familiar with this technique, use it to solve for the gains in this example. Then compare the results with those obtained here using PSpice.

You will not need to replace R_e in the PSpice analysis. The circuit, using the *h*-parameter model, is shown in Fig. 3.17. The input file is as follows:

```
Common-Emitter Amplifier with Emitter Resistor
VS 1 0 1mV
```

Fig. 3.16 *CE* amplifier with emitter resistor.

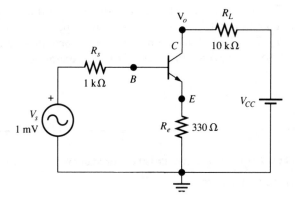

```
VO 3 3A 0
E 3A 4 5 4 2.5E-4
F 5 4 VO 50
RS 1 2 1k
RI 2 3 1.1k
RO 5 4 40k
RL 5 0 10k
RE 4 0 330
.OP
.TF V(5) VS
.END
```

Run the analysis, obtain a printed copy, and verify that the overall voltage gain $A_V = V(5)/VS = -25.74$, $R_i' = 15.44$ kΩ, and $R_o' = 9.752$ kΩ. Calculate and verify that A_V (at the base) $= -27.5$, $A_I = I_L/I_S = -39.7$, $R_i = 14.44$ kΩ (at the base), and $R_o = 393$ kΩ (without R_L).

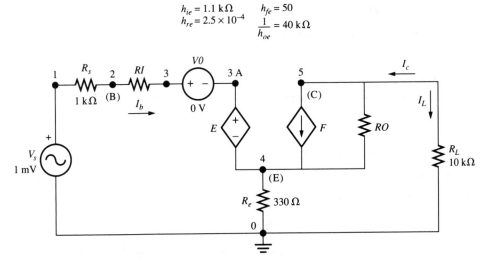

Fig. 3.17 PSpice *CE* amplifier with emitter resistor.

The effect of the emitter resistor on the input and output resistances is of particular interest in this analysis. The input resistance increases by the factor $(1 + h_{fe})R_e$. The voltage gain is usually approximated by the expression $-R_L/R_e$. Check to see how accurate this is in the example. Figure 3.18 shows the output file.

```
Common-Emitter Amplifier with Emitter Resistor

****        CIRCUIT DESCRIPTION

VS 1 0 1mV
V0 3 3A 0
E 3A 4 5 4 2.5E-4
F 5 4 V0 50
RS 1 2 1k
RI 2 3 1.1k
RO 5 4 40k
RL 5 0 10k
RE 4 0 330
.OP
.TF V(5) VS
.END

  NODE    VOLTAGE      NODE    VOLTAGE      NODE    VOLTAGE      NODE    VOLTAGE

(    1)      .0010  (    2) 935.2E-06  (    3) 864.0E-06  (    4) 870.6E-06
(    5)    -.0257  (   3A) 864.0E-06

    VOLTAGE SOURCE CURRENTS
    NAME            CURRENT

    VS             -6.477E-08
    V0              6.477E-08

    TOTAL POWER DISSIPATION   6.48E-11   WATTS

**** VOLTAGE-CONTROLLED VOLTAGE SOURCES

NAME          E
V-SOURCE    -6.651E-06
I-SOURCE     6.477E-08

**** CURRENT-CONTROLLED CURRENT SOURCES

NAME          F
I-SOURCE     3.239E-06

   ****      SMALL-SIGNAL CHARACTERISTICS

      V(5)/VS = -2.574E+01

      INPUT RESISTANCE AT VS =  1.544E+04

      OUTPUT RESISTANCE AT V(5) =   9.752E+03
```

Fig. 3.18

COMMON-COLLECTOR CIRCUIT WITH COLLECTOR RESISTOR

Another circuit of interest is a slight variation of the usual common-collector circuit. It contains an external collector resistor, added to protect the transistor from a short circuit across the emitter resistor. This modified circuit is shown in Fig. 3.19, and the PSpice model is shown in Fig. 3.20. If you want to analyze this by conventional pencil-and-paper techniques, the presence of R_c presents a problem that might call for the application of the dual of Miller's theorem. The derivation of formulas is tedious and adds little insight to the operation of the circuit. Look at the input file; then compare the results to the amplifier without R_c.

```
Common-Collector Circuit with Collector Resistor
VS 1 0 1mV
VO 3 3A 0
E 3A 4 5 4 1
F 5 4 VO -51
RS 1 2 1k
RI 2 3 1.1k
RC 4 0 1k
```

Fig. 3.19 *CC circuit with collector resistor.*

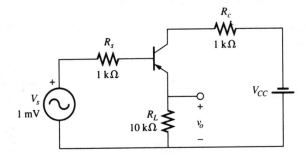

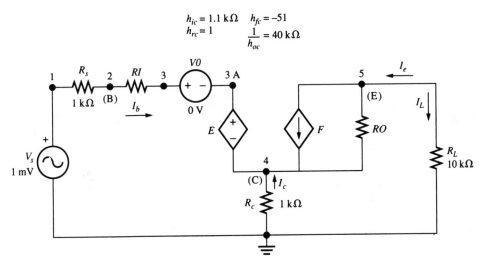

Fig. 3.20 *CC circuit model with collector resistor.*

```
RO 5 4 40k
RL 5 0 10k
.OP
.TF V(5) VS
.END
```

Run the analysis and compare the results to those obtained for the simple *CC* amplifier. You will see that the voltage gain is almost identical in both cases and that the input and output resistances changed very little. We conclude that the addition of R_c to the basic circuit has minimal effect on the operation of the circuit.

HIGH-INPUT-RESISTANCE AMPLIFIER

When you need an amplifier with high input resistance, you might choose the Darlington circuit of Fig. 3.21. The circuit consists of a pair of common-collector transistors, often produced in a single package. Note that the first stage can be thought of as having an infinite external emitter resistor $R_{e1} = \infty$. Using the *h*-parameter model for the cascaded *CC* stages results in Fig. 3.22, from which the input file on page 132 is created:

Fig. 3.21　Darlington circuit for high input resistance.

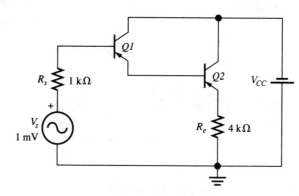

$$h_{ic} = 1.1\ \text{k}\Omega \qquad h_{fc} = -51$$
$$h_{rc} = 1 \qquad \frac{1}{h_{oc}} = 40\ \text{k}\Omega$$

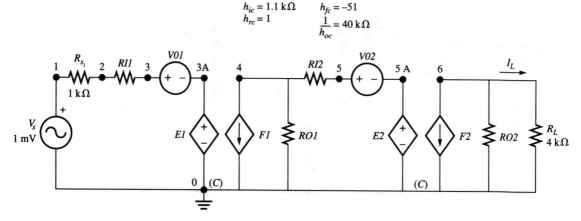

Fig. 3.22　Darlington *CC* h-parameter circuit model.

```
        Darlington-Pair (High-Input-Resistance) Amplifier

        ****        CIRCUIT DESCRIPTION

        VS 1 0 1mV
        V01 3 3A 0
        V02 5 5A 0
        E1 3A 0 4 0 1
        E2 5A 0 6 0 1
        F1 4 0 V01 -51
        F2 6 0 V02 -51
        RS 1 2 1k
        RI1 2 3 1.1k
        RO1 4 0 40k
        RI2 4 5 1.1k
        RO2 6 0 40k
        RL 6 0 4k
        .OP
        .TF V(6) VS
        .END

        NODE    VOLTAGE      NODE    VOLTAGE      NODE    VOLTAGE      NODE    VOLTAGE

        (   1)     .0010  (    2) 999.4E-06  (    3) 998.8E-06  (    4) 998.8E-06
        (   5) 992.9E-06  (    6) 992.9E-06  (   3A) 998.8E-06  (   5A) 992.9E-06

            VOLTAGE SOURCE CURRENTS
            NAME           CURRENT

            VS           -5.946E-10
            V01           5.946E-10
            V02           5.354E-09

            TOTAL POWER DISSIPATION   5.95E-13  WATTS

        **** VOLTAGE-CONTROLLED VOLTAGE SOURCES

        NAME           E1           E2
        V-SOURCE       9.988E-04    9.929E-04
        I-SOURCE       5.946E-10    5.354E-09

        **** CURRENT-CONTROLLED CURRENT SOURCES

        NAME           F1           F2
        I-SOURCE      -3.032E-08   -2.730E-07

         ****        SMALL-SIGNAL CHARACTERISTICS

            V(6)/VS =  9.929E-01

            INPUT RESISTANCE AT VS =  1.682E+06

            OUTPUT RESISTANCE AT V(6) =  2.224E+01
```

Fig. 3.23

```
Darlington-Pair (High-Input-Resistance) Amplifier
VS 1 0 1mV
VO1 3 3A 0
VO2 5 5A 0
E1 3A 0 4 0 1
E2 5A 0 6 0 1
F1 4 0 VO1 -51
F2 6 0 VO2 -51
RS 1 2 1k
RI1 2 3 1.1k
RO1 4 0 40k
RI2 4 5 1.1k
RO2 6 0 40k
RL 6 0 4k
.OP
.TF V(6) VS
.END
```

Run the analysis and verify that the voltage gain $V(6)/VS = 0.9929$, $R_i' = 1.682$ MΩ, and $R_o' = 22.24$ Ω. From your calculations show that $R_i = 1.681$ MΩ at the base of the first transistor $Q1$ and that $R_o = 22.36$ Ω with R_L removed. Also find $A_I = I_L/I_b = 417.5$, which is much higher than for the single-stage CC amplifier. This analysis has assumed that the h parameters for both stages are the same. In reality, the quiescent currents of the first stage are less than those of the second stage. Figure 3.23 shows the output file.

TWO-STAGE AMPLIFIERS

The treatment of two-stage amplifiers is much simplified using SPICE rather than wading through the usual formula calculations, which are tedious and error prone. If you understand the basic principles of amplifier analysis, you should not hesitate to use PSpice for multistage analysis. As an illustration, consider the CE-CC amplifier of Fig. 3.24. The input is to the base of the first transistor. The output from its collector goes directly to the base of the second transistor, with the output taken on the emitter at R_{e2}. The h parameters are shown in Fig. 3.25, differing

Fig. 3.24 Two-stage amplifier; CE and CC stages.

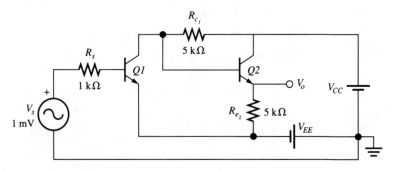

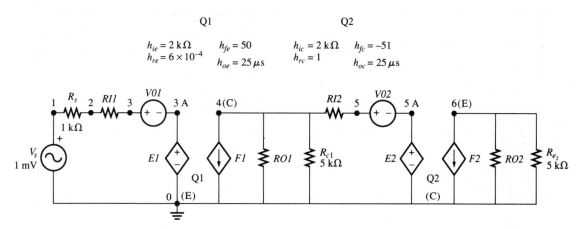

Fig. 3.25 Two-stage amplifier; *CE* and *CC* stages.

slightly from those of previous examples. The input file is

```
Two-Stage Amplifier; CE and CC Stages
VS 1 0 1mV
VO1 3 3A 0
VO2 5 5A 0
E1 3A 0 4 0 6E-4
F1 4 0 VO1 50
E2 5A 0 6 0 1
F2 6 0 VO2 -51
RS 1 2 1k
RI1 2 3 2k
RO1 4 0 40k
RC1 4 0 5k
RI2 4 5 2k
RO2 6 0 40k
RE2 6 0 5k
.OP
.TF V(6) VS
.END
```

After running the analysis, you should verify that V(6)/VS = −75.31 is the overall voltage gain. From your calculations, show that $A_I = I_L/I_{b1} = -43.2$, $R_i = 1.869$ kΩ (at the base of *Q1*), and $R_o = 130$ Ω. Figure 3.26 shows the output file.

SIMPLIFIED *h*-PARAMETER MODEL

The examples that have been used so far were based on the full *h*-parameter model for the transistor that is normally used for small-signal, low-frequency analysis. Another model that is often used for certain BJT circuits is the simplified *h*-parameter model. In this model h_{fe} and h_{ie} are used and the other *h* parameters are omitted. There is little justification for using the simplified model with SPICE. It is often in error by 10% or more. It is given here for reference, along with another

```
   Two-stage Amplifier; CE and CC stages

   ****        CIRCUIT DESCRIPTION

VS 1 0 1mV
V01 3 3A 0
V02 5 5A 0
E1 3A 0 4 0 6E-4
F1 4 0 V01 50
E2 5A 0 6 0 1
F2 6 0 V02 -51
RS 1 2 1k
RI1 2 3 2k
RO1 4 0 40k
RC1 4 0 5k
RI2 4 5 2k
RO2 6 0 40k
RE2 6 0 5k
.OP
.TF V(6) VS
.END

  NODE    VOLTAGE      NODE    VOLTAGE      NODE    VOLTAGE      NODE    VOLTAGE

 (    1)      .0010  (    2) 651.5E-06  (    3)-45.58E-06  (    4)     -.0760
 (    5)    -.0753  (    6)    -.0753  (   3A)-45.58E-06  (   5A)     -.0753

     VOLTAGE SOURCE CURRENTS
     NAME           CURRENT

     VS           -3.485E-07
     V01           3.485E-07
     V02          -3.322E-07

     TOTAL POWER DISSIPATION   3.49E-10   WATTS

**** VOLTAGE-CONTROLLED VOLTAGE SOURCES

NAME         E1           E2
V-SOURCE    -4.558E-05   -7.531E-02
I-SOURCE     3.485E-07   -3.322E-07

**** CURRENT-CONTROLLED CURRENT SOURCES

NAME         F1           F2
I-SOURCE     1.743E-05    1.694E-05

  ****        SMALL-SIGNAL CHARACTERISTICS

      V(6)/VS = -7.531E+01

      INPUT RESISTANCE AT VS =  2.869E+03

      OUTPUT RESISTANCE AT V(6) =  1.267E+02
```

Fig. 3.26

Fig. 3.27 Simplified *h*-parameter model.

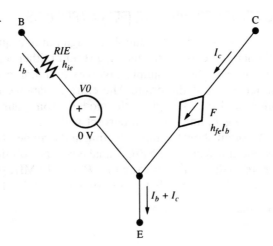

look at the common-emitter amplifier. Remember that the values for h_{ie} and h_{fe} are used for all three configurations, *CE, CB*, and *CC*.

The *CE* Amplifier Using the Simplified *h*-Parameter Model

Figure 3.27 shows the simplified model, using PSpice notation, and Fig. 3.28 shows the common-emitter circuit using the model. The input file for the analysis is given here.

```
Simplified h-Parameter Analysis
VS 1 0 1mV
V0 3 3A 0V
F 4 0 V0 50
RS 1 2 1k
RI 2 3 1.1k
RL 4 0 10k
.OP
.TF V(4) VS
.END
```

You can easily predict the results of this analysis using pencil-and-paper methods. Compare your predictions with the PSpice answers.

Fig. 3.28 *CE* amplifier using simplified *h*-parameter model.

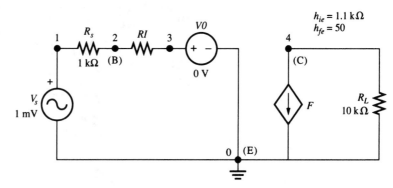

FIELD-EFFECT TRANSISTOR (FET) AMPLIFIERS

The FET amplifier is often simple enough not to require computer analysis. In the cases where an extra resistance is present (either R_d or R_s), the situation is more interesting. Our first example involves a common-source FET with the output taken across R_d at the drain. The extra resistance is R_s. Figure 3.29 shows the amplifier, and Fig. 3.30 gives the model. From your study of FETs, predict what the voltage gain and the load current will be.

    We often show the input floating in the model (the gate is not connected), but this would not work in a SPICE analsyis. The solution is to place a large resistor from gate to drain. In our example, $R_{GD} = 10$ MΩ, $g_m = 2$ mS, $r_d = 40$ kΩ, $R_L = 2$ kΩ, $R_s = 500$ Ω, and the input voltage is 1 mV. The input file becomes

```
Common-Source FET with RS
VI 1 0 1mV
G 2 3 1 3 2mS
RD 2 3 40k
RL 2 0 2k
```

Fig. 3.29 Common-source FET with R_s.

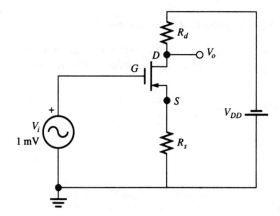

Fig. 3.30 FET model and common-source amplifier.

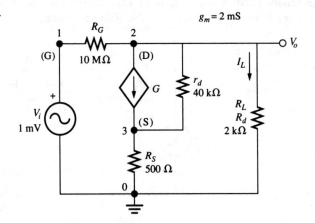

```
RS 3 0 500
RG 1 2 10MEG
.OP
.TF V(2) VI
.END
```

After running the analysis, verify that $V(2)/VI = -1.939$ and $R'_o = 1.95 \text{ k}\Omega$. From your calculations, show that $R_o = 79.6 \text{ k}\Omega$ and $I_L = -950$ nA.

Common-Drain FET with External Drain Resistor

In Fig. 3.29 if the output is taken across R_s, this is a common-drain amplifier. Assuming that R_d is still in place, you are interested in the analysis. The SPICE model shown in Fig. 3.31 produces this input file:

```
Common-Drain FET with Drain Resistor
VI 1 0 1mV
G 3 2 1 2 2mS
RD 2 3 40k
RD1 3 0 1k
RS 2 0 2k
RG 1 3 10MEG
.OP
.TF V(2) V
.END
```

From your study of FETs, predict what the voltage gain for this source-follower circuit should be; then run the PSpice analysis and verify that $V(2)/VI = 0.7882$ and $R'_o = 403.9 \ \Omega$. Calculate $R_o = 506 \ \Omega$ and $I_L = 394$ nA.

Frequency Response of FET Amplifiers

When you use the FET amplifier over a wide range of frequencies, you need to account for the internal node capacitances. Figure 3.32 shows a common-source amplifier model including C_{gd}, C_{gs}, and C_{ds}. Usually these internal capacitances are small. In our example, we choose $C_{gs} = 3$ pF, $C_{ds} = 1$ pF, and $C_{gd} = 2.8$ pF. Other values include $g_m = 1.6$ mS, $r_d = 44 \text{ k}\Omega$, along with $R_s = 1 \text{ k}\Omega$ and $R_L = 100 \text{ k}\Omega$. For the SPICE analysis, we will choose a frequency range of 100 Hz to

Fig. 3.31 Common-drain FET with drain resistor.

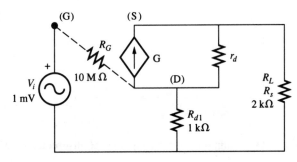

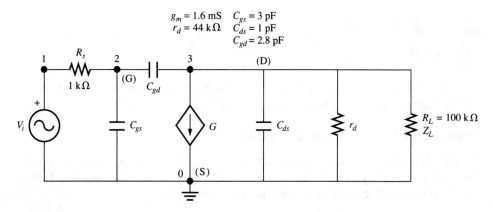

Fig. 3.32   Common-source amplifier; high-frequency model.

100 kHz. The capacitances will be of interest only at high frequencies. The input file will become the following:

```
Common-Source Amplifier; High-Frequency Model
VI 1 0 AC 1mV
G 3 0 2 0 1.6mS
RD 3 0 44k
```

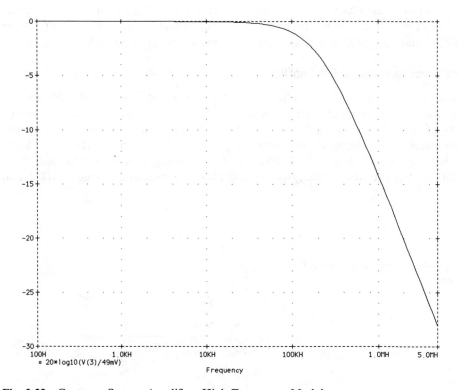

Fig. 3.33   Common-Source Amplifier; High-Frequency Model.

```
RL 3 0 100k
RS 1 2 1k
CGS 2 0 3pF
CGD 2 3 2.8pF
CDS 3 0 1pF
.AC DEC 20 100 10MEG
.PROBE
.END
```

Run the analysis and obtain a printed output of the Probe results using a log-frequency X-axis, with V(3) on the Y-axis. Can you identify at what frequency the output becomes seriously attenuated? Since this is not quite a Bode plot, it is difficult to be specific. Remove this trace and add a new logarithmic trace. The trace which you call for should be

$$20*\log10(V(3)/49mV)$$

The plot is now in standard Bode form. In the expression, 49 mV represents the midfrequency gain as shown on the first plot. We used this value to normalize the plot. The vertical axis now shows 0 at the top followed by -5, -10, and so forth. Adjust the X-axis to show the frequency range 100 Hz to 5 MHz. Use the cursor to verify that the -3 dB point is at $f = 619$ kHz. Obtain a printed copy of this, and draw lines tangent to both linear portions of the curve. The point where these lines intersect indicates where the frequency response is down 3 dB. Figure 3.33 shows the plot.

HIGH-FREQUENCY MODEL OF THE BIPOLAR-JUNCTION TRANSISTOR

For the CE circuit, we often use the hybrid-π model. Figure 3.34 shows this model along with V_s, R_s, and R_L, the external components. In this model an extra node B' is required to account for the behavior at high frequencies. Parameters used in this

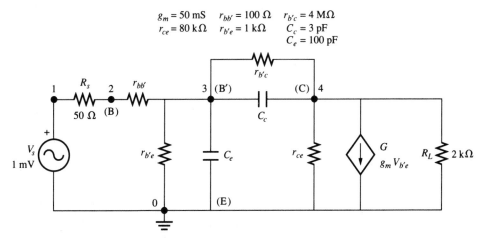

Fig. 3.34 Hybrid-π model for BJT, common-emitter circuit.

model are resistors r_{ce}, $r_{bb'}$, $r_{b'e}$, and $r_{b'c}$ and capacitors C_c and C_e. The gain is represented by a voltage-dependent current source $g_m V_{b'e}$. Figure 3.34 shows the values used in this example. These lead to the following input file:

```
High-Frequency Model of Bipolar-Junction Transistor
VS  1  0  AC  1mV
G  4  0  3  0  50mS
RS  1  2  50
RBB  2  3  100
RBE  3  0  1k
RBC  3  4  4MEG
RCE  4  0  80k
RL  4  0  2k
CE  3  0  100pF
CC  3  4  3pF
.AC  DEC  50  100k  10MEG
.PROBE
.END
```

Run the analysis and determine the midfrequency voltage output. Verify that it is approximately 85 mV. Then obtain a plot of

$$20*\log 10(V(4)/85mV)$$

This will allow you to find the 3 dB point. Verify that it is at $f = 2.85$ MHz. Figure 3.35 shows the plot.

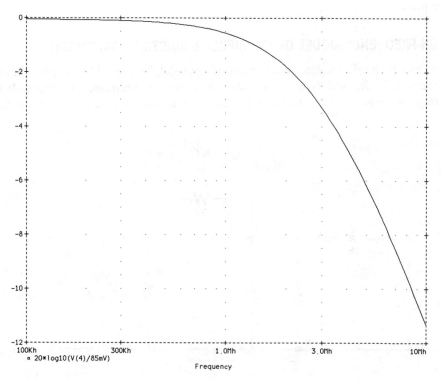

Fig. 3.35 High-Frequency Model of Bipolar-Junction Transistor.

Without the aid of a powerful tool such as PSpice, the equations needed to correctly solve problems such as this become very difficult to derive and employ. The circuit has four independent nodes, and the elements are complex. Often for the mathematical treatment simpler models are used instead.

EMITTER FOLLOWER AT HIGH FREQUENCIES

We now present a variation of the high-frequency analysis. This circuit includes a load impedance Z_L, consisting of R_L and C_L. The amplifier has low output resistance and is used as a driver for a capacitive load. Figure 3.36 shows the circuit with the hybrid-π model. Note that the current arrow on G still points toward the emitter node. The input file is

```
Emitter Follower High-Frequency Model
VS 1 0 AC 1mV
G 0 4 3 4 50mS
RS 1 2 50
RBB 2 3 100
RBE 3 4 1k
RBC 3 0 4MEG
RL 4 0 2k
CL 4 0 3nF
CC 3 0 3pF
CE 3 4 100pF
.AC DEC 50 100k 10MEG
.PROBE
.END
```

Run the analysis; then plot V(4). Note that the gain is slightly less than unity, as expected for the emitter follower. To obtain the Bode plot, use the trace for

$$20*\log10(V(4)/0.99\text{mV})$$

Then using the cursor mode, verify that the 3 dB point is at $f = 2.7$ MHz. Add a second graph which is a plot of the phase shift of V(4). To do this simply plot

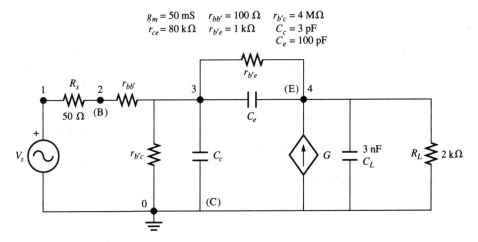

Fig. 3.36 Hybrid-π model, emitter follower with capacitive load.

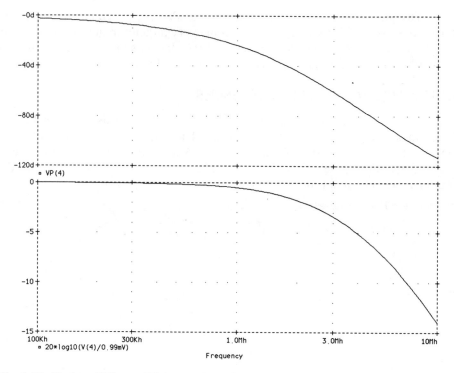

Fig. 3.37 Emitter Follower High-Frequency Model.

VP(4). See that the phase shift at the -3 dB frequency is about $-55.8°$. Note that at the lowest frequency of the plot, 100 kHz, there is already a phase shift of a few degrees due to the capacitive nature of the load. Figure 3.37 shows the phase-shift and magnitude plots for this circuit.

DC SENSITIVITY

Variations in component values can cause circuits to operate improperly. In some cases, expected voltages and currents fall outside acceptable values. In other cases, an improper biasing condition might lead to distortion problems, and so forth. Using PSpice, the sensitivity of an output variable may be determined by including the .SENS statement.

For example, in a series circuit as shown in Fig. 3.38, the resistor R_2 represents the load resistance. The voltage across this resistor is 1.25 V. The input file will contain a statement to determine the sensitivity of this voltage with respect to the other elements in the circuit. The file is

```
Sensitivity of Load Voltage in Series Circuit
Vs 1 0 5V
R1 1 2 300
R2 2 0 100
.sens V(2)
.end
```

Fig. 3.38 Circuit to illustrate sensitivity.

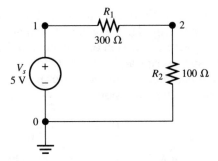

The output from this analysis is shown in Fig. 3.39. The dc sensitivities of the output voltage V_2 are shown in relation to the various elements in the circuit. The first element listed is R_1, with its value of 300 Ω. Its sensitivity is given as $-3.125E-03$ volts per unit. Since R_1 is a resistor, the *unit* is the ohm. The sensitivity is then $-3.125E-03$ V/Ω. The sensitivity of V_2 to changes in the R_2 value of 9 Ω is 9.375E$-$03 V/Ω. Finally, the sensitivity of V_2 to changes in V_s is 0.25 V/V. We would like to discover how these values are found and what they mean.

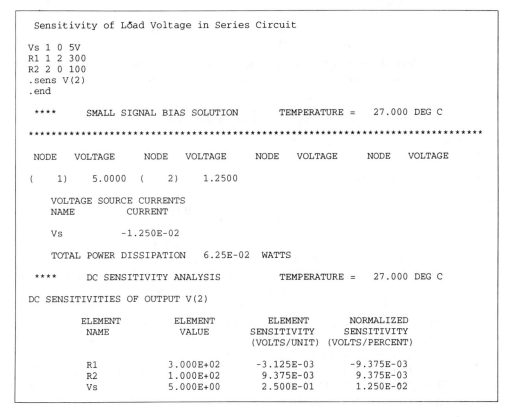

Fig. 3.39

The concept of sensitivity was introduced by Bode in *Network Analysis and Feedback Amplifier Design*. He was interested in how a transfer function T would change when there were changes in one element in the system. The symbol S was introduced, with superscript and subscript, to represent sensitivity. The superscript is the output parameter, and the subscript is the element in question. In our series circuit, using voltage division,

$$V_2 = V_s \frac{R_2}{R_1 + R_2}$$

First consider that R_1 becomes the variable while the other elements are held constant. R_1 might increase by a small amount, ΔR_1, bringing about a small change in V_2 which will be ΔV_2. In the limit, $\Delta R_1 \to \partial R_1$ and $\Delta V_2 \to \partial V_2$. The element sensitivity for element R_1 is defined as

$$\frac{V_2}{R_1} S_{R_1}^{V_2} = \frac{V_2 \, \partial \ln V_2}{R_1 \, \partial \ln R_1} = \frac{\partial V_2}{\partial V_1}$$

$$\frac{\partial V_2}{\partial R_1} = \frac{\partial V_s}{\partial R_1} \frac{R_2}{R_1 + R_2} = \frac{V_s(-R_2)}{(R_1 + R_2)^2}$$

In our example

$$\frac{\partial V_2}{\partial R_1} = \frac{-5(100)}{(400)^2} = -0.003125$$

which is in agreement with the element sensitivity shown in the output file.

The sensitivity of V_2 with respect to R_1 is defined as

$$\frac{V_2}{R_2} S_{R_2}^{V_2} = \frac{V_2 \, \partial \ln V_2}{R_2 \, \partial \ln R_2} = \frac{\partial V_2}{\partial R_2}$$

$$\frac{\partial V_2}{\partial R_2} = \frac{\partial V_s}{\partial R_2} \frac{R_2}{R_1 + R_2} = \frac{V_s R_1}{(R_1 + R_2)^2}$$

In our example

$$\frac{\partial V_2}{\partial R_2} = \frac{5(300)}{(400)^2} = 0.009375$$

which is also in agreement with the element sensitivity shown in the output file.

Normalized values are shown in the last column of the PSpice analysis of Fig. 3.39. These are found as the product of element value and element sensitivity.

Now that we have seen how the sensitivity values are found, our next step is to determine what they mean. Suppose that there is an incremental change in the value of R_1. For example, let R_1 increase by 1%. This gives $R_1 = 303 \ \Omega$ and $\Delta R_1 = 3 \ \Omega$. Based on the value $\partial V_2/\partial R_1 = -0003125$, $\Delta V_2 = 3(-0.003125) = -0.009375$, and the new value of $V_2 = 1.240625$ V.

In like manner, suppose that there is an increase in the value of R_2 by 1%. This gives $R_2 = 101 \ \Omega$ and $\Delta R_2 = 1 \ \Omega$. Based on the value of $\partial V_2/\partial R_2 = 0.009375$, $\Delta V_2 = 0.009375$ V, and the new value of $V_2 = 1.259375$ V.

But wait: could we not find the new value of V_2 in each case by the voltage-division formula? Thus when $R_1 = 303\ \Omega$,

$$V_2 = V_s \frac{R_2}{R_1 + R_2} = 5\,\frac{100}{403} = 1.240695\ \text{V}$$

On the other hand, when $R_2 = 101\ \Omega$,

$$V_2 = V_s \frac{R_2}{R_1 + R_2} = 5\,\frac{101}{401} = 1.25935\ \text{V}$$

Careful comparison of the two methods for predicting the new value of V_2 shows that they are not in exact agreement. In fact, had we used changes greater than 1%, the disagreement would have been greater. It now becomes obvious that sensitivity values are not to be used for predicting actual values of the output voltage. Instead, they are to be used in determining which elements are more critical in maintaining a stable circuit.

When the normalized sensitivities for our circuit are compared, we look at the larger values and conclude that these are the more critical. For example, the largest value shown in the last column of Fig. 3.39 is the value for V_s, which is 0.0125. Thus V_s becomes the most critical element. The normalized sensitivities for R_1 and R_2 are the same in magnitude. Thus they are equally sensitive elements.

DC SENSITIVITY OF BIAS CIRCUIT

The circuit of Fig. 3.40 is our model for a BJT bias circuit. In this example the transistor has $V_{BE} = 0.7$ V and $h_{FE} = 80$. The output voltage is taken as the collector-to-emitter voltage V(3,2). This voltage will be the subject of the sensitiv-

Fig. 3.40 Simple BJT model for sensitivity.

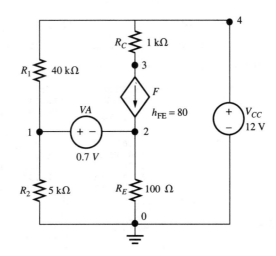

ity analysis. Here is the input file:

```
Sensitivity of Model Transistor Bias Circuit
VCC 4 0 12V
VA 1 2 0.7V
F 3 2 VA 80
R1 4 1 40k
R2 1 0 5k
RC 4 3 1k
RE 2 0 100
.SENS V(3,2)
.END
```

Run the analysis and study the listings for element sensitivity. This output is shown in Fig. 3.41. Verify that for a 0.12-V increase in V_{CC}, V(3,2) will increase

```
Sensitivity of Model Transistor Bias Circuit

VCC 4 0 12V
VA 1 2 0.7V
F 3 2 VA 80
R1 4 1 40k
R2 1 0 5k
RC 4 3 1k
RE 2 0 100
.SENS V(3,2)
.END


****     SMALL SIGNAL BIAS SOLUTION        TEMPERATURE =   27.000 DEG C

 NODE    VOLTAGE      NODE    VOLTAGE      NODE    VOLTAGE      NODE    VOLTAGE

(    1)    1.1089  (     2)     .4089  (     3)    7.9610  (     4)   12.0000

     VOLTAGE SOURCE CURRENTS
     NAME            CURRENT

     VCC         -4.311E-03
     VA           5.049E-05


     TOTAL POWER DISSIPATION   5.17E-02   WATTS

****     DC SENSITIVITY ANALYSIS           TEMPERATURE =   27.000 DEG C

DC SENSITIVITIES OF OUTPUT V(3,2)

            ELEMENT        ELEMENT        ELEMENT        NORMALIZED
            NAME           VALUE          SENSITIVITY    SENSITIVITY
                                          (VOLTS/UNIT)   (VOLTS/PERCENT)

            R1             4.000E+04      2.125E-04       8.499E-02
            R2             5.000E+03     -1.385E-03      -6.923E-02
            RC             1.000E+03     -4.039E-03      -4.039E-02
            RE             1.000E+02      2.463E-02       2.463E-02
            VCC            1.200E+01      2.197E-01       2.636E-02
            VA             7.000E-01      7.023E+00       4.916E-02
```

Fig. 3.41

by 0.02636 V. Also verify that of the various resistors in the circuit, R_1 is the most sensitive in determining V(3,2). As an exercise, use the element sensitivity values to approximate V(3,2) when first R_1 increases by 1% and then when R_2 increases by 1%. Remember that the results obtained by this method will be approximate and will apply only to incremental changes in the element values.

SENSITIVITY OF LIBRARY BJT CIRCUIT

When the PSpice model of a BJT is used in a sensitivity analysis, sensitivity of the output variable to changes in transistor parameter values are obtained, as well as the information previously given. For example, the circuit shown in Fig. 3.42 uses one of the BJTs in the evaluation library. It is our desire to obtain the sensitivity of V(3,4), the collector-to-emitter voltage. The input file is

```
Sensitivity of BJT Biasing Circuit
VCC 2 0 12V
R1 2 1 40k
R2 1 0 3.3k
RC 2 3 7.7k
RE 4 0 220
Q1 3 1 4 Q2N2222A
.SENS V(3,4)
.LIB EVAL.LIB
.END
```

The output file, shown in Fig. 3.43, gives dc sensitivities of V(3,4) not only to the external components but also to the transistor parameters as well. In the latter category, note that under *Q1* in the sensitivity analysis, *RB* is the (internal) base resistance, *RC* is the collector ohmic resistance, *RE* is the emitter ohmic resistance, and so forth. Of particular interest is the sensitivity with respect to the dc beta, *BF*.

Fig. 3.42 Built-in model for sensitivity.

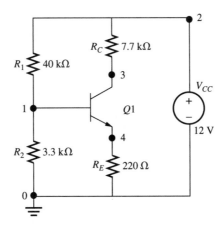

```
    Sensitivity of BJT Biasing Circuit
VCC 2 0 12V
R1 2 1 40k
R2 1 0 3.3k
RC 2 3 7.7k
RE 4 0 220
Q1 3 1 4 Q2N2222A
.SENS V(3,4)
.LIB EVAL.LIB
.END

    ****        BJT MODEL PARAMETERS
                Q2N2222A
                NPN
            IS   14.340000E-15
            BF   255.9

    ****        SMALL SIGNAL BIAS SOLUTION        TEMPERATURE =    27.000 DEG C

    NODE    VOLTAGE        NODE    VOLTAGE      NODE    VOLTAGE      NODE    VOLTAGE
    (   1)     .8926     (   2)   12.0000     (   3)    3.4815     (    4)     .2450

       VOLTAGE SOURCE CURRENTS
       NAME            CURRENT
       VCC            -1.384E-03

       TOTAL POWER DISSIPATION   1.66E-02  WATTS

    ****        DC SENSITIVITY ANALYSIS          TEMPERATURE =    27.000 DEG C

DC SENSITIVITIES OF OUTPUT V(3,4)
            ELEMENT            ELEMENT        ELEMENT        NORMALIZED
            NAME              VALUE        SENSITIVITY    SENSITIVITY
                                          (VOLTS/UNIT) (VOLTS/PERCENT)
            R1          4.000E+04        6.263E-04        2.505E-01
            R2          3.300E+03       -7.395E-03       -2.440E-01
            RC          7.700E+03       -1.086E-03       -8.363E-02
            RE          2.200E+02        3.186E-02        7.009E-02
            VCC         1.200E+01       -1.274E+00       -1.529E-01
Q1
            RB          1.000E+01        2.127E-04        2.127E-05
            RC          1.000E+00        2.020E-05        2.020E-07
            RE          0.000E+00        0.000E+00        0.000E+00
            BF          2.559E+02       -1.586E-03       -4.059E-03
            ISE         1.434E-14        2.022E+13        2.899E-03
            BR          6.092E+00        3.790E-11        2.309E-12
            ISC         0.000E+00        0.000E+00        0.000E+00
            IS          1.434E-14       -6.888E+13       -9.878E-03
            NE          1.307E+00       -4.250E+00       -5.555E-02
            NC          2.000E+00        0.000E+00        0.000E+00
            IKF         2.847E-01       -1.831E-02       -5.213E-05
            IKR         0.000E+00        0.000E+00        0.000E+00
            VAF         7.403E+01        6.382E-04        4.725E-04
            VAR         0.000E+00        0.000E+00        0.000E+00
```

Fig. 3.43

SUMMARY OF NEW PSPICE STATEMENTS USED IN THIS CHAPTER

E[*name*] *<+node> <−node> <+controlling node>*
                 <−controlling node> <gain>

For example,

```
E 6 5 2 1 18
```

means that a voltage-controlled voltage source is connected between nodes *6* and *5*. It is dependent on the voltage between nodes *2* and *1*, and it has a voltage gain of 18. This statement (along with the *F* and *G* statements) was given in Chapter 1; it is repeated here since it is a fundamental amplifier equation. Some of the examples and problems in this chapter require its use. Like other dependent sources, its form could also involve a POLY expression. Note that the gain is a dimensionless voltage ratio.

F[*name*] *<+node> <−node> <controlling V device name> <gain>*

For example,

```
F 4 3 VA 80
```

means that a current-controlled current source is connected between nodes *4* and *3*. The current arrowhead is at node *3*. The current through the dependent source is greater than the current through VA by a factor of 80. The voltage source *VA* may be an actual source or a dummy source of zero volts. The dummy source is often needed to specify the path for the controlling current.

    The *h*-parameter model of the transistor requires the use of the *F* statement. The gain is h_{fe} and is dimensionless. Other transistor models involving beta also require the *F* statement.

G[*name*] *<+node> <−node> <+controlling node> <−controlling node>*
                 <transconductance>

For example,

```
G 8 7 5 3 20mS
```

means that a voltage-controlled current source is connected between nodes *8* and *7*. The current arrowhead is at node *7*. The current through the dependent source is a function of the voltage between nodes *5* and *3* as specified by the transconductance of 20 mS. This means, for example, that if $v_{53} = 10$ mV, then $i_{87} = (10$ mV$) \cdot (20$ mS$) = 200$ μA.

DOT COMMANDS USED IN THIS CHAPTER

.TF *<output variable> <input source>*

For example,

```
.TF V(4) VS
```

When used with the *h*-parameter model as introduced in this chapter, this statement will give the small-signal gain V_4/V_S. This is possible when we are using ac voltages in circuits where the passive components are purely resistive. As far as PSpice is concerned, the analysis could be ac or dc.

.LIB *<file name>*

For example,

```
.LIB EVAL.LIB
```

means that the library *EVAL.LIB* will be searched for the models used in the input file. In the example based on the BJT biasing circuit, the transistor Q_1 was used. This is based on the model for the *Q2N2222A*. This model is found in the library *EVAL.LIB* that comes with the evaluation version of PSpice.

.SENS *<output variable>*

For example,

```
.SENS V(2)
```

means that the dc sensitivities of the output voltage V(2) will be computed in relation to the various elements in the circuit.

PROBLEMS

3.1 A biasing circuit for a silicon transistor with $h_{FE} = 100$ is shown in Fig. 3.44. Assume that $V_{BE} = 0.7$ V in your PSpice model. Find the currents I_B and I_C. Find the bias voltage V_{CE}. Your results should show $I_B = 21.5$ μA, $I_C = 2.15$ mA, and $V_{CE} = 3.55$ V. Is the transistor operating in the active region?

Fig. 3.44

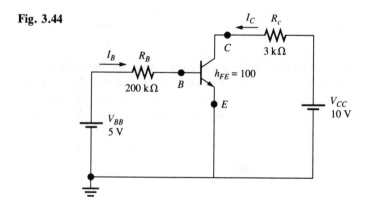

3.2 Change the value of R_B in Problem 3.1 to 50 kΩ. Assume that all other values remain the same, and use PSpice to find I_B, I_C, and V_{CE}. Study your results carefully, and explain why the values are incorrect. *Hint:* Recall that relatively large values of base current may put the transistor into saturation.

3.3 Using the PSpice bias model with $V_{BE} = 0.7$ V, solve for I_B, I_C, and V_{CE} in Fig. 3.45. Is the transistor operating in the active region?

Fig. 3.45

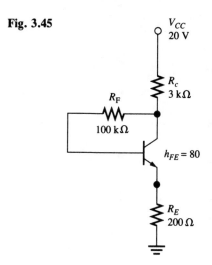

3.4 From the PSpice bias model with $V_{BE} = 0.7$ V, determine I_B, I_C, and V_{CE} in Fig. 3.46.

Fig. 3.46

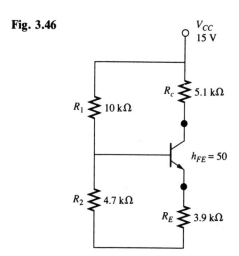

Note: Use the *h* parameters given in Fig. 3.7 for each of the following problems where *h*-parameter analysis is required.

3.5 Use the PSpice model based on the full set of *h* parameters to solve this problem. In Fig. 3.47 find $A_I = I_o/I_i$, $A_V = V_c/V_b$, and $A_{V_s} = V_c/V_s$. *Hint:* For small-signal, low-frequency analysis, the capacitor can be replaced by a short circuit.

Fig. 3.47

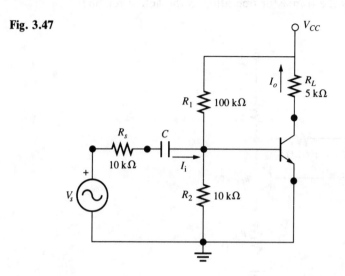

3.6   Use the same PSpice model as in Problem 3.5. For the circuit shown in Fig. 3.48 with R_E added to the circuit, solve for A_I, A_V, and A_{V_s}.

Fig. 3.48

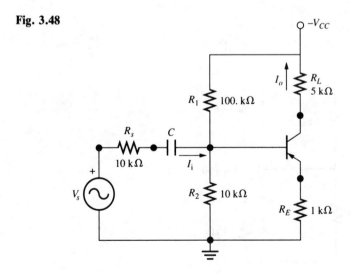

3.7   For each of the amplifiers of Problems 3.5 and 3.6, find the input resistance as seen by the source using PSpice.

3.8 Using the full *h*-parameter model, find A_I, A_V, and R_i for the circuit of Fig. 3.49.

Fig. 3.49

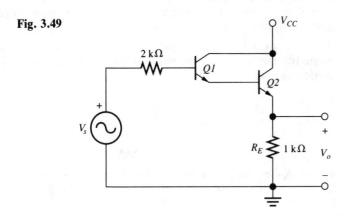

3.9 A *CS* FET amplifier is shown in Fig. 3.50. When the input voltage $V_i = 4$ mV, what will be the output voltage from drain to ground? What is the voltage gain of the amplifier? Is the gain positive or negative? What does this mean?

Fig. 3.50

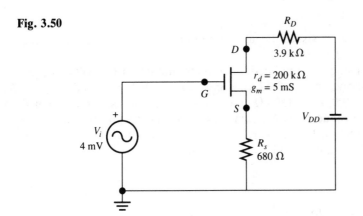

3.10 When the output voltage of the amplifier in Problem 3.9 is taken from the source terminal, it becomes a *CD* amplifier. Using the same values as previously given, what will be the output voltage from source to ground? What is the voltage gain of the amplifier? Is the gain positive or negative?

3.11 A *CS* FET amplifier is to be used over a wide range of frequencies. Given: $R_s = 1$ kΩ, $C_{gs} = 2$ pF, $C_{gd} = 3$ pF, $C_{ds} = 1.5$ pF, $R_L = 48$ kΩ, $g_m = 3$ mS, and $r_d = 100$ kΩ. Run a PSpice analysis and obtain a plot of frequency response for the amplifier. Find the upper 3 dB frequency. What is the midfrequency gain of the amplifier?

3.12 A common-emitter amplifier is shown in Fig. 3.51. It has the following parameters: $g_m = 70$ mS, $R_{ce} = 100$ kΩ, $r_{bb'} = 120$ Ω, $r_{b'e} = 1100$ Ω, $r_{b'c} = 2$ MΩ, $C_c = 2.5$ pF, and $C_e = 80$ pF. In the external circuit $R_s = 1050$ Ω, $R_L = 2.4$ kΩ, and $V_s = 5$ mV. Run a PSpice analysis to determine the frequency response. Determine the midfrequency output voltage and the voltage gain. Find the upper 3 dB frequency.

Fig. 3.51

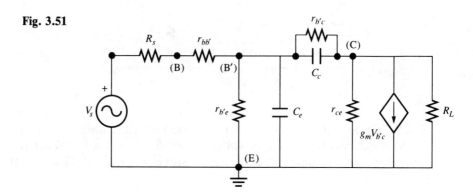

3.13 Run a Probe analysis for the amplifier of Problem 3.12 to determine the input impedance at $f = 50$ kHz.

3.14 Instead of using the simplified h-parameter model for the BJT, an equivalent model is shown in Fig. 3.52(a). Clearly, $r_{bb'} + r_{b'e} = h_{ie}$ and $g_m v_{b'e} = h_{fe} i_b$. Given $h_{fe} = 100$, $h_{ie} = 1200$ Ω, and $r_{bb'} = 100$ Ω, and using the model shown, find the midfrequency gain (from source to collector) of the amplifier shown in Figure 3.52(b). In the figure $R_1 = 20$ kΩ, $R_2 = 10$ kΩ, $R_c = 4.8$ kΩ, and $R_e = 800$ Ω. Treat C_1 and C_2 as short circuits.

Fig. 3.52

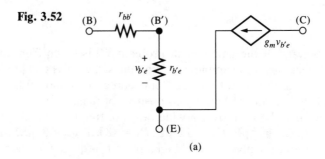

(a)

Fig. 3.52 (continued).

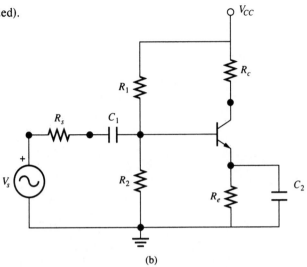

(b)

Multistage Amplifiers, Frequency Response, and Feedback

This chapter covers a variety of topics based on frequency response. We will look at how frequency affects Bode plots, decibel notation, and high-frequency models for the BJT and the FET. We will also present the effects of feedback on single-stage and multistage amplifiers.

LOW-PASS FILTER

For an introduction and a bit of a review, consider the low-pass RC circuit shown in Fig. 4.1(a). Values are $R = 100$ kΩ, $C = 1$ nF, and $V = 1\underline{/0°}$ V. The output is across the capacitor at V(2). The input file for this circuit will allow for a Probe investigation extending from 1 Hz to 1 MHz.

```
High-Frequency Response of Simple Filter
V 1 0 AC 1V
R 1 2 100k
C 2 0 1nF
.AC DEC 20 1HZ 1MEG
.PROBE
.END
```

Run the analysis; then spend some time looking at various aspects of the response. First plot V(2) and look at the shape of the curve. The output level varies from 1 V at $f = 1$ Hz to almost 0 V at $f = 1$ MHz. When the frequency is low, the value of X is large, allowing most of the source voltage of 1 V to appear across node 2. As the frequency increases, X becomes smaller and V(2) diminishes. When $|V_R| = |V_C|$, what will be the value of each voltage? Remember that

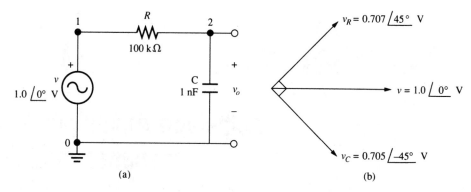

Fig. 4.1   (a) Low-pass filter. (b) Phasor diagram for low-pass filter.

you are dealing with phasors and that these two voltages are always 90° apart as shown in Fig. 4.1(b). When the two voltages are equal in magnitude, $v_C = 0.707\underline{/-45°}$ V.

On the Probe screen, use the cursor to find the frequency that gives V(2) = 0.707 V. Verify that this is at $f = 1.591$ kHz. In ac analysis, the formula is easily shown to be $f_H = 1/(2\pi RC)$, agreeing with your results.

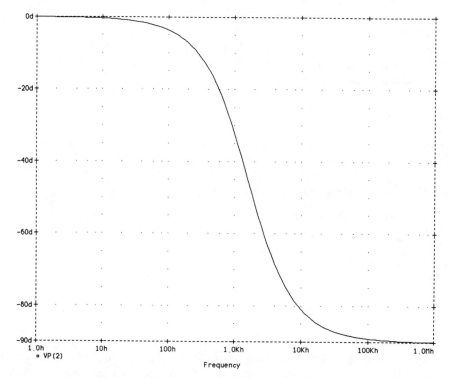

Fig. 4.2   High-Frequency Response of Simple Filter.

Add a plot of VP(2) and verify that when $f = 1.591$ kHz, $\theta = -44.99°$. This would be exactly $\theta = 45°$ if more points were used in the plot. Change the Y-axis to go from $-90°$ to $0°$, and look for the midway point ($-45°$) on the Y-axis. Note that again $f = 1.591$ kHz and that this is the point of inflection on the phase angle plot. Figure 4.2 shows the Bode phase plot.

LOW-FREQUENCY RESPONSE OF HIGH-PASS *RC* NETWORK

The counterpart of the low-pass filter of the previous example is the high-pass network shown in Fig. 4.3. The circuit is again an *RC* combination, but here the

Fig. 4.3 High-pass filter.

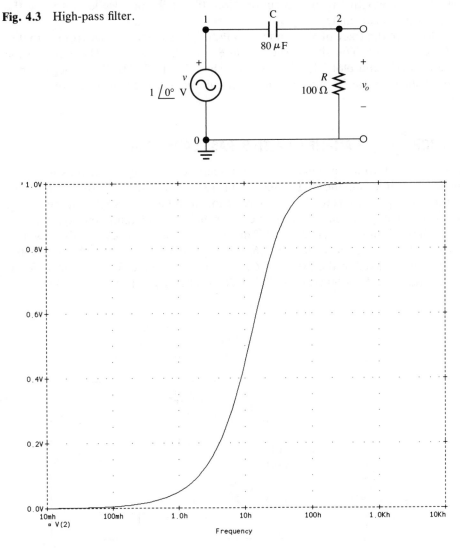

Fig. 4.4 Low-Frequency Response of Simple Filter.

output is taken across R. Values are $R = 100\ \Omega$, $C = 80\ \mu\text{F}$, and $V = 1.0\underline{/0°}$ V. The input file is

```
Low-Frequency Response of Simple Filter
V 1 0 AC 1V
R 2 0 100
C 1 2 80µF
.AC DEC 20 0.01Hz 10kHz
.PROBE
.END
```

Run the analysis; then plot V(2). Using the cursor mode, find the frequency at which the output is down 3 dB. Verify that when V(2) = 0.7081 V, f = 19.95 Hz. Figure 4.4 shows this response curve. Note that it is not a Bode plot since the Y-axis is not logarithmic.

Remove the trace of V(2) and plot VP(2). Using a phase range from 0° to 90°, find the 45° point. You should see that at $\theta = 44.92°$, f = 19.95 Hz. Remove the trace of VP(2) and plot I(R). Verify that when f = 19.95 Hz, $I = 7.08\underline{/45°}$ mA. These values are easily checked with ac circuit theory, and you are encouraged to do so.

COMMON-EMITTER AMPLIFIER WITH BYPASS CAPACITOR

It is customary to use a bypass capacitor such as C_e in Fig. 4.5 across R_e in the common-emitter amplifier. This allows for a larger voltage gain than if C_e were omitted. The problem is to choose a large enough value for C_e so that at the lowest usable frequency the gain will not be down below 3 dB (and consequently the phase shift will not be greater than 45° due to the Z_e value). The ac analysis will be based on the model shown in Fig. 4.6. The h-parameter values used here are the same as those used in the previous CE transistor examples of Chapter 3. Additional values are $R_s = 50\ \Omega$, $R_1 = 50\ \text{k}\Omega$, $R_2 = 8\ \text{k}\Omega$, $R_e = 1\ \text{k}\Omega$, $R_c = 2\ \text{k}\Omega$, $C_b =$

Fig. 4.5 CE amplifier with bypass capacitor.

$h_{fe} = 50$ $h_{re} = 2.5E - 4$
$h_{ie} = 1.1\ \text{k}\Omega$ $\dfrac{1}{h_{oe}} = 40\ \text{k}\Omega$

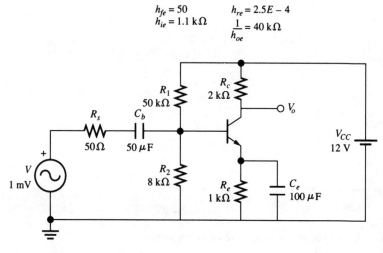

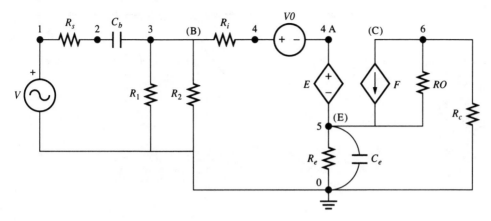

Fig. 4.6 Circuit model for *CE* amplifier with C_e and C_b.

50 μF, $C_e = 100 \, \mu$F, and $V = 1$ mV. The analysis extends from 0.01 Hz to 10 kHz, requiring this input file:

```
Common-Emitter Amplifier with Emitter-Bypass Capacitor
V  1  0  AC  1mV
VO  4  4A  0
E  4A  5  6  5  2.5E-4
F  6  5  VO  50
RS  1  2  50
R1  3  0  50k
R2  3  0  8k
RI  3  4  1.1k
RE  5  0  1k
RO  6  5  40k
RC  6  0  2k
CB  2  3  50uF
CE  5  0  100uF
.AC  DEC  20  0.01Hz  10kHz
.PROBE
.END
```

Run the analysis and in Probe plot V(6), the output voltage. The curve should look like that obtained from the example of the *RC* high-pass filter. Use the cursor mode to determine the midfrequency output voltage. Verify that this is 83.89 mV. Now, look at the *Y*-axis in terms of decibels. Remove this trace and replace it with

$$20*log10(V(6)/84mV)$$

Suddenly, the plot takes on a strange appearance. The information that was essentially hidden in the linear plot now adds more detail to the dB plot. Refer to Fig. 4.7 for this plot. Where are the two curved portions located, and what causes them? The answers will require further investigation.

Set the *Y*-axis for −20 to 0 range, and the *X*-axis for 1 Hz to 10 kHz. Use the cursor to locate the −3 dB point. Verify that this is at about $f = 72$ Hz. This

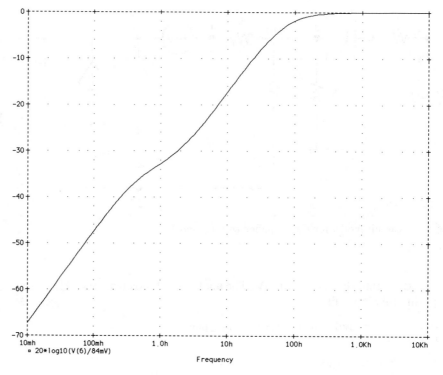

Fig. 4.7   Common-Emitter Amplifier with Emitter-Bypass Capacitor.

frequency is called a pole frequency, but because the circuit also has another capacitor C_b, it will produce a secondary pole at a lower frequency.

In order to concentrate on the effects of C_e alone, modify your input file to eliminate C_b. This is easily done by changing the CB statement to become

```
RB  2  3  0.001
```

Make this change and run the analysis again. In Probe, plot

$$20*\log10(V(6)/84mV)$$

as before. Verify that close to the -3 dB point, $f = 70.8$ Hz. Thus the presence of C_b does not greatly change the location of the first pole. There is also a zero in this circuit, indicated by the frequency at which the response is up by 3 dB from its lowest level. Using the cursor mode, see that on the far left on the plot, the response is -32.91 dB. Verify that when 3 dB is added to this (giving -29.91 dB), the frequency $f = 1.585$ Hz. Thus the zero is located at a frequency of about 1.6 Hz.

If you have interest in seeing what happens when C_e is not responsible for the first pole, simply change the value of R_e to 0.001 Ω and rerun the analysis. This gives a single pole at 3.255 Hz.

TWO-STAGE AMPLIFIER AT HIGH FREQUENCIES

A two-stage CE amplifier based on a simplified hybrid-π model is shown in Fig. 4.8. The values are $V = 1$ mV, $R_s = 50\ \Omega$, $R_{L1} = R_{L2} = 2\ k\Omega$, $r_{bb'} = 100\ \Omega$, $r_{b'e} = 1\ k\Omega$, $g_m = 50$ mS, $C_e = 100$ pF, and $C_c = 3$ pF. The input file becomes

```
Two-Stage CE Amplifier at High Frequencies
V  1  0  AC  1mV
G1 4 0 3 0 50mS
G2 6 0 5 0 50mS
RS 1 2 50
RBB1 2 3 100
RBE1 3 0 1k
RL1 4 0 2k
RBB2 4 5 100
RBE2 5 0 1k
RL2 6 0 2k
CE1 3 0 100pF
CC1 3 4 3pF
CE2 5 0 100pF
CC2 5 6 3pF
.AC DEC 20 100Hz 1MEG
.PROBE
.END
```

Run the analysis and in Probe plot V(6). Using the cursor, verify that at midfrequencies, V(6) = 2.805 V. Remove this trace; then plot

$$20*\log10(V(6)/2.806V)$$

Use the cursor mode to show that the -3 dB point is at $f = 541.2$ kHz. You will notice that the plot does not show an exact straight line region needed to find the -3 dB point by the Bode technique. This is because the amplifier has more than one pole frequency. There is a pole frequency for each capacitor, hence a total of 4 poles. The poles are usually not close together in networks of this type. When one pole is dominant, it will be close to the -3 dB point. From a practical

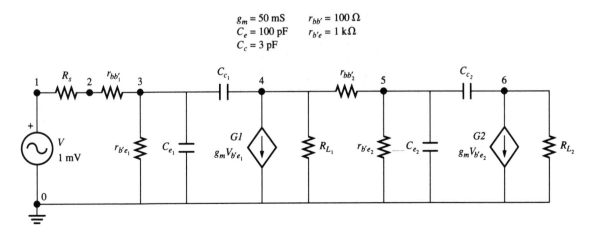

Fig. 4.8 Two-stage CE amplifier at high frequencies.

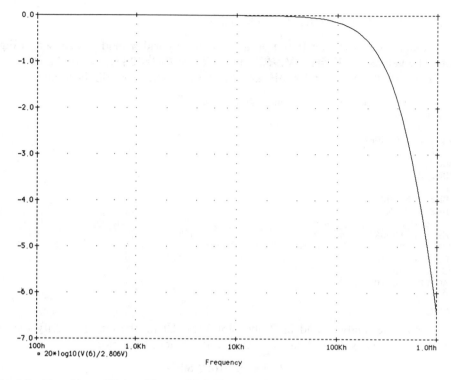

Fig. 4.9   Two-Stage *CE* Amplifier at High Frequencies.

standpoint, finding the -3 dB frequency is more important than locating the frequencies of all of the poles. Refer to Fig. 4.9 for this Bode plot.

Next look at VP(6) and show that at $f = 541.2$ kHz, $\theta = -48°$, completing the analysis.

TWO-STAGE *CE* AMPLIFIER WITH VOLTAGE-SERIES FEEDBACK

Using conventional feedback development, the circuit shown in Fig. 4.10 is analyzed only with a degree of difficulty. Keeping the full set of *h* parameters in the analysis leads to a complicated set of formulas. On the other hand, with the help of SPICE, the nodal analysis is greatly simplified. In the small-signal study of the circuit, assume that all capacitors have been selected to represent short circuits for the chosen range of frequencies. This gives the circuit model shown in Fig. 4.11. Find the voltage gain, R_i, and R_o. Carefully label each node; then create the input file. Compare your file with the one given here:

```
Small-signal Model Voltage-series Feedback, CE Pair
V  1  0  1mV
VO1  3  3A  0
VO2  6  6A  0
E1  3A  4  5  4  2.5E-4
E2  6A  0  7  0  2.5E-4
F1  5  4  VO1  50
```

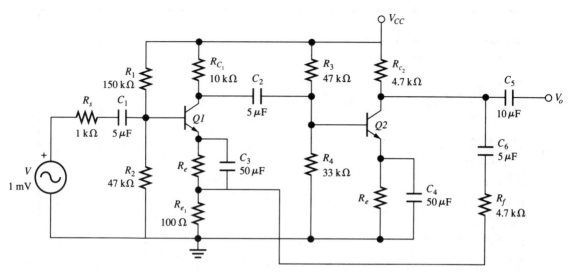

Fig. 4.10 Voltage-series feedback with two CE stages.

```
F2  7  0  V02  50
RS  1  2  1k
R1  2  0  150k
R2  2  0  47k
RI1  2  3  1.1k
RE1  4  0  100
RO1  5  4  40k
RC1  5  0  10k
R3  5  0  47k
R4  5  0  33k
RI2  5  6  1.1k
RO2  7  0  40k
RC2  7  0  4.7k
RF  7  4  4.7k
.TF  V(7)  V
.OP
.END
```

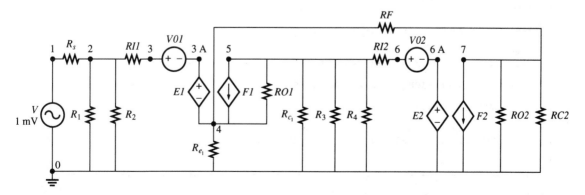

Fig. 4.11 Small-signal, low-frequency model voltage-series feedback, CE.

```
    Small-signal Model Voltage-series Feedback, CE Pair

V 1 0 1mV
V01 3 3A 0
V02 6 6A 0
E1 3A 4 5 4 2.5E-4
E2 6A 0 7 0 2.5E-4
F1 5 4 V01 50
F2 7 0 V02 50
RS 1 2 1k
R1 2 0 150k
R2 2 0 47k
RI1 2 3 1.1k
RE1 4 0 100
RO1 5 4 40k
RC1 5 0 10k
R3 5 0 47k
R4 5 0 33k
RI2 5 6 1.1k
RO2 7 0 40k
RC2 7 0 4.7k
RF 7 4 4.7k
.TF V(7) V
.OP
.END


   NODE    VOLTAGE       NODE    VOLTAGE       NODE    VOLTAGE       NODE    VOLTAGE

(     1)     .0010  (     2) 963.4E-06  (     3) 952.7E-06  (     4) 953.0E-06
(     5)-416.6E-06  (     6) 10.89E-06  (     7)     .0436  (    3A) 952.7E-06
(    6A) 10.89E-06


     VOLTAGE SOURCE CURRENTS
     NAME            CURRENT

     V              -3.664E-08
     V01             9.719E-09
     V02            -3.886E-07


     TOTAL POWER DISSIPATION   3.66E-11  WATTS

**** VOLTAGE-CONTROLLED VOLTAGE SOURCES

NAME           E1          E2
V-SOURCE    -3.424E-07   1.089E-05
I-SOURCE     9.719E-09  -3.886E-07

**** CURRENT-CONTROLLED CURRENT SOURCES

NAME           F1          F2
I-SOURCE     4.860E-07  -1.943E-05

   ****      SMALL-SIGNAL CHARACTERISTICS

       V(7)/V =  4.358E+01

       INPUT RESISTANCE AT V =  2.729E+04
       OUTPUT RESISTANCE AT V(7) =  1.486E+02
```

Fig. 4.12

```
    Small-signal Model Voltage-series Feedback, CE Pair (With RF removed)
    V 1 0 1mV
    V01 3 3A 0
    V02 6 6A 0
    E1 3A 4 5 4 2.5E-4
    E2 6A 0 7 0 2.5E-4
    F1 5 4 V01 50
    F2 7 0 V02 50
    RS 1 2 1k
    R1 2 0 150k
    R2 2 0 47k
    RI1 2 3 1.1k
    RE1 4 0 100
    RO1 5 4 40k
    RC1 5 0 10k
    R3 5 0 47k
    R4 5 0 33k
    RI2 5 6 1.1k
    RO2 7 0 40k
    RC2 7 0 4.7k
    .TF V(7) V
    .OP
    .END

    NODE    VOLTAGE      NODE    VOLTAGE      NODE    VOLTAGE      NODE    VOLTAGE

    (    1)    .0010  (     2) 838.3E-06  (     3) 686.3E-06  (     4) 688.0E-06
    (    5)   -.0061  (     6) 305.9E-06  (     7)   1.2235  (    3A) 686.3E-06
    (   6A) 305.9E-06

       VOLTAGE SOURCE CURRENTS
       NAME          CURRENT

       V            -1.617E-07
       V01           1.382E-07
       V02          -5.818E-06

       TOTAL POWER DISSIPATION   1.62E-10   WATTS

    **** VOLTAGE-CONTROLLED VOLTAGE SOURCES

    NAME        E1           E2
    V-SOURCE   -1.695E-06   3.059E-04
    I-SOURCE    1.382E-07  -5.818E-06

    **** CURRENT-CONTROLLED CURRENT SOURCES

    NAME        F1           F2
    I-SOURCE    6.911E-06  -2.909E-04

     ****      SMALL-SIGNAL CHARACTERISTICS

         V(7)/V =  1.223E+03

         INPUT RESISTANCE AT V =   6.186E+03

         OUTPUT RESISTANCE AT V(7) =   4.236E+03
```

Fig. 4.13

Run the analysis and produce a printed copy of the output file. With careful editing of unnecessary lines, you should be able to print the results on a single page. Refer to Fig. 4.12 for comparison with your results. The analysis shows that $V(7)/V = 43.58$. This is the overall voltage gain. From the results $R_o' = 148.6\ \Omega$, verify that $R_o = 153.5\ \Omega$ with the load removed. From the results $R_i' = 27.29\ k\Omega$, verify that $R_i = 99.01\ k\Omega$ at the base of the first transistor.

How do the results of the analysis using feedback compare to those obtained when R_f is removed? Simply run the analysis again with the RF statement omitted. The results show $V(7)/V = 1223$, $R_o = 42.9\ k\Omega$, and $R_i = 6.06\ k\Omega$. See Fig. 4.13.

TWO-POLE AMPLIFIER MODEL WITH FEEDBACK

As a continuation of the PSpice analysis of frequency response and related topics, consider the circuit of Fig. 4.14. The circuit consists of the resistance, inductance, and capacitance such that it may be used to illustrate important properties of a two-pole feedback amplifier. Although the circuit does not physically contain the active devices associated with amplifiers, it nevertheless has the same frequency, phase, and transient response as an amplifier. Look closely at this circuit, because many of the terms associated with frequency response and transient response may be better understood from a simple circuit such as this.

We will begin our analysis by using the following elements: $V = 1$ V, $R_s = 1\ \Omega$, $L = 20$ mH, $R = 333.33\ \Omega$, and $C = 0.5\ \mu F$. The undamped resonant frequency of this circuit is given by

$$f_o = \frac{1}{2\pi\sqrt{LC}} = 1.59\ \text{kHz}$$

The angular frequency is

$$\omega_o = 2\pi f_o = 10\ \text{krad/s}$$

Other quantities of interest are $Q = R/(\omega_o L)$ and $k = 1/2Q$ (the damping factor). Later, you will see the effects of changing k by changing R; however, in the beginning analysis use $R = 333.33\ \Omega$ and $k = 0.3$. It is interesting to look at the frequency response of this two-pole circuit, keeping in mind that it behaves like an

Fig. 4.14 Two-pole circuit model, amplifier with feedback.

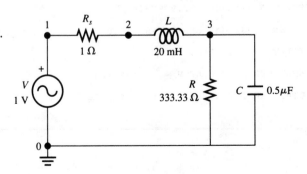

amplifier with feedback. The input file is

```
Two-Pole Circuit Model for Amplifier with Feedback
V 1 0 AC 1
RS 1 2 1
L 2 3 20mH
R 2 0 333.33 Ω
C 3 0 0.5uF
.AC DEC 50 100 10kHz
.PROBE
.END
```

Run the analysis and plot V(3) over the range 100 Hz to 5 kHz. The vertical axis shows that over a certain range of frequencies, the output voltage V(3) exceeds the input voltage of 1 V. From the transfer function you find that the peak occurs at $\omega = \omega_o \sqrt{1 - 2k^2}$. Also the peak value is

$$V_p = \frac{1}{2k\sqrt{1 - k^2}}$$

Calculate what these values should be for this example; then use the cursor mode in Probe to verify these values. The results should indicate a peak at $f = 1.436$ kHz, with $V_p = 1.73$ V. Refer to Fig. 4.15 for this plot.

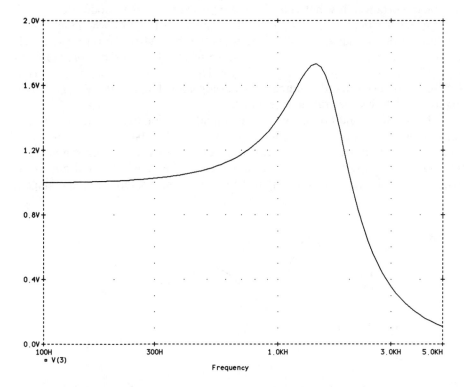

Fig. 4.15 Two-Pole Circuit Model for Amplifier with Feedback.

The next portion of the analysis involves using a step function for the input voltage to see the degree of ringing, or overshoot, associated with this value of k. For this step input voltage, the V statement is changed to include a pulse, *PWL*. The values in parentheses are ordered pairs for time and voltage. Thus at the beginning (0, 0) means that at zero time, the voltage is zero. Then (0.01ms, 1) means that at 0.01 ms, the voltage is 1 V. The rise is taken to be linear in the time interval. The voltage remains at 1 V until $t = 2$ ms. The *.TRAN* statement uses two values, the first of which deals with the print interval (for printing and plotting) and can be ignored for the Probe analysis. The second value represents the final time, 1 ms. Thus the input file is

```
Transient Response of Two-Pole Circuit Model for Amplifier with Feedback
V 1 0 PWL(0,0 0.01ms,1 2ms,1)
RS 1 2 1
R 3 0 333.33
L 2 3 20mH
C 3 0 0.5uF
.TRAN 0.05ms 1.0ms
.PROBE
.END
```

Run the analysis and in Probe plot V(3). Note that the X-axis represents time, because you called for a transient analysis. It extends to 1.5 ms. The Y-axis shows the overshoot of the circuit with its damped oscillatory response. There are several important times that you will find by using the cursor mode. Refer to Fig. 4.16 for the designation of these times. The time $t_{0.1}$ is the time when the response reaches 0.1 of its final value. The time $t_{0.5}$ is the time when the response reaches 0.5 of its final value (the delay time), and so forth. Using the cursor, verify that $t_{0.1} = 52$ μs, $t_{0.5} = 124$ μs, and $t_{0.9} = 187$ μs. This gives a rise time of $(t_{0.9} - t_{0.1}) = 135$ μs. Also verify that the voltage reaches a peak value of 1.368 V at $t = 334$ μs. Figure 4.17 shows the transient response.

Thus PSpice and Probe have given you information that would be very time-consuming to obtain from pencil-and-paper methods. Using such methods, you could hope to find only a few critical points of the plots without enormous effort.

Of equal importance, you can now change the value of k and quickly run another analysis. Return to the input file for frequency response, and change the

Fig. 4.16 Step response of a two-pole network.

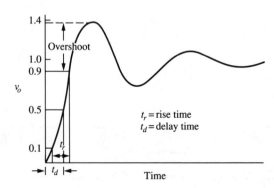

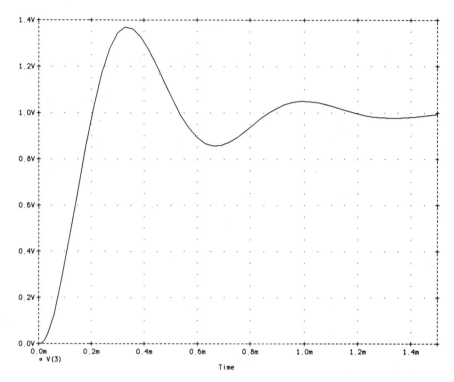

Fig. 4.17 Transient Response of Two-Pole Circuit Model for Amplifier with Feedback.

value of resistance to $R = 141.41\ \Omega$. This is in keeping with the fact that when $2k^2 > 1$, there will be no peak in the frequency response. When $R = 141.41\ \Omega$, $k = 0.707$. Run the frequency analysis with this value of R, and confirm that the response does not peak but begins to drop at a lower frequency. Use other values of k if you would like to continue this analysis. Remember that larger values of R (smaller values of k) will give peaks in the response. Suggested values are $k = 0.4$ and $k = 0.6$.

The transient analysis for each value of k should also be investigated. We have used $k = 0.3$ in the transient analysis. When $k = 0.707$, although there is no peaking in the frequency plot, demonstrate that there is still some overshoot and consequent ringing in the transient response to the step input voltage. According to theory, when $k = 1$, critical damping will be reached and overshoot will be eliminated. This will also mean that the frequency response will show more attenuation for lower frequencies. Run the analysis with $k = 1$, and verify that $t_{0.1} = 57\ \mu s$, $t_{0.5} = 173\ \mu s$, and $t_{0.9} = 403\ \mu s$. Also show that the response is down 3 dB (to 0.707 V) at $f = 1.015$ kHz.

In summary, we have looked at the frequency and transient responses of a two-pole circuit that has the characteristics of a feedback amplifier. Study the results until you have a clear picture of the roles played by Q, k, R, L, and C in the circuit.

Fig. 4.18 *CE* amplifier with voltage-shunt feedback.

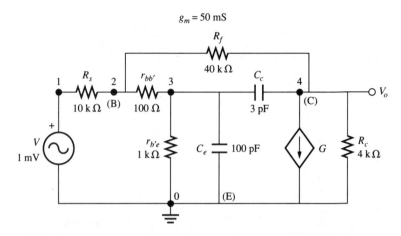

CE AMPLIFIER WITH VOLTAGE-SHUNT FEEDBACK

For an actual amplifier example, Fig. 4.18 shows the simplified hybrid-π model for a *CE* amplifier with voltage-shunt feedback. Since we are interested in the high-frequency response, we will use an ac analysis and a frequency range extending from 1 kHz to 10 MHz. The input file is

```
CE Amplifier with Voltage-Shunt Feedback
V 1 0 AC 1mV
G 4 0 3 0 50mS
RS 1 2 10k
RBB 2 3 100
RBE 3 0 1k
RF 2 4 40k
RC 4 0 4k
CE 3 0 100pF
CC 3 4 3pF
.AC DEC 40 1kHz 10MEGHz
.PROBE
.END
```

Run the analysis; then verify using the cursor mode that V(4) = 3.199 mV in the midfrequency range. With that knowledge, remove the trace and plot

$$20*log10(V(4)/3.2mV)$$

Use the cursor to find the 3 dB point at f = 1.37 MHz.

To demonstrate the effect of R_f on the circuit, remove the input-file statement for RF and run the analysis again. Verify that with R_f removed, V(4) = 18.02 mV at midfrequencies and that the 3 dB point is at f = 243 kHz. As you expect from your study of feedback, R_f stabilizes the circuit, producing a lower voltage gain and a larger bandwidth.

CURRENT-SHUNT FEEDBACK TWO-STAGE CE AMPLIFIER

To further illustrate the effects of feedback over a range of frequencies, Fig. 4.19 shows a current-shunt feedback pair of *CE* stages. The hybrid-π simplified model

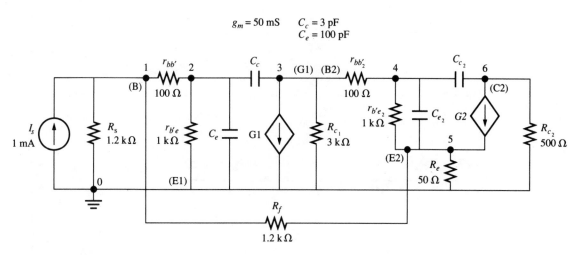

Fig. 4.19 Current-shunt feedback pair, *CE* amplifier.

is again chosen, and $R_f = 1.2$ kΩ is used between the emitter of Q_2 and the base of Q_1. Use this input file to run the analysis:

```
Current-Shunt Feedback Pair
I  0  1  AC  1mA
G1  3  0  2  0  50mS
G2  6  5  4  5  40mS
RS  1  0  1.2k
RBB  1  2  100
RBE  2  0  1k
RC1  3  0  3k
RBB2  3  4  100
RBE2  4  5  1k
RE  5  0  50
RC2  6  0  500
RF  5  1  1.2k
CE  2  0  100pF
CC  2  3  3pF
CE2  4  5  100pF
CC2  4  6  3pF
.AC  DEC  40  10kHz  100MEGHz
.PROBE
.END
```

Verify in Probe that I(RC2) = 22.82 mA (which is 27.16 dB above I_s) at midfrequencies with a current peak of 26.35 mA at $f = 6.6$ MHz. Then use

$$20*\log10(I(RC2)/22.82mA)$$

to obtain the dB plot of the output current. In order to see the peak more clearly, let the *X*-axis extend from 10 kHz to 20 MHz, and let the *Y*-axis go from −5 to 5. Use the cursor to verify that the −3 dB point is at $f = 11.73$ MHz. The plot should look like Fig. 4.20.

Run the analysis with R_f removed from the circuit to show that I(RC2) = 598.9 mA at midfrequencies without the feedback path.

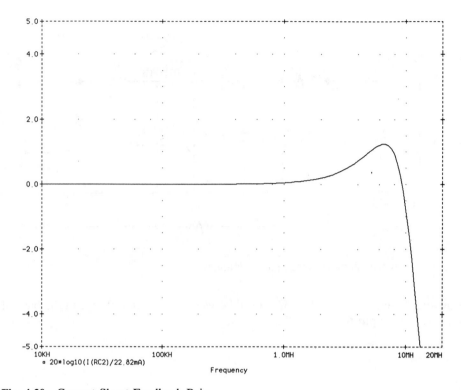

Fig. 4.20   Current-Shunt Feedback Pair.

THREE-STAGE *CE* AMPLIFIER FREQUENCY RESPONSE

We will now look at a case study of a three-stage *CE* amplifier. The circuit is too difficult to analyze without the help of a computer. We will find that SPICE comes to our aid, allowing for an in-depth analysis of the circuit with various parameters. First, we will look at the frequency response of the amplifier without feedback. After that we will introduce a feedback resistor, connected between the collector of the last stage and the base of the first stage. Finally, we will see what needs to be done to correct a problem with severe peaking of the feedback amplifier.

The circuit is shown in Fig. 4.21. Again the simplified hybrid-π model is used for each transistor. For simplicity, the load resistor of each stage is chosen as 2 kΩ. Actually, the load resistor of each of the first two stages represents the parallel combination of biasing and collector resistors. The source is $V = 0.1$ mV, with $R_s = 50\ \Omega$. The input file is

```
Three-Stage CE Amplifier Frequency Response
V 1 0 AC 0.1mV
G1 4 0 3 0 50mS
G2 6 0 5 0 50mS
G3 8 0 7 0 50mS
RS 1 2 50
RBB1 2 3 100
```

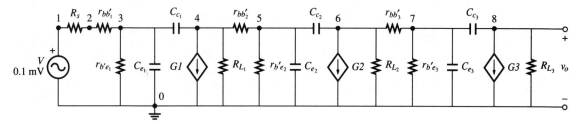

Fig. 4.21 Three-stage *CE* amplifier frequency response.

```
RBE1  3 0 1k
RL1   4 0 2k
RBB2  4 5 100
RBE2  5 0 1k
RL2   6 0 2k
RBB3  6 7 100
RE3   7 0 1k
RL3   8 0 2k
CE1   3 0 100pF
CC1   3 4 3pF
CE2   5 0 100pF
CC2   5 6 3pF
CE3   7 0 100pF
CC3   7 8 3pF
.AC DEC 20 100kHz 1MEGHz
.PROBE
.END
```

Run the analysis, showing that V(8) = 9.046 V at midfrequencies. Then plot

$$20*\log 10(V(8)/9.05V)$$

and check to see that the −3 dB point is located at f = 420 kHz. The gain is probably considerably greater than needed, and the frequency response is somewhat limited compared to what it might be with feedback. Refer to Fig. 4.22 for comparison with your results.

Effects of Circuit Modifications

The second portion of the analysis will be performed with a slightly modified circuit. Change the voltage source to a current source, based on the Norton equivalent, and change the load resistor to let R_{L3} = 50 Ω. The modifications give

```
I  2 0 AC 2uA
RS 2 0 50
RL3 8 0 50
.AC DEC 20 10kHz 10MEGHz
```

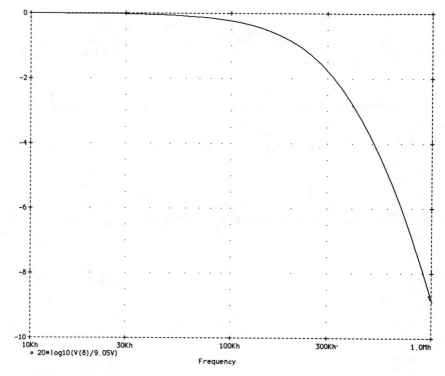

Fig. 4.22 Three-Stage *CE* Amplifier Frequency Response.

Note that the *RS* statement was changed because node *1* has been eliminated. Refer to Fig. 4.23 for this detail. Run the analysis and verify that I(RL3) =
4.52 mA at midfrequencies. Then using

$$20*\log10(I(RL3)/4.53mA)$$

to obtain the dB plot, verify that the response is down 3 dB at f = 771.8 kHz. This
simply demonstrates that the smaller value of *RL* extends the bandwidth.

Fig. 4.23 Modified three-stage
CE amplifier.

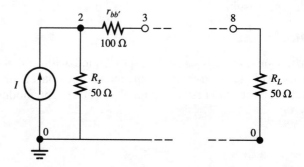

Three-Stage Amplifier with Voltage-Shunt Feedback

Next, consider a more significant change. Insert a feedback resistor, $R_f = 5$ kΩ, between nodes 8 and 2, that is, from the collector of stage 3 back to the base of stage 1. This produces voltage-shunt feedback, which you have seen in previous examples. Modify the input file by adding the statement for R_f:

```
RF  8  2  5k
```

Then run the analysis, verifying that I(RL3) = 191 μA at midfrequencies. Adjust the X-axis to display a range from 100 kHz to 20 MHz, and adjust the Y-axis to show a range of -20 to 20. Verify that the current peak is 17.85 dB at $f_p = 7.94$ MHz. Also show that the -3 dB point is at $f = 11.68$ MHz. Figure 4.24 shows this plot.

The sudden, steep rise in the output response is, of course, undesirable. Consider eliminating this feature by placing a suitable capacitor across the feedback resistor, R_f. The capacitor will introduce another zero into the gain expression. The obvious choice for the location of this zero is at f_{peak}. This is accomplished using $C_f = 1/2\pi R_f f_p$. Using $f_p = 8$ MHz gives $C_f = 4$ pF. Insert the statement

```
CF  8  2  4pF
```

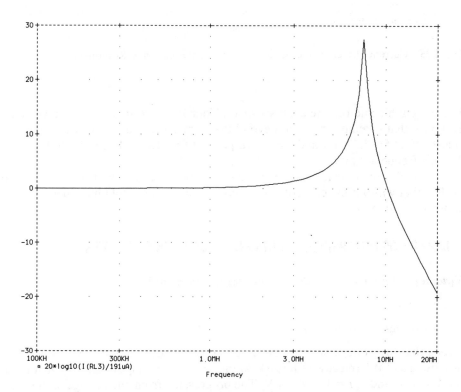

Fig. 4.24 Current-Driven, Three-Stage *CE* Amplifier Frequency Response.

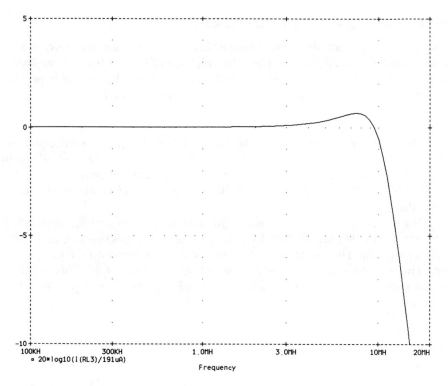

Fig. 4.25   Current-Driven Three-Stage CE Amplifier Frequency Response.

in the input file and run the analysis again. Your Probe results should show a gain response that is almost flat well beyond the former peak location, with a peak of 0.6695 dB at $f = 7.5$ MHz and the -3 dB point at $f = 11.67$ MHz. Refer to Fig. 4.25 for this frequency plot.

Spend some time looking over the results of these studies. It is important to realize that without the computer as a tool, the analyses would have been far too difficult to attempt.

SUMMARY OF NEW PSPICE TREATMENT USED IN THIS CHAPTER

V[*name*] *<+node> <−node>* [*transient specification*]

For example,

```
V 1 0 PWL(0us 0V 1us 1V 1s 1V)
```

means that the voltage source connected between nodes *1* and *0* is a waveform described as *PWL* (piecewise linear). At time $= 0$, the voltage is zero; then at $t = 1$ μs, $V = 1$ V, and at $t = 1$ s, $V = 1$ V. The progression from one voltage to the next is linear.

Various Forms of Transient Specification

Several forms of transient specifications are available in PSpice for describing independent voltage or current sources. These are useful when the source is not simply dc or ac and you want to run a transient analysis. We will now describe these sources in detail, including simple examples of each.

The Exponential Source

This transient specification has the form

$$\text{exp}(<v1> <v2> <td1> <\tau1> <td2> <\tau2>)$$

where

$v1$ = initial voltage

$v2$ = peak voltage

$td1$ = rise delay time

$\tau1$ = rise time constant

$td2$ = fall delay time

$\tau2$ = fall time constant

As an example, consider the following input file:

```
The Exponential Source
V 1 0 exp(2V 12V 2s 1s 7s 1s)
R 1 0 1
.tran 0.1s 12s
.probe
.end
```

Figure 4.26 shows the Probe output of v(1). The trace shows $V = 2$ V as the initial value; then at $t = 2$ s the voltage begins an exponential rise toward 12 V with a rise time constant of $\tau1 = 1$ s. At $t = 7$ s the voltage begins to fall exponentially toward its initial voltage with a fall time constant of $\tau2 = 1$ s. Note that $td1$ and $td2$ are specified with respect to $t = 0$.

The Pulse Source

This transient specification has the form

$$\text{pulse}(<v1> <v2> <td> <tr> <tf> <pw> <per>)$$

where

$v1$ = initial voltage

$v2$ = pulsed voltage

td = delay time

tr = rise time

tf = fall time

pw = pulse width

per = the period if the wave recurs

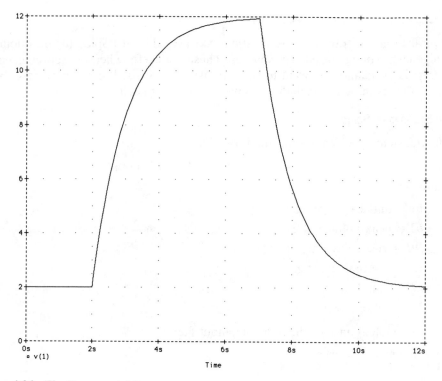

Fig. 4.26 The Exponential Source.

For example, look at this input file:

```
The Pulse Source
V 1 0 pulse(0 5V 0.5ms 0.1ms 0.1ms 0.8ms 2ms)
R 1 0 1
.tran 0.02ms 4ms
.probe
.end
```

Figure 4.27 shows the Probe output of v(1). The trace shows V = 0 V until 0.5 ms (the delay time); then the voltage rises toward 5 V with a rise time of 0.1 ms. The pulse width is 0.8 ms, followed by a fall time of 0.1 ms. The period is 2 ms, after which the pulse repeats. Note the slope on both the leading and the trailing edges of the pulse; this is due to the rise and fall times of 0.1 ms.

The Piecewise-Linear Source

This transient specification has the form

$$PWL(<t1> <v1> <t2> <v2> \cdots <tn><vn>)$$

where $t1$ = time associated with voltage $v1$, $t2$ = time associated with voltage $v2$, and so forth. The progression from one voltage level to another is linear, as you

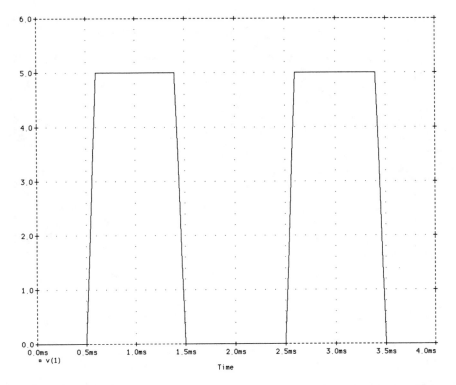

Fig. 4.27   The Pulse Source.

would get by "connecting the dots." As an example, consider this input file:

```
The Piecewise-Linear Source
V 1 0 PWL(0s 0V 0.2s 3V 0.4s 5V 0.6s -5V 0.8s -3V 1s 0V)
R 1 0 1
.tran 0.01s 1s
.probe
.end
```

Figure 4.28 shows the Probe output of v(1). Note that in the *PWL* statement, times are given first, followed by the corresponding voltages. Successive times must be given; the voltages can be either positive or negative values.

The Frequency-Modulated Source

This transient specification has the form

$$\text{SFFM}(<vo> <va> <fc> <m> <fs>)$$

where

    vo = offset voltage

    va = amplitude of voltage

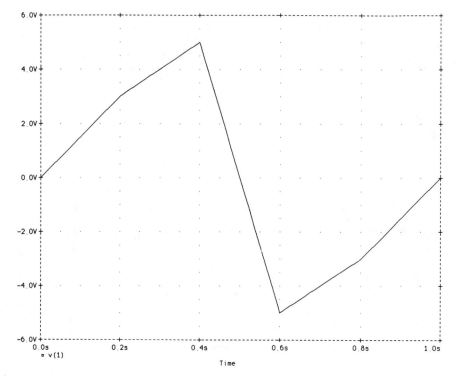

Fig. 4.28   The Piecewise-Linear Source.

fc = carrier frequency

m = index of modulation

fs = signal frequency

For example, look at this input file:

```
Single-Frequency FM Source
V 1 0 sffm(0V 5V 10kHz 3 1kHz)
R 1 0 1
.tran 0.005ms 1ms
.probe
.end
```

Figure 4.29 shows the Probe output of v(1). Since the carrier frequency f_c = 10 kHz, the time display on the X-axis shows 10 cycles of the carrier in the time of 1 ms. The carrier is modulated at the rate determined by the signal frequency and the index of modulation. Note the larger spacing of the waves toward the center of the trace. When a small value of m is used, the shifting of the carrier is less noticeable. When a larger value of m is used, the shifting is more pronounced.

As an exercise, run the *sffm* analysis using $m = 6$, and compare the results with those shown here.

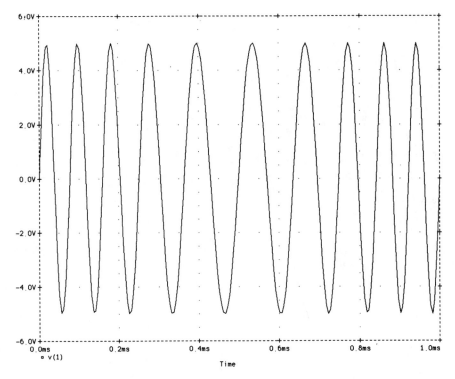

Fig. 4.29 Single-Frequency FM Source.

The Sine-Wave Source

This transient specification has the form

$$\text{sin}(<vo> <va> <f> <td> <df> <phase>)$$

where

   vo = offset voltage

   va = voltage amplitude

   f = frequency

   td = time delay

   df = damping factor

   $phase$ = the phase of the sine wave

An example will clarify the use of this transient specification.

```
The Sine-Wave Source
V 1 0 sin(0.3V 1V 500Hz 0 500 0)
R 1 0 1
.tran 0.06ms 6ms
.probe
.end
```

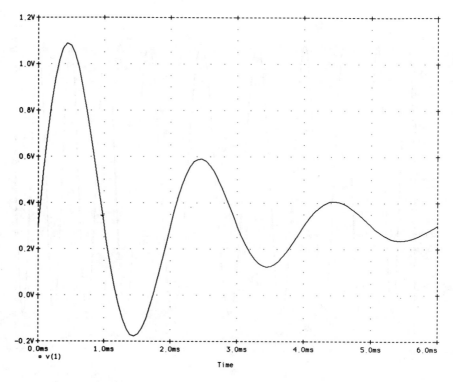

Fig. 4.30   The Sine-Wave Source.

Figure 4.30 shows the results. The Probe output of v(1) is for 6 ms, representing 3 cycles of the damped sine wave. The damping is of the form

$$e^{-at}$$

where a is the damping factor, which in our example is 500. Note that when $t = 2$ ms, this represents

$$e^{-1}$$

It is obvious that smaller values of a will produce less damping and that when $a = 0$, the wave is undamped.

In summary, the transient specifications for independent sources (voltages or currents) may be shown in a variety of ways. These specifications are designed to be used with transient analyses, requiring the use of the .tran statement in the input file.

THE STIMULUS EDITOR

PSpice comes with a program *STMED.EXE* that is intended to allow the user to create and modify various stimuli. The stimuli are the exponential, pulse, piecewise-linear, frequency-modulated, and sine-wave sources that were introduced in this chapter.

A stimulus in PSpice is an independent-current or independent-voltage source. Therefore its description must be either *I[name]* or *V[name]*.

The main feature of the Stimulus Editor is that it allows the user to customize the source in an editing environment. In the evaluation version, the program is limited to sine-wave stimuli and is therefore of limited value.

An Example Using the Stimulus Editor

Create the input file with the name *stim.cir* for a circuit with a sine-wave source in series with a coil as shown in Fig. 2.1. The input file is

```
Stimulus Circuit
V1 1 0 sin(0.2 1.0 60 0 0 0)
R 1 2 1.5
L 2 0 5.3mH
.tran 0.5ms 20ms
.end
```

In this example, the input sine wave has these parameters: voffset = 0.2, vamplitude = 1.0, frequency = 60, tdelay = 0, dampfactor = 0, and phase = 0. Run the PSpice analysis and plot v(1), v(2), and v(1,2). The traces are shown in Fig. 4.31.

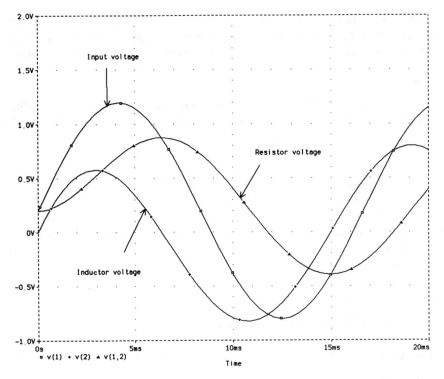

Fig. 4.31 Stimulus Circuit.

If you would like to modify the sine-wave parameters, the most direct way to do this is in the input file. But the Stimulus Editor can also be used to accomplish the job as follows: At the DOS prompt, type

```
stmed stim
```

Since the input file has the name *stim.cir,* it will be loaded into the Stimulus Editor, and the stimulus V1 will be displayed along with the main menu. Select *Modify_stimulus*, in order to make changes in the sine-wave parameters. Next select *Transient_parameters*, which will display this menu:

1)VOFF 2)VAMPL 3)FREQ 4)TD 5)DF 6)PHASE

Assume that we want to change the value of voltage offset to zero, the damping factor to 10, and the phase angle to 45°. Select these items (*1, 5,* and *6,* respectively) and make the changes, following the prompts. Exit after the changes have been completed, and notice the new appearance of V1 on the screen. Since this is an interactive program, other changes can be made if necessary. Finally, choose *Exit* again, which gives you the opportunity to produce hard copy, and *Exit* (twice) again to leave the program.

When you examine the input file you will notice this entry for V1:

```
V1 1 0
+ SIN( 0 1 60 0 10 45)
```

When the Stimulus Editor is used to modify or create a stimulus, the stimulus specifications will appear on a separate line. Now you may run the analysis again to compare the traces of v(1), v(2), and v(1,2) with those of the previous run.

Adding a New Stimulus

The procedure for adding a new stimulus is similar to that for modifying a stimulus. If you want to add a second voltage to the series loop of the *stim.cir* input file, go into the stimulus editor by typing

```
stmed stim
```

Then from the main menu choose *Add_stimulus*.

Let us add another sine wave of double the frequency, with an amplitude of 1.5 V. Select *Transient_parameters* and enter the values as prompted. Now when you exit, you will see a display of both sine waves on the screen, allowing for further changes as necessary. When you exit from the program and look at the input file of *stim.cir*, you will see the following addition:

```
V2 PNode NNode
+ SIN( 0 1.5 120 0 0 0)
```

Since the nodes *PNode* and *NNode* have not been identified in the input file, you must determine where the new stimulus goes in the circuit. If you desire to place V2 in series with V1, the following changes should be made:

```
Stimulus Circuit
V1 1 0
```

```
+ SIN ( 0 1 60 0 10 45)
R 1a 2 1.5
L 2 0 5.3mH
V2 1a 1
+ SIN (0 1.5 120 0 0 0)
.tran 0.5ms 20ms
.end
```

Save the input file with a new name if desired and run the analysis.

The Stimulus Editor is available for use under the shell *ps* also. Its operation is menu driven and requires no further explanation.

PROBLEMS

4.1   As an extension of the low-pass filter shown in Fig. 4.1, Fig. 4.32 shows a circuit with two resistors and two capacitors. Using a PSpice analysis, produce a graph showing magnitude and phase of the output voltage. Identify the 3 dB frequency.

Fig. 4.32

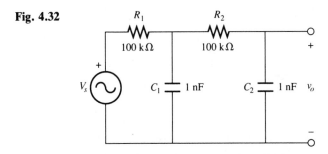

4.2   Analyze the *CE* amplifier of Fig. 4.5 using the simplified *h*-parameter model. Compare the results with those obtained using the full *h*-parameter model.

4.3   Use the simplified *h*-parameter model with $h_{ie} = 1.1$ kΩ and $h_{fe} = 80$ for each of the two stages of the amplifier shown in Fig. 4.33. Note that the first stage is *CE* and the second stage is *CC*. Given $R_s = 100$ Ω, $R_c = 4$ kΩ, and $R_e = 2$ kΩ, and with $V_s = 2$ mV, find the output voltage at midfrequencies.

Fig. 4.33

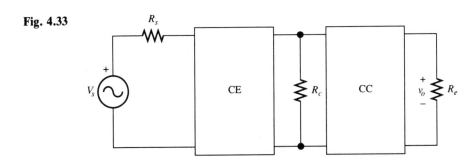

4.4 A cascode amplifier is often used for high-frequency applications. Assume in Fig. 4.34 that both FETs have $g_m = 5$ mS, $r_d = 50$ kΩ, $C_{gs} = 5$ pF, $C_{gd} = 4$ pF, $C_{ds} = 0.5$ pF, $C_1 = C_2 = C_3 = C_4 = 50$ μF, $R_1 = 470$ kΩ, $R_2 = 100$ kΩ, $R_3 = 180$ kΩ, $R_s = 800$ Ω, and $R_d = 4$ kΩ. Find the midfrequency gain of the amplifier and the upper 3 dB frequency based on a PSpice analysis.

Fig. 4.34

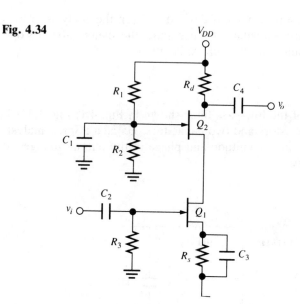

4.5 Use the high-frequency model of the FET with $C_{gs} = 5$ pF, $C_{gd} = 3$ pF, $C_{ds} = 0.4$ pF, $g_m = 6$ mS, and $r_d = 500$ kΩ. A source follower is shown in Fig. 4.35. Using $V_s = 1$ mV, obtain a plot of frequency showing the upper 3 dB point when (a) $R_s = 2$ kΩ and (b) $R_s = 10$ kΩ.

Fig. 4.35

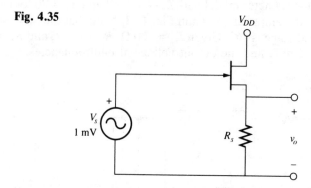

4.6 Use the hybrid-π model for the current-series feedback amplifier shown in Fig. 4.36. Given: $g_m = 50$ mS, $r_{bb'} = 100$ Ω, $r_{b'e} = 1$ kΩ, $C_c = 4$ pF, $C_e =$

80 pF, $r_{ce} = 80$ kΩ. With $R_s = 500$ Ω, $V_s = 1$ mV, and $R_L = 4$ kΩ, plot the frequency response when (a) $R_e = 300$ Ω and (b) $R_e = 500$ Ω.

Fig. 4.36

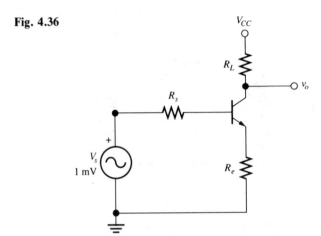

4.7 Figure 4.37 shows an example of voltage-shunt feedback. Use the simplified h-parameter model with $h_{fe} = 100$ and $h_{ie} = 1.1$ kΩ. Given: $R_s = 500$ Ω and $R_e = 4$ kΩ. Find the midfrequency voltage gain and the input and output resistances with $V_s = 1$ mV when (a) $R_f = 27$ kΩ and (b) $R_f = 40$ kΩ.

Fig. 4.37

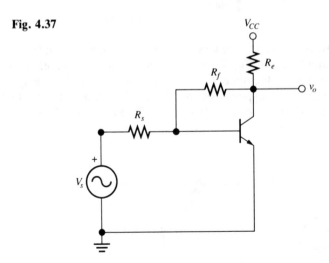

4.8 A *CE* amplifier uses a shunt-peaking coil as shown in Fig. 4.38(a) to enhance its high-frequency response. The Miller-approximation model of the circuit is shown in Fig. 4.38(b). Given: $R_L = 500$ Ω, $R_1 = 100$ Ω, $R_i = 1.1$ kΩ, $L = 5$ μH, $g_m = 0.2$ mS, and $r_b = 100$ Ω. Using $C_c = 5$ pF and $C_e = 100$ pF, verify that $C_i = 605$ pF. Create a PSpice input file to determine the frequency response of the circuit. Find the midfrequency gain of the amplifier. Compare the upper 3 dB frequencies with and without the peaking coil in the circuit.

Fig. 4.38

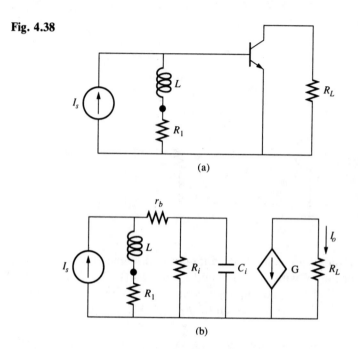

(a)

(b)

4.9 Wiring often introduces series inductance which has an effect on frequency response in an amplifier. In Fig. 4.39 a simplified amplifier model is shown which includes series inductance $L = 0.5$ mH. Run a PSpice analysis to determine the frequency response of the circuit. Given $v = 1$ mV, find v_o over the frequency range from 100 Hz to 10 MHz. For comparison, assume that L is neglected, and run the analysis again.

Fig. 4.39

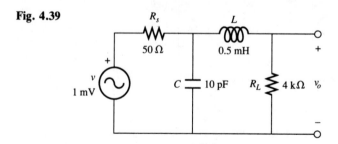

The Operational Amplifier

The operational amplifier (op amp) is an integrated-circuit device widely used in electronics. The actual circuitry is complex and has many features which may or may not be needed in our model. Therefore we will begin with only the essential components and refer to this model as the ideal op amp.

THE IDEAL OPERATIONAL AMPLIFIER

The ideal operational amplifier will be modeled for SPICE as an amplifier with high input resistance, zero output resistance, and high voltage gain. Typical values of these parameters are shown in Fig. 5.1, where $R_i = 1$ GΩ, $A = 200,000$, and $v_o = A(v_2 - v_1)$. Note that v_1 is an inverting input and v_2 is a noninverting input. This model will serve for dc and low-frequency analysis. We will add other features to the model as needed.

Although you do not need SPICE in the analysis of simple op amp circuits, it is desirable to see what information you can obtain even in these situations. There are also some limitations that deserve your attention.

The circuit of Fig. 5.2(a) shows the op amp being used in a negative feedback connection. The feedback resistor, R_2, is connected from the output to v_1, the inverting *minus* input terminal. The noninverting input is grounded. Figure 5.2(b) shows the PSpice version of the circuit. The input file is

```
Ideal Operational Amplifier
VS 1 0 1V
E 3 0 0 2 200E3
RI 2 0 1G
R1 1 2 1k
```

```
R2  3  2  10k
. OP
. TF  V(3)  VS
. END
```

Run the analysis and look at the results in the output file. Verify that V(3)/VS = −9.999. The gain is very close to −10 and is sometimes approximated

Fig. 5.1 The ideal operational amplifier.

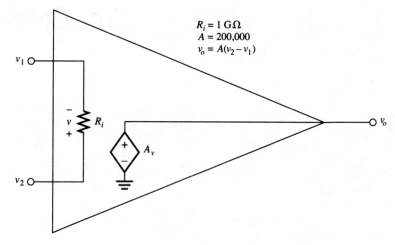

$$R_i = 1\,\text{G}\Omega$$
$$A = 200{,}000$$
$$v_o = A(v_2 - v_1)$$

Fig. 5.2 The ideal inverting op amp.

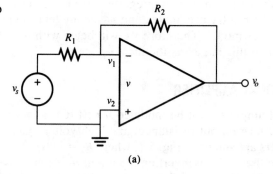

(a)

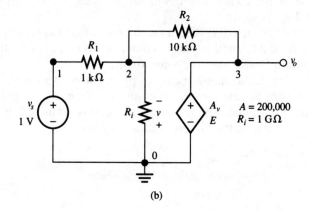

$$A = 200{,}000$$
$$R_i = 1\,\text{G}\Omega$$

(b)

as $v_o/v_s = -R_2/R_1$. Using nodal analysis, write the equations required to solve for the ratio v_o/v_s. Demonstrate that the results depend on the value of A and that only when A becomes infinite is the approximation equation correct.

Your analysis should show $R_{in} = 1\,k\Omega$. Can you explain this? Remember that the inputs to the operational amplifier can be thought of as a virtual ground. This means that the input resistance is seen as R_1.

NONINVERTING IDEAL OPERATIONAL AMPLIFIER

Figure 5.3 shows another simple circuit for the op amp. This circuit has v_s connected to the noninverting (+) input. Figure 5.4 shows the model along with the components. The input file is

```
Ideal Operational Amplifier, Noninverting
VS 1 0 1V
E 3 0 1 2 200E3
RI 1 2 1G
R1 2 0 1k
R2 3 2 9k
.OP
.TF V(3) VS
.END
```

Fig. 5.3 Noninverting ideal op amp.

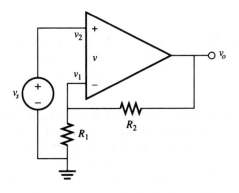

Fig. 5.4 Noninverting ideal op amp model.

$$A = 200,000$$
$$R_1 = 1\,k\Omega$$
$$R_2 = 9\,k\Omega$$

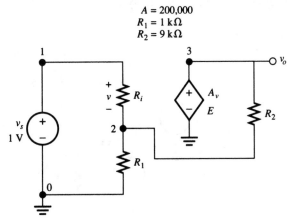

Verify that V(3)/VS = 10, in keeping with the formula $v_o/v_s = 1 + R_2/R_1$. Also verify that $R_{in} = 2.0E13$. Why is the input resistance so large a value? Since the ideal draws virtually no current, the source v_s sees what appears to be almost an open circuit in this case.

OP AMP GIVING VOLTAGE DIFFERENCE OUTPUT

If the op amp has inputs to both its plus and minus terminals, it may be used to find the difference of these two voltages, with appropriate amplification. To keep the analysis simple, assume that $R_1 = R_3 = 5\ k\Omega$ and $R_2 = R_4 = 10\ k\Omega$ in Fig. 5.5. The SPICE model for the ideal op amp and its external components is shown in Fig. 5.6. The input file is

```
Op Amp Giving Voltage Difference Output
VA 1 0 3V
VB 4 0 10V
E 5 0 3 2 200E3
RI 2 3 1G
R1 1 2 5k
R2 5 2 10k
R3 4 3 5k
R4 3 0 10k
.OP
.TF V(5) VB
.END
```

The analysis will show that V(5) = 14 V. Using nodal analysis on the ideal op amp, you should verify that

$$v_o = \frac{R_2(v_b - v_a)}{R_1}$$

which agrees with our results. A few more pencil-and-paper calculations will give insight. Begin by finding the voltage at the plus input of the op amp. This is easily done when you recall that ideally the op amp inputs draw no current. The voltage v_b divides (using voltage division) to give $v+ = 6.667$ V. This means that $v-$ is also 6.667 V (actually PSpice gives this as 6.666 V). Using this voltage, the current through R_1 and R_2 can easily be found. The output file is shown in Fig. 5.7.

Fig. 5.5 Op amp giving voltage difference output.

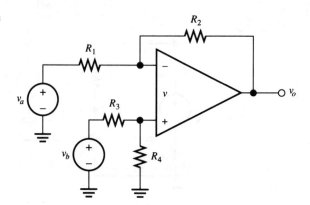

Fig. 5.6 Op amp model for voltage difference.

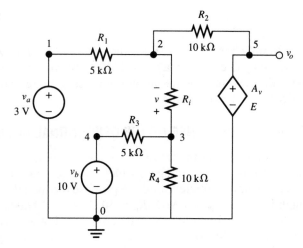

```
   Opamp Giving Voltage Difference Output

   ****      CIRCUIT DESCRIPTION

   VA 1 0 3V
   VB 4 0 10V
   E 5 0 3 2 200E3
   RI 2 3 1G
   R1 1 2 5k
   R2 5 2 10k
   R3 4 3 5k
   R4 3 0 10k
   .OP
   .TF V(5) VB
   .END

   NODE    VOLTAGE    NODE    VOLTAGE    NODE    VOLTAGE    NODE    VOLTAGE

   (    1)    3.0000  (    2)    6.6666  (    3)    6.6667  (    4)   10.0000
   (    5)   14.0000

      VOLTAGE SOURCE CURRENTS
      NAME            CURRENT

      VA              7.333E-04
      VB             -6.667E-04

      TOTAL POWER DISSIPATION    4.47E-03   WATTS

   **** VOLTAGE-CONTROLLED VOLTAGE SOURCES

   NAME          E
   V-SOURCE      1.400E+01
   I-SOURCE     -7.333E-04

   ****      SMALL-SIGNAL CHARACTERISTICS

       V(5)/VB =  2.000E+00

       INPUT RESISTANCE AT VB =  1.500E+04

       OUTPUT RESISTANCE AT V(5) =  0.000E+00
```

Fig. 5.7

Remember that SPICE should not be used simply as a way to get numerical results in problems such as this. Hopefully, you will ask yourself some questions about the results, which will help you learn more about circuit analysis and the devices being investigated.

FREQUENCY RESPONSE OF THE OPERATIONAL AMPLIFIER

When you consider the frequency response of an op amp, you must use a model that accounts for the rolling off which will occur as the frequency increases. Using typical op amp characteristics, we propose the model of Fig. 5.8. Study the model, which includes $R_{in} = 1$ MΩ, $R_o = 50$ Ω, $A_o = 100,000$, and $f_c = 10$ Hz. The low-frequency or dc open-loop gain is represented by A_o. The symbol f_c represents the frequency at which the open-loop response is down by 3 dB.

For an example of how to look at the frequency response, refer to Fig. 5.9. The external resistors are $R_1 = 10$ kΩ, and the feedback resistor is $R_2 = 240$ kΩ. In creating the input file for SPICE, note carefully the power choice of the *plus* and *minus* input terminals, shown as p and m, relating to the *EG* statement. The input file becomes

```
Op Amp Model with 3 dB Frequency at 10 Hz for Open-Loop Gain
VS 2 0 AC 1mV
EG 3 0 2 1 1E5
E 6 0 4 0 1
RI1 3 4 1k
RO 6 5 50
R1 0 1 10k
R2 5 1 240k
RIN 1 2 1MEG
C 4 0 15.92uF
.AC DEC 40 100 1MEG
.PROBE
.END
```

Run the file; then use this statement to find the frequency response:

$$20*\log10(V(5)/V(2))$$

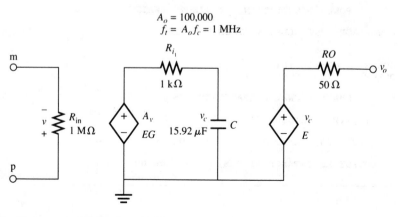

Fig. 5.8 Op amp model for $f_c = 10$ Hz.

Fig. 5.9 Frequency response of op amp model.

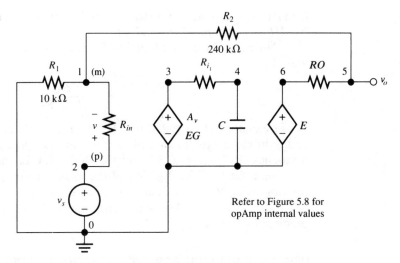

Verify that at midfrequencies the gain $A_{\text{mid}} = 27.96$ dB and that the frequency where the gain is down by 3 dB is $f = 39.15$ kHz. In order to check the accuracy of these values, recall that $f_t = A_o f_c$, representing the unity-gain frequency. The model assumes that $f_t = 1$ MHz, which is a typical value for the unity-gain

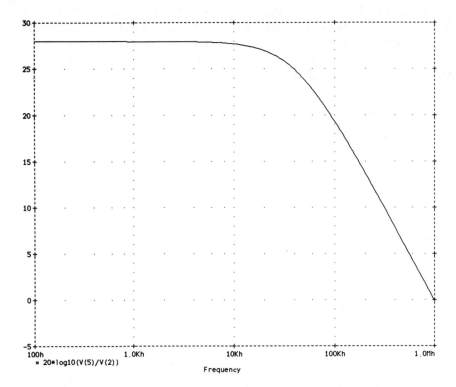

Fig. 5.10 Opamp model with 3 dB frequency at 10 Hz for open-loop gain.

frequency. It also assumes that $f_c = 10$ Hz, giving $A_o = 1E5$. The value of f_c is set by $RII = 1$ kΩ and $C = 15.92$ μF.

Note that the closed-loop bandwidth is given approximately by $CLBW = f_t\beta$ and

$$\beta = \frac{R_1}{R_1 + R_2}$$

In our example, $\beta = 10/250 = 0.04$, and $f_t\beta = 40$ kHz. This is an approximation and is in close agreement with our model that gave $f = 39.15$ kHz as the 3 dB frequency. Refer to Fig. 5.10 for the Bode plot for this example.

As a further investigation of the model, change the value of the feedback resistor, using $R_2 = 15$ kΩ, and run the analysis again. Verify that $A_{\text{mid}} = 7.959$ dB and $f_{3dB} = 393.6$ kHz. Using the approximate formula and the new β, what is the predicted value of f_{3dB}?

USING A SUBCIRCUIT FOR THE OPERATIONAL AMPLIFIER

The model we have used for the op amp in the previous example contains enough elements to make it a good candidate for use in a subcircuit. This will also serve as an introduction to the concept of subcircuits. The model is shown in Fig. 5.11. Observe that the nodes are shown using lower-case letters. This is not necessary, since PSpice is not case-sensitive. That is, upper- and lower-case can be used interchangeably. However, to make the subcircuit and its elements easier to iden-tify, we chose lower-case node labels. We chose letters rather than numbers, so that the nodes will not be confused with elements external to the subcircuit. The subcircuit will be given as an independent portion of an input file, but it is not a complete input file in itself. The statements in the subcircuit will be

```
.subckt opamp m p vo
   eg 1 0 p m 1e5
   e c 0 b 0 1
   rin m p 1meg
   ril a b 1k
   c b 0 15.92uf
   ro c vo 50
.ends
```

Fig. 5.11 Subcircuit for op amp with nodes designated.

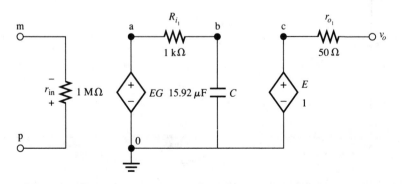

Each subcircuit begins with a *.subckt* statement. The first item in its list is the subcircuit name, which is *opamp* in this example. This is followed by a set of nodes, which link the subcircuit to the rest of the input file. You can think of these as externally available nodes. In this example they are m, p, and v_o. The reference node is always given as node *0*, and this need not be included in the node list.

Identify the elements in the subcircuit in the usual manner. Since this is a subcircuit and not a complete input file, it does not matter that some of the nodes appear to be floating. The element statements are indented to make them easy to identify, but this is not a requirement. Finally, the statement *.ends* marks the end of the subcircuit.

Now you are ready to look at a new version of the analysis of the op amp using the subcircuit. The complete circuit is shown in Fig. 5.9 and is repeated as Fig. 5.12. After you have more experience, you may want to represent the subcircuit simply by using a box or a triangle. In the figure you will see that the nodes m, p, and v_o have new designations. These are called nodes *1*, *2*, and *3*, respectively. In order to use the subcircuit, the main circuit file must contain a statement such as this:

```
X 1 2 3 opamp
```

The X designates a subcircuit call. The nodes *1*, *2*, and *3* are in the proper order to conform to nodes m, p, and v_o in the subcircuit. This allows the subcircuit to receive the node designation being passed from the main circuit file. The statement also contains the subcircuit name, *opamp*. Now look at the entire input file:

```
Op Amp Analysis Using Subcircuit
VS 2 0 AC 1mV
R1 1 0 10k
R2 3 1 240k
X 1 2 3 opamp
.AC DEC 40 100 1MEG
.PROBE
```

Fig. 5.12 Model showing subcircuit to be called by main circuit.

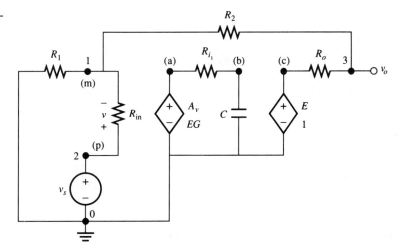

```
.subckt opamp m p vo
  eg 1 0 p m 1e5
  e c 0 b 0 1
  rin m p 1meg
  ril a b 1k
  c b 0 15.92uf
  rol c vo 50
.ends
.END
```

Run the analysis and verify that it gives the same result as the previous analysis where the subcircuit was not used.

OP AMP DIFFERENTIATOR CIRCUIT

A differentiator circuit based on an ideal op amp is shown in Fig. 5.13. With the inverting input at a potential of zero volts, $v_c = v$. It is easily shown that with $R = 0.5 \ \Omega$,

$$v_o = \frac{-dv}{dt}$$

Thus when the input voltage is a triangle as shown in Fig. 5.13(b), the output should be a square wave. Use this input file to test the conclusion:

```
Differentiator Circuit
v 1 0 PWL (0, 0 1s, 1V 2s, 0)
C 1 2 2
R 2 3 0.5
X 2 0 3 iop
.subckt iop m.p vo
  ri m p 1meg
  e vo 0 p m 2e5
.ends
.TRAN 0.05s 2s
.PROBE
.END
```

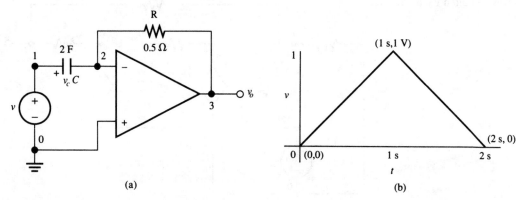

Fig. 5.13 Differentiator using ideal op amp.

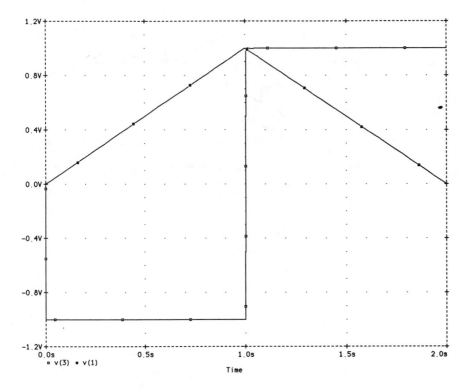

Fig. 5.14 Differentiator circuit.

Run the analysis and verify that the output is a square wave alternating from −1.0 V to 1 V. The polarity of the output voltage indicates the inversion that takes place in the op amp also. Plot v(3) along with v(1). Refer to Fig. 5.14 for the results. Note that the input-file statement for *C* should *not* be given as

```
C 1 2 2F
```

where the *F* is intended for farads. It will be taken as a prefix, giving a value of 2 fF (femtofarads). If you like to use unit symbols whenever possible, you might use this alternate form:

```
C 1 2 2E6UF
```

OP AMP INTEGRATOR CIRCUIT

The counterpart of the differentiator is the integrator. In the circuit of Fig. 5.15(a) the positions of *R* and *C* are interchanged when compared to Fig. 5.13. The new circuit is an (inverting) integrator. In order to test its properties, use the waveshape shown in Fig. 5.15(b) and this input file:

```
Integrator Circuit
v 1 0 PWL (0, 0 0.01ms, −1V 1s, −1V 1000.01ms, 0V 2s, 0V 2000.01ms, 1V
3s, 1V)
```

Fig. 5.15 Integrator using ideal op amp.

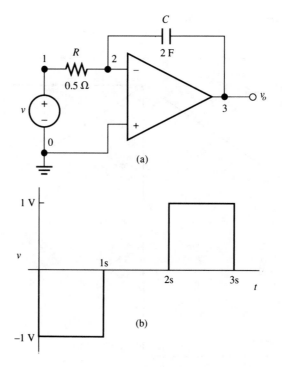

(a)

(b)

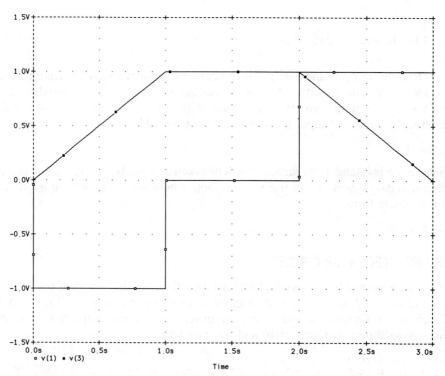

Fig. 5.16 Integrator circuit.

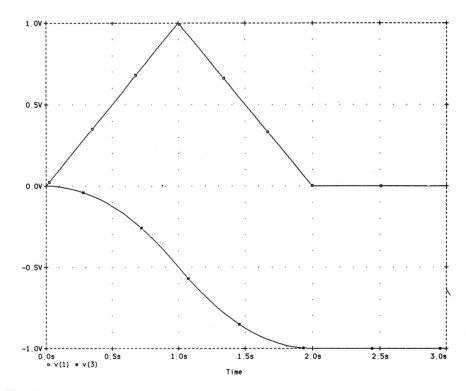

Fig. 5.17 Integrator circuit.

```
R 1 2 0.5
C 2 3 2
X 2 0 3 iop
.subckt iop m p vo
  ri m p 1meg
  e vo 0 p m 2e5
.ends
.tran 0.05s 3s
.probe
.end
```

Run the analysis and plot v(1) along with v(3). Verify that the output starts out as a ramp, reaching a peak value of 1 V, then begins to fall back to zero between 2 s and 3 s. Refer to Fig. 5.16 for the results.

As an additional exercise, use the input waveform from the differentiator example and see what output voltage you obtain. Verify its parabolic shape with a final value of −1 V. Figure 5.17 shows this plot.

RESPONSE TO UNIT STEP FUNCTION

A unit step function is shown in Fig. 5.18(b). By definition, it remains at zero volts until $t = 0$, and from that time forward it is 1 V. The circuit shown in Fig. 5.18(a) has for its external components $R = 2\ \Omega$, $R_1 = 1\ \Omega$, and $C = 0.125$ F. An analysis

Fig. 5.18 Response of first-order circuit to unit step function.

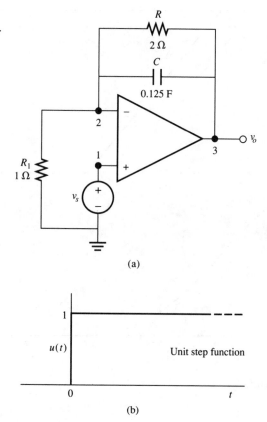

(a)

(b)

of the circuit shows that

$$v_o(t) = (3 - 2e^{-4t})u(t)$$

You may want to sketch this response before beginning the PSpice analysis, in order to know what you will be looking for. The input file is

```
Response to Unit Step Function
vs 1 0 PWL (0, 0 1us, 1V 5s, 1V)
C 2 3 0.125
R 2 3 2
R1 2 0 1
X 2 1 3 iop
.subckt iop m p vo
  ri m p 1meg
  e vo 0 p m 2e5
.ends
.TRAN 0.05s 3s
.PROBE
.END
```

When you run the Probe analysis verify, using the cursor, that at $t = 0.5$ s, $v_o = 2.729$ V. This is in agreement with the equation given for this circuit. The results are shown in Fig. 5.19.

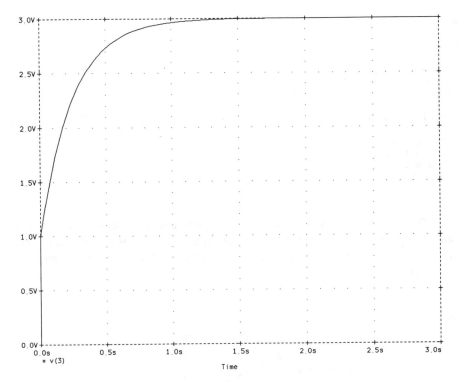

Fig. 5.19 Response to unit step function.

DOUBLE OP AMP CIRCUIT

Unless there is a need to use a device such as the op amp more than once in an input file, there is little to gain by setting up the subcircuit. But in some situations, there may be several of the same devices. In these cases, it is much easier to work with the subcircuit. Suppose that you would like to compare the frequency responses of the two op amp circuits we looked at previously (in the section "Frequency Response of the Operational Amplifier"). Recall that the circuits were alike except that in the first case R_2 = 240 kΩ, while in the second case R_2 = 15 kΩ. By looking at their frequency characteristics on the same graph, you can get a better comparison.

In order to do this, simply extend the circuit so that both cases are covered at the same time. We will define the op amp in the subcircuit and use Fig. 5.20 to provide for easy identification of the nodes. Note that *Op1* and *Op2* are shown merely as triangles, but because you are already familiar with their model, there is no need to repeat the internal details. Now your input file is easily obtained:

```
Double Op Amp Circuit for Gain-Bandwidth Analysis
VS1 2 0 AC 1mV
R1 1 0 10k
R2 3 1 240k
```

```
X1 1 2 3 opamp
VS2 5 0 AC 1mV
R3 4 0 10k
R4 6 4 15k
X2 4 5 6 opamp
.AC DEC 40 100 10MEG
.PROBE
.subckt opamp m p vo
  eg 1 0 p m 1e5
  e c 0 b 0 1
  rin m p 1meg
  ri1 a b 1k
  c b 0 15.92uf
  ro c vo 50
.ends
.END
```

The subcircuit is described as before. Once you develop the subcircuit, you can merely copy it into any input file where it is needed. It is called twice, first by the *X1* statement, then next by the *X2* statement. The list of nodes used in each case is in keeping with Fig. 5.20.

Fig. 5.20 Double op amp circuit.

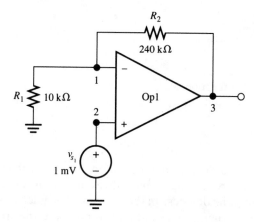

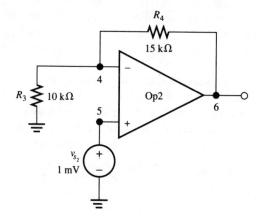

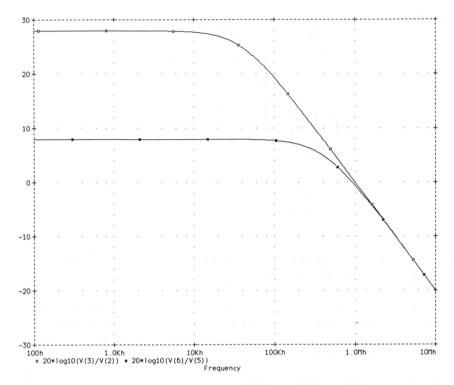

Fig. 5.21 Double op amp circuit for gain-bandwidth analysis.

Run the analysis; then plot

$$20*\log10(V(3)/V(2))$$

and

$$20*\log10(V(6)/V(5))$$

Use the cursor mode to find the 3 dB point of the first trace. Note that the first trace is automatically selected when you use the cursor mode. Verify that $A_{mid} = 27.96$ dB and $f_{3dB} = 39.8$ kHz.

Now follow the second trace with the cursor. Press [*Ctrl*] [*Rt arrow*] to activate the cursor for the second trace. Then move along the second trace until you find the desired information. Note that the second trace shows $A_{mid} = 7.96$ dB, which is down 20 dB from the first trace. The frequency to look for will correspond to a gain of 4.96 dB (7.96 − 3.00). Verify that this gives $f_{3dB} = 398$ kHz. These results are in agreement with the previous examples. Refer to Fig. 5.21 for this double plot.

ACTIVE FILTERS

Active filters can be used to provide low-pass, high-pass, and band-pass filters with improved cutoff properties when compared to simple single-pole filters,

Fig. 5.22 Second-order Butterworth filter, low pass.

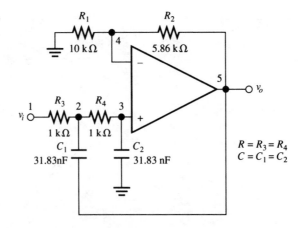

which contain only a single capacitor, for example. The Butterworth filter serves as a classic example of an active filter.

We often use the operational amplifier in the construction of active filters because it is available with high gain-bandwidth products. We will not attempt to include the theory of filters in this discussion. If you are studying active filters for the first time, refer to a good reference to better understand the elegance and simplicity of these circuits.

Second-Order Butterworth Low-Pass Filter

Using a table of normalized Butterworth polynomials, we find these factors for the second-order filter:

$$s^2 + 1.414s + 1$$

The second-order filter is shown in Fig. 5.22. For an introductory example, we would like to find the elements R_1, R_2, R, and C for a Butterworth filter with a cutoff frequency at f_c = 5 kHz. As usual, the cutoff frequency is taken as the frequency at which the response is down by 3 dB. According to theory, the low-frequency gain is given by

$$A_{vo} = 3 - 2k$$

where k represents the damping factor, defined as one-half the coefficient of s in the quadratic given from the Butterworth polynomial table.* For this example, $k = 0.707$ and

$$A_{vo} = 3 - 1.414 = 1.586$$

Let R_1 = 10 kΩ. Since

$$A_{vo} = \frac{R_1 + R_2}{R_1}$$

---

* See Hillburn and Johnson, *Manual of Active Filter Designs*, McGraw-Hill, 1973.

Fig. 5.23 Ideal op amp subcircuit.

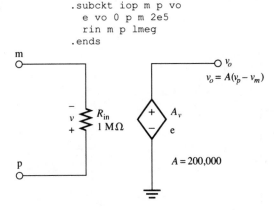

```
.subckt iop m p vo
  e vo 0 p m 2e5
  rin m p 1meg
.ends
```

$v_o = A(v_p - v_m)$

$A = 200,000$

then $R_2 = 5.86$ kΩ. If you let $R = 1$ kΩ, then since $f_c = 1/(2\pi RC)$, you find $C = 31.83$ nF. To test the Butterworth theory, use the ideal model of the op amp as a subcircuit, as shown in Fig. 5.23. Now construct the input file as follows:

```
Second-Order Butterworth Filter
VI 1 0 AC 1mV
R3 1 2 1k
R4 2 3 1k
R1 4 0 10k
R2 5 4 5.86k
C1 2 5 31.83nF
C2 3 0 31.83nF
X 4 3 5 iop
.AC DEC 40 1 100kHz
.PROBE
.subckt iop m p vo
  e vo 0 p m 2e5
  rin m p 1meg
.ends
.END
```

Run the analysis and plot V(5)/V(1). Check to see that $A_{vo} = 1.586$, in agreement with our prediction. Then remove the trace and plot

$$20*\log 10(V(5)/(V(1)*1.587V))$$

to confirm that $f_c = 5$ kHz. This second-order filter should have about twice the attenuation rate as a first-order filter. Recall that a first-order filter has an attenuation rate of 20 dB/decade. Verify that at $f = 10$ khz, A_v is down 12.31 dB, and at $f = 100$ kHz, A_v is down by 52.05 dB. This is approximately 40 dB/decade. This plot is shown in Fig. 5.24.

Fourth-Order Butterworth Low-Pass Filter

For another example, consider a fourth-order Butterworth filter designed to give $f_c = 1$ kHz. From the polynomial table, we find these factors:

$$(s^2 + 0.765s + 1)(s^2 + 1.848s + 1)$$

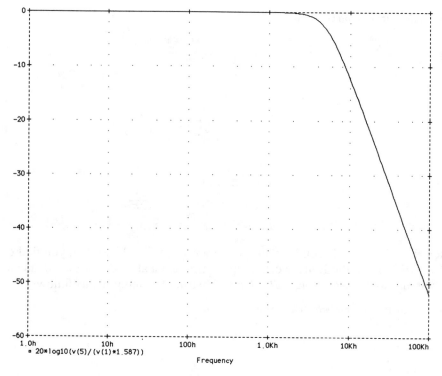

Fig. 5.24   Second-order butterworth filter.

The damping factor k is one half the coefficient of s in each quadratic factor, giving $k_1 = 0.383$ and $k_2 = 0.924$.

$$A_{v1} = 3 - 2k_1 = 3 - 0.765 = 2.235 \quad \text{and} \quad A_{v2} = 3 - 2k_2 = 1.152$$

For the first stage, let $R_1 = 10$ kΩ, and using

$$A_{v1} = \frac{R_1 + R_2}{R_1}$$

we find that $R_2 = 12.35$ kΩ. For the second stage, let $R_1 = 10$ kΩ, giving $R_2 = 1.52$ kΩ. For $f_c = 1$ kHz, if you let $R = 1$ kΩ, then $C = 0.16$ μF. The circuit is shown in Fig. 5.25. Note that since each element must have a unique designation, the R and C values shown are expanded from those given in this paragraph. The input file becomes

```
Fourth-Order Butterworth Filter
VI 1 0 AC 1mV
R3 1 2 1k
R4 2 3 1k
R1 4 0 10k
R2 5 4 12.35k
R7 5 6 1k
R8 6 7 1k
R5 8 0 10k
```

```
R6 9 8 1.52k
C1 2 5 0.16uF
C2 3 0 0.16uF
C3 6 9 0.16uF
C4 7 0 0.16uF
.AC DEC 40 1 10kHz
.PROBE
.subckt iop m p vo
  e vo 0 p mm 2e5
  rin m p 1meg
.ends
X1 4 3 5 iop
X2 8 7 9 iop
.END
```

Run the analysis; then make traces together for V(5)/V(1), V(9)/V(5), and V(9)/V(1). These represent the gain of the first stage, the gain of the second stage, and the overall gain, respectively. Since these are not decibel plots, you should be able to easily verify that $A_{v1} = 2.235$, $A_{v2} = 1.152$, and A_v (overall) $= A_{v1}A_{v2} = 2.575$. You can find these values by using the cursor mode at low frequencies. Refer to Fig. 5.26 for these plots.

Obtain a printed copy of this graph including all three traces for further study. Note the interesting peak in the A_{v1} graph. This is compensated in the graph of A_{v2} so that the trace of overall gain is flat over almost all of its pass band of frequencies; then it drops off steeply near the frequency of 1 kHz.

The rate of attenuation can be determined more easily from the decibel plot. Use the method involving 20*log10(V(9)/V(1)), and so forth, to replace the three traces with logarithmic traces. Verify that for the overall circuit, $f_c = 1$ kHz. Also observe the rate of attenuation for each of the three traces. You should be able to show that for each of the two stages, the rate of attenuation is about 10 dB/decade compared to about 20 dB/decade for the combined stages. Looking at the results of this example, you will feel some of the delight that comes from seeing the main ideas conveyed in such a graphical manner. You should also appreciate how much

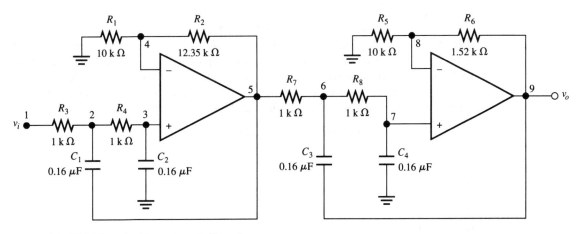

Fig. 5.25 Fourth-order Butterworth filter, low pass.

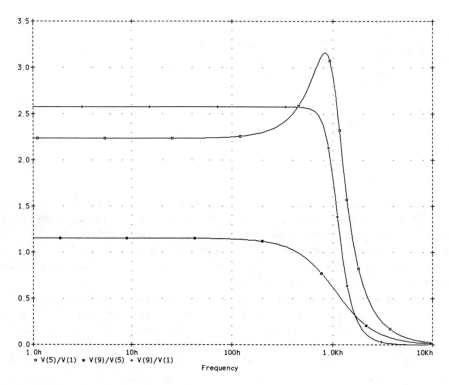

Fig. 5.26   Fourth-order Butterworth filter.

time and effort are spared by using such a powerful computational tool as PSpice. Refer to Fig. 5.27 for these plots.

   We can show one additional feature of the Butterworth filter by using a slight modification of the previous input file. Compare the two-stage filter with the four-stage filter. A few calculations will be necessary because you do not have data for a two-stage filter with $f_c = 1$ kHz.

   The low-frequency gain will be the same as before for the two-stage filter. This is, $A_v = 1.586$. Let $R_1 = 10$ kΩ, giving $R_2 = 5.86$ kΩ. Using $R = 1$ kΩ, find that $C = 0.159$ μF. The circuit extension for this filter is shown in Fig. 5.28. Note that the extension involves an extension of node numbering more than anything else; this filter has its own input and is not physically linked with the four-stage filter. When this information is added to the original input file, it becomes

```
Fourth-Order Butterworth Filter Compared to Second-Order
VI 1 0 AC 1mV
R3 1 2 1k
R4 2 3 1k
R1 4 0 10k
R2 5 4 12.35k
R7 5 6 1k
R8 6 7 1k
R5 8 0 10k
R6 9 8 1.52k
```

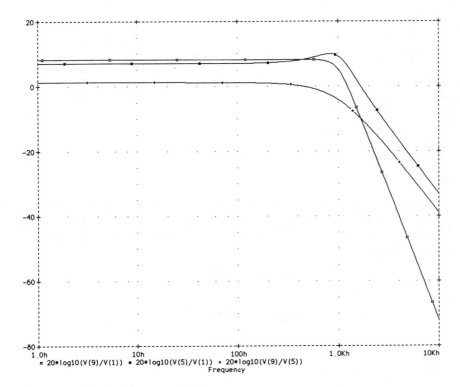

Fig. 5.27   Fourth-order Butterworth filter.

```
C1  2  5  0.16uF
C2  3  0  0.16uF
C3  6  9  0.16uF
C4  7  0  0.16uF
VI1 10 0  AC 1mV
R9  13 0  10k
R10 14 13 5.86k
```

Fig. 5.28   Circuit extension to
include second-order filter.

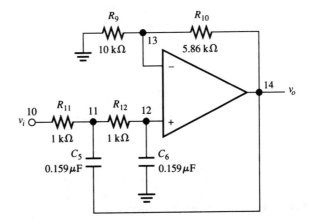

```
R11 10 11 1k
R12 11 12 1k
C5 11 14 0.159uF
C6 12 0 0.159uF
X1 4 3 5 iop
X2 8 7 9 iop
X3 13 12 14 iop
.AC DEC 40 1 10kHz
.PROBE
.subckt iop m p vo
  e vo 0 p m 2e5
  rin m p 1meg
.ends
.END
```

Run the analysis and plot decibel traces of V(9)/V(1) for the fourth-order filter and V(14)/V(10) for the second-order filter. You should find that $A_v = 4.006$ dB (second order) and $A_v = 8.214$ dB (fourth order). We want to show these on a comparable basis, so plot

$$20*\log10(V(14)/V(10))$$

and

$$20*\log10(V(9)/V(1)) - 4.208$$

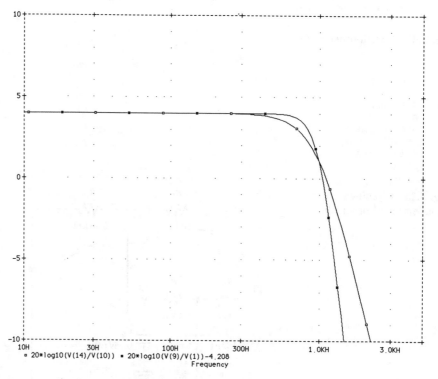

Fig. 5.29 Fourth-order Butterworth filter.

The value 4.208 represents the offset of the second trace from the first, normalizing the second trace with respect to the first. This graph with its two overlapping traces in the low-frequency range clearly shows that both Butterworth filters have the same f_c. This will apply to Butterworth filters of all orders. See Fig. 5.29 for these plots.

ACTIVE RESONANT BAND-PASS FILTER

A simple resonant circuit can give steep cutoff characteristics by making use of the resonant properties of an *RLC* combination. Figure 5.30 shows an input loop containing V_s, R, L, and C. We will choose the component values to provide a particular bandwidth B and quality factor Q as outlined here.

The center frequency is taken as the frequency at which L and C resonate, which is given by

$$f = \frac{1}{2\pi\sqrt{LC}}$$

The quality factor Q is defined by $Q = \omega_o L/R$. In this type of filter circuit, $B = f_0/Q = R/(2\pi L)$. For example, we will choose $Q = 2$, $f_0 = 11$ kHz, and $R = 10$ kΩ. This leads to $L = 0.289$ H, and $C = 0.724$ nF. To complete the circuit, $R_1 = 10$ kΩ and $R_2 = 10$ kΩ are chosen for feedback and to produce a suitable A_v for this noninverting amplifier. The input file is

```
Active Resonant Band-Pass Filter
VS 1 0 AC 1mV
R 3 0 10k
R1 4 0 10k
R2 5 4 10k
L 1 2 0.289H
C 2 3 0.724nF
.AC DEC 40 1kHz 100kHz
.PROBE
```

Fig. 5.30 Active resonant band-pass filter with $Q = Z$.

for $Q = 2$ $L = 0.289$ H
      $R = 10$ kΩ $C = 0.724$ nF

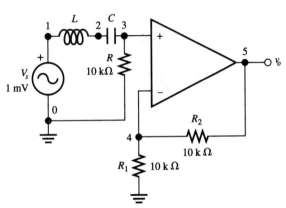

```
.subckt iop m p vo
  e vo 0 p m 2e5
  rin m p 1meg
.ends
.END
```

Run the analysis and look at the ratio of output to source voltage V(5)/V(1) on a logarithmic scale. Verify the predicted center frequency and the bandwidth. Values to look for at the 3 dB points are about f_1 = 8.5 kHz and f_2 = 14.1 kHz, giving B = 5.6 kHz. You will see that the center frequency turns out to be at about 11.2 kHz.

Also obtain a trace of VP(5) to observe the phase shift changes near the resonant frequency.

It is interesting to compare two circuits of this type that have different values of Q, the quality factor. We have seen the results when Q = 2, and now we will add another circuit with Q = 5. Refer to Fig. 5.31, which shows the added circuit. The bandwidth B = 2.2 kHz, and retaining R = 10 kΩ gives L = 0.723 H, and C = 0.289 nF. The nodes are numbered in such a way that the original input file can be appended. This will allow us to look at the response of both circuits at the same time in Probe.

Add the following statements to the input file shown above:

```
VS1 6 0 AC 1mV
R11 8 9 10k
R3 9 0 10k
R4 10 9 10k
L1 6 7 0.723H
C1 7 8 0.289nF
X1 9 8 10 iop
```

Run the analysis and plot together

$$20*\log10(V(5)/V(1))$$

and

$$20*\log10(V(10)/V(6))$$

Fig. 5.31 Circuit extension for band-pass filter with Q = 5.

for Q = 5 L_1 = 0.723 H
       R_{11} = 10 kΩ C_1 = 0.289 nF

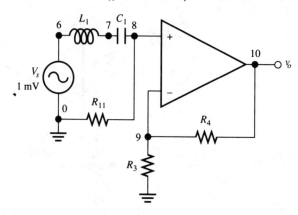

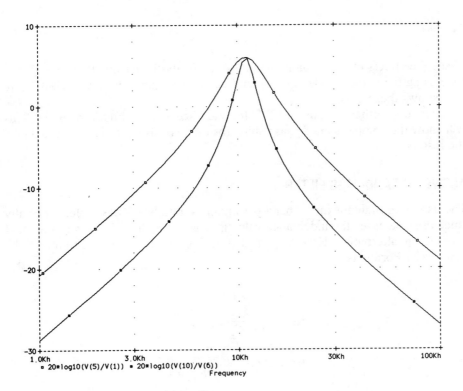

Fig. 5.32 Active resonant band-pass filter.

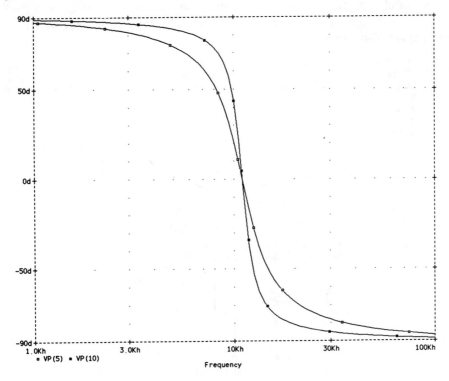

Fig. 5.33 Active resonant band-pass filter.

to see the effects of $Q = 5$ along with $Q = 2$. Verify the bandwidth of the $Q = 5$ case using the cursor mode. This should be almost exactly $B = 2.2$ kHz. Figure 5.32 shows these curves.

Obtain another plot, using VP(5) for one trace and VP(10) for the other. This will show the comparison of phase shifts for the two cases. Refer to Fig. 5.33 for the results.

ACTIVE *RC* BAND-PASS FILTER

The use of an inductor in the band-pass filter is not always desirable, especially since in some cases the inductance value is large. The circuit shown in Fig. 5.34 provides an alternative. Here only capacitors and resistors are used to obtain the band pass. Formulas are

$$R_1 = \frac{Q}{A_o \omega_o C_1}$$

$$R_3 = \frac{Q}{\dfrac{\omega_o C_1 C_2}{C_1 + C_2}}$$

$$R_p = R_1 \| R_2 = \frac{1}{\omega_o^2 R_3 C_1 C_2}$$

For an example, we will choose $A_o = 50$, $f_o = 160$ Hz, and $B = 16$ Hz. Convenient values can also be chosen for the capacitors, so we will let $C_1 = C_2 = 0.1 \ \mu F$. Solve for $Q = f_o/B$. Also find R_1, R_2, and R_3. Check your answers with those shown below. The input file is

```
Active RC Band-Pass Filter
VS 1 0 AC 1mV
R1 1 2 2k
R2 2 0 667
R3 4 3 200k
C1 2 4 0.1uF
C2 2 3 0.1uF
```

Fig. 5.34 Active *RC* band-pass filter.

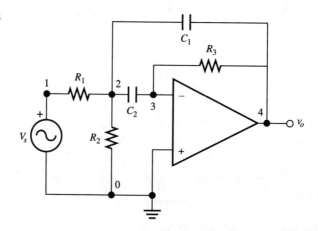

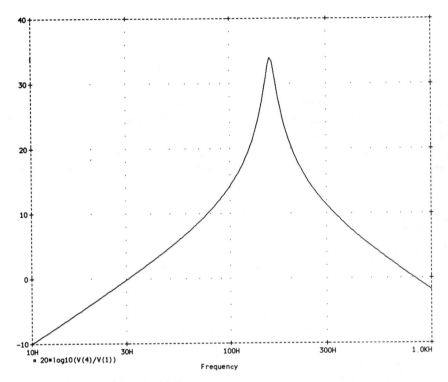

Fig. 5.35 Active RC band-pass filter.

```
.AC DEC 100 1 1 kHz
.PROBE
.subckt iop m p vo
 e vo 0 p m 2e5
 rin m p 1meg
.ends
.END
```

Run the analysis and plot V(4)/V(1) to show that $A_o = 50$ at $f_o = 160$ Hz; then remove this trace and plot the same voltage ratio on a logarithmic scale to find the bandwidth. You should verify that $f_1 = 151$ Hz and $f_2 = 167$ Hz, giving $B = 16$ Hz. Figure 5.35 shows the results.

NEW PSPICE STATEMENT USED IN THIS CHAPTER

X[*name*] [<*node*>]* <*sname*>

For example,

```
X1 9 8 10 iop
```

means that a subcircuit is connected at nodes *9, 8,* and *10* in the (main) circuit. The name of the subcircuit is *iop*. The input file contains a description of the subcircuit. It might look like this, for example:

```
.subckt iop 1 2 3
. . . .
. . . .
.ends
```

where the name *iop* identifies the subcircuit, and the nodes *1, 2,* and *3* refer to nodes *9, 8,* and *10,* respectively, of the *X* statement. The line with *.ends* shows the end of the subcircuit description.

The use of subcircuits is most convenient when it is necessary to use a device, model, or group of elements more than once in an input file. For example, *X1, X2,* and *X3* could all refer to the same device *iop.*

PROBLEMS

5.1 An ideal inverting op amp as shown in Fig. 5.2 has $R_1 = 2$ kΩ, $R_2 = 15$ kΩ, $A = 100,000$ and $R_i = 1$ MΩ. Run a PSpice analysis to determine the voltage gain, the input resistance, and the output resistance. The value of $R_i = 1$ MΩ is a practical value. When $R_i = 1$ GΩ is used in the PSpice analysis, what differences are obtained in the results?

5.2 Design an ideal noninverting op amp as shown in Fig. 5.3 to have a voltage gain of 20. Select values for R_1 and R_2, and run a PSpice analysis to verify your design.

5.3 An ideal op amp as shown in Fig. 5.5 is to be used with inputs $v_a = 3$ V and $v_b = 10$ V. When $R_1 = 5$ kΩ, $R_2 = 10$ kΩ, $R_3 = 10$ kΩ, and $R_4 = 5$ kΩ, find the output voltage using PSpice. Compare the results to those obtained in the text example where $R_1 = R_3$ and $R_2 = R_4$. Define the role of R_3 and R_4 in determining the voltage gain.

5.4 The op amp model in Fig. 5.8 assumes that $f_t = 1$ MHz and $f_c = 10$ Hz. Revise the model to allow for $f_t = 2$ MHz and $f_c = 10$ Hz. Use $R_1 = 10$ kΩ, and $R_2 = 240$ kΩ. Find the midfrequency gain and the upper 3 dB frequency. Compare your results with those given in the text example.

5.5 In Fig. 5.13 the product of R and C equals 1 (second). Using more practical values $C = 50$ μF and $R = 20$ kΩ with the same input should produce the same results as those given in the text example. Demonstrate that this is the case. Then using $C = 50$ μF and $R = 10$ kΩ, run the analysis again. Explain the difference in these and the previous results.

5.6 Using the circuit of Fig. 5.15 with $C = 50$ μF and $R = 20$ kΩ, run the PSpice analysis with the same input as shown in the figure. Compare your results with Fig. 5.16. Then using $C = 50$ μF and $R = 10$ kΩ, run the analysis again. Explain the difference in these and the previous results.

5.7 Figure 5.36 shows a first-order op amp circuit, which has

$$v_s = 4 - 4u(t)$$

where $u(t)$ represents a unit step function. Analysis shows that

$$v_c(t) = 10e^{-4t} \text{ V}$$

Fig. 5.36

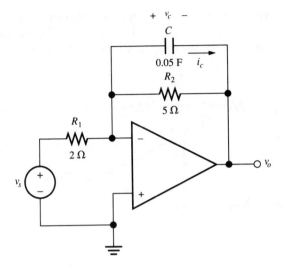

and

$$v_o(t) = -v_c(t)$$

for $t \geq 0$. Run the PSpice analysis to verify the predicted results.

5.8   Figure 5.37 shows an op amp circuit where

$$v_s(t) = 3 - 3u(t) \text{ V}$$

find (a) $v_o(0)$, (b) $i_c(0)$, (c) $i_o(0)$, and (d) the plot of $v_o(t)$ using PSpice analysis.

Fig. 5.37

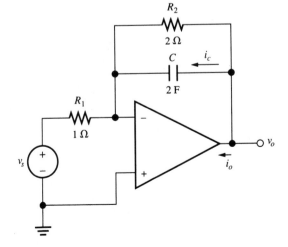

5.9   Design a first-order, low-pass filter such as that shown in Fig. 5.38 with a cut-off frequency $f_o = 5$ kHz. Use $R = R_1 = 1$ kΩ and solve for C. Find the midfrequency gain; then use Probe to verify your design.

Fig. 5.38

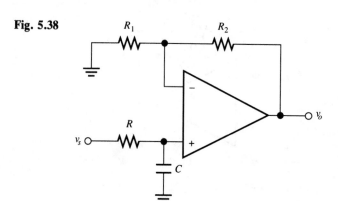

Transients and the Time Domain

One of the important features in PSpice is its transient analysis capability. Transients can be tedious from the mathematical point of view. Lengthy derivations often involve differential equations with specified boundary conditions and assumptions for the solutions. PSpice will allow you to gain additional insight, especially when you need plots of voltages and currents as a function of time.

SWITCH CLOSING IN AN *RL* CIRCUIT

Every circuit, when first energized, will experience at least a brief period of transient conditions. As an example, Fig. 6.1 shows a circuit with a 1-V source (a battery), a switch (to be closed at the beginning of the time measurements), a resistor R, and an inductor L. See what happens immediately after the switch closes. From your study of circuit analysis, you know that the current will not immediately reach its final value V/R, but that it will rise exponentially. The time constant $\tau = L/R$ represents the time required for the current to reach 63.2% of its final value. After about 5τ, the current will have reached its final value within less than 1%.

In PSpice, we will investigate this response using the PWL (piecewise linear) function. It will be used in the statement describing the applied voltage, as follows:

```
V 1 0 PWL(0,0 10us, 1V 10ms, 1V)
```

The statement shows that the voltage is between nodes *1* and *0* and that its form is given as piecewise linear. The arguments in parentheses represent time,

Fig. 6.1 Switch closing in an
RL circuit.

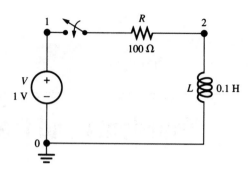

voltage pairs. In this example at $t = 0$, $V = 0$; then at $t = 10$ μs, $V = 1$ V; then at $t = 10$ ms, $V = 1$ V. If you move from point to point between successive pairs with a straight line between, you will see how the voltage looks as a function of time. Now you are ready to look at the complete input file:

```
Switch Closing in RL Circuit
V 1 0 PWL(0,0 10us,1V 10ms,1V)
R 1 2 100
L 2 0 0.1H
.TRAN 1ms 10ms
.PROBE
.END
```

Incidentally, the first time shown in the .TRAN statement is a print-step value (make it about one tenth of the second value), and the second value indicates the length of the time for the analysis.

Run the analysis and plot I(R). Note that the current begins to build exponentially as expected, reaching a final value of 10 mA. Use the cursor mode to determine the initial rate of change of current $\Delta i/\Delta t$. You might choose to do this for a time interval of about 50 μs. Verify that this initial $\Delta i/\Delta t = 10$ A/s. If the current continued to increase at this linear rate, when would it reach its final value of 10 mA?

As you know, after 1 time constant the current should reach 0.632 of its final value. Verify on the graph that this value, 6.32 mA, is reached at $t = 1$ ms. Refer to Fig. 6.2 for this graph.

If you are just learning about time constants, make another graph to help you with this concept. Remove the current trace and plot three curves V(1), V(2), and V(1,2). The voltage V(1,2) is the same as V(1) − V(2). If you change the time scale to run from −1 ms to 10 ms, you can see the initial switch closing more readily. What do the curves represent? V(1) is the applied voltage, which rises suddenly from zero to 1 V; V(2) is the inductor voltage, which begins at 1 V when $t = 0$. Can you apply Kirchhoff's voltage law to explain why? V(1,2) is the voltage drop across the resistor. This plots in the same fashion as the current, obviously, since $v_R = Ri$. Since $v_R + v_L$ must equal V (the applied voltage) at all times, v_R and v_L appear to be mirror images. Figure 6.3 shows these relationships.

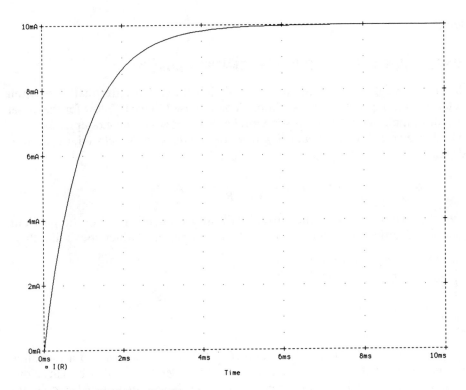

Fig. 6.2 Switch Closing in *RL* Circuit.

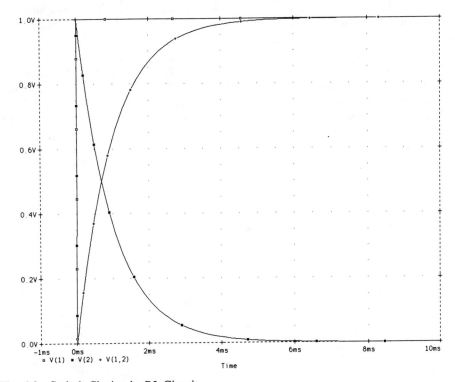

Fig. 6.3 Switch Closing in *RL* Circuit.

NONZERO INITIAL CURRENT IN THE TRANSIENT ANALYSIS

The circuit of Fig. 6.4 contains an open switch before $t = 0$. After that, the switch is closed, creating a transient condition. You can use PSpice to allow for the initial current in the circuit if you do some preliminary work. As an example, let $R_1 = 15\,\Omega$, $R = 5\,\Omega$, $L = 0.5$ mH, and $V = 10$ V. Before the switch is closed, the current is

$$i(0) = \frac{V}{R_1 + R} = 0.5 \text{ A}$$

After the switch is closed, the current will rise exponentially as in the previous example. Use the initial current of 0.5 A in the input file, which looks like this:

```
Transient with Nonzero Initial Current
V 1 0 PWL(0,2.5V 1us,10V 1ms,10V)
R 1 2 5
L 2 0 0.5mH IC=0.5A
.TRAN 10us 1ms
.PROBE
.END
```

Note that the L statement contains $IC = 0.5$ A. This is a *guess* for the initial current. In reality it is more than a guess, but it is called that in SPICE. This alone will not make the analysis run correctly, however. Notice that our *PWL* contains $0, 2.5$ V. Where did that come from? Before the switch was closed, with $i = 0.5$ A, the voltage drop across R is given by $v_R = Ri = 0.5 \cdot 5 = 2.5$ V. You must help the analysis along by using this as the initial voltage. Recall that R_1 drops out of the picture during the PSpice analysis, and there is no node for R_1. With only R and L in the loop along with V, initially all of V appears across R.

Now you are ready to run the analysis and look at the current. Plot I(R) and verify that the initial current is 0.5 A and that the final current is 2 A. After one time constant, the current should reach what value? The total change is 1.5 A; in one time constant the current should reach $0.632 \cdot 1.5 = 0.948$ of that change. Add this to 0.5 A, giving the current $i = 1.448$ A. Verify this on the plot, using the cursor mode. Refer to Fig. 6.5 for this plot.

Fig. 6.4 Circuit with nonzero initial current.

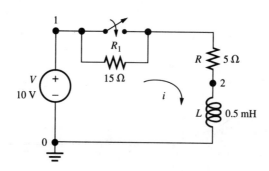

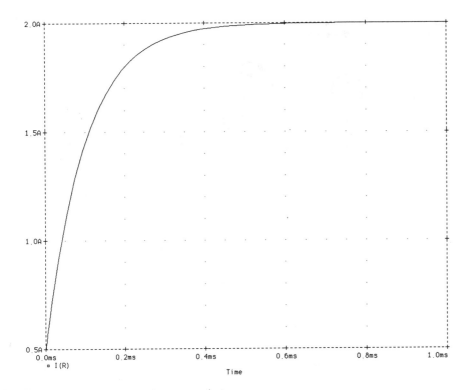

Fig. 6.5 Transient with Nonzero Initial Current.

RESISTOR AND CAPACITOR IN THE TRANSIENT ANALYSIS

If a capacitor is used in series with a resistor, as shown in Fig. 6.6, there will be an initial inrush of current when the switch is closed. Our analysis is based on $\tau = 1$ ms and $C = 0.1$ μF, giving $R = \tau/C = 10$ kΩ. The input file is simply

```
Switch Closing in RC Circuit
V 1 0 PWL(0,0 10us, 1V 10ms,1V)
R 1 2 10k
C 2 0 0.1uF
.TRAN 1ms 10ms
.PROBE
.END
```

Run the analysis and plot I(R). What is the value of current at the instant the switch is closed? What will be the value of current when $t = \tau$? If the current continued to drop at its initial rate, when would it become zero? Refer to Fig. 6.7 for this trace.

Remove the current trace and plot V(1), V(2), and V(1,2)—the applied voltage, the capacitor voltage, and the resistor voltage, respectively. Note the exponential rise in capacitor voltage as the resistor voltage decreases exponentially. Figure 6.8 shows these features.

Fig. 6.6 Switch closing in *RC* circuit.

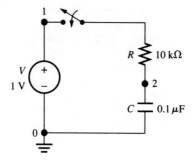

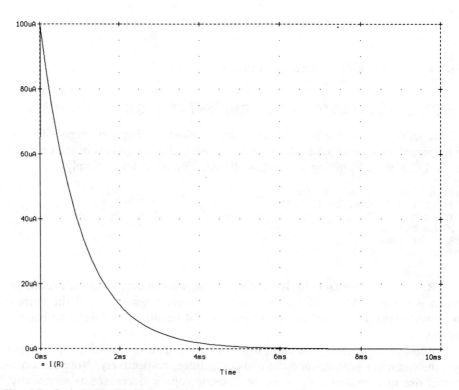

Fig. 6.7 Switch Closing in *RC* Circuit.

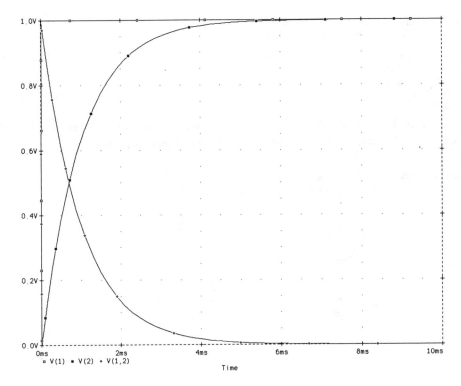

1.0V
0.8V
0.6V
0.4V
0.2V
0.0V

Oms   2ms   4ms   6ms   8ms   10ms

□ V(1) ■ V(2) ◆ V(1,2)

Time

Fig. 6.8   Switch Closing in *RC* Circuit.

A DOUBLE-ENERGY CIRCUIT

A double-energy circuit is one that contains an inductor and a capacitor in addition to one or more resistors.

When a circuit contains *R*, *L*, and *C* in series, the transient response is classically divided into three categories. The first is for overdamping, the second for critical damping, and the third for underdamping or oscillatory conditions. We will begin with the overdamped case.

Overdamped *RLC* Series Circuit

Figure 6.9 shows the circuit with a 12-V source. The switch is to be closed just after $t = 0$, causing the transient conditions to begin. Values are $C = 1.56 \, \mu\text{F}$, $L = 10 \, \text{mH}$, and $R = 200 \, \Omega$. The value of *R* will later be adjusted to allow for the other conditions. With $R = 200 \, \Omega$, the circuit is overdamped. A time of 1 ms will be sufficient to allow the current to rise to a peak value and then exponentially decay. A mathematical analysis of this circuit shows that there are two exponential components of current, which combine to give the actual current which you will see on the graph. The input file is

```
Double-Energy Circuit, Overdamped
V 1 0 PWL(0,0 10us,12V 10ms,12V)
R 1 2 200
L 2 3 10mH
C 3 0 1.56uF
.TRAN 10us 1ms
.PROBE
.END
```

Run the analysis; then plot I(R). Verify that the current peaks at $i = 47.4$ mA when $t = 115$ μs. Figure 6.10 shows the overdamped response.

Fig. 6.9 A double-energy circuit, overdamped.

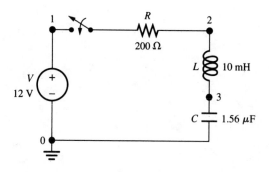

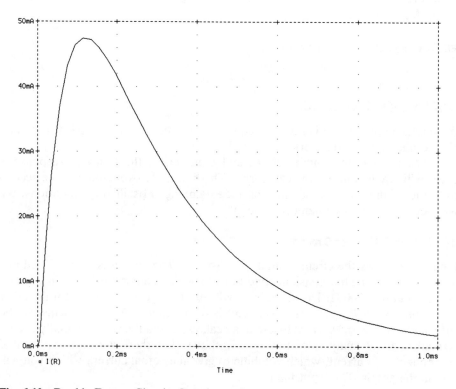

Fig. 6.10 Double-Energy Circuit, Overdamped.

Now you can make an interesting discovery relating to the voltage components in the circuit. Remove the current trace and plot V(1), V(3), V(2,3), and V(1,2). These are readily identified in Fig. 6.9. Observe that the resistor voltage reaches a peak of $v_R = 9.49$ V at $t = 115$ μs. Also observe that the inductor voltage appears to suddenly rise to nearly $v_L = 10.9$ V and then passes through zero to a negative peak of $v_L = -1.206$ V at $t = 235$ μs.

The Critically Damped *RLC* Circuit

Use the same circuit as in Fig. 6.9. Analysis shows that for critical damping,

$$R^2 = \frac{4L}{C}$$

Keep the former values of L and C. When $R = 160$ Ω, the critical condition exists. To see the results, simply change the value of R in the input file to the required value and run the analysis again.

After running the analysis, verify that the current reaches a peak value of $i = 55.36$ mA at $t = 135$ μs. Remove the current trace and plot the various voltages as in the previous analysis. These curves will have the same appearance as those of the former case involving overdamping. Refer to Fig. 6.11 for these plots.

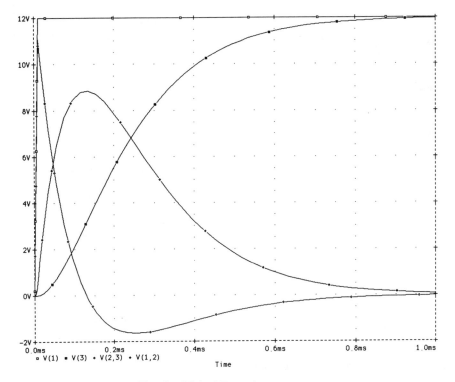

Fig. 6.11 Double-Energy Circuit, Critical Damping.

The Underdamped *RLC* Circuit

To show the effects of underdamping, you need to reduce the resistance to less than the critical value of 160 Ω. Run the analysis using $R = 60$ Ω. Change this value in the input file and look at the plot of current using I(R). Verify that the current peaks at $i = 92.7$ mA when $t = 174$ μs. Also observe that the current goes negative, then positive once again. This oscillatory pattern is typical for the underdamped case. Of course, smaller values of R produce a more extended period of oscillation. Refer to Fig. 6.12 for the oscillatory current trace. You may want to try several other smaller values of resistance and observe the effect on the transient response.

After looking at the current plot, remove the trace and plot V(1), V(3), V(2,3), and V(1,2). It is interesting to see that the capacitor voltage rises above the applied voltage of 12 V, reaching its peak as the inductor voltage is rising from its negative peak. If you try other values of R, you will see a variation of the ways in which the component voltages interact, always in keeping with Kirchhoff's voltage law, of course. Figure 6.13 shows these plots.

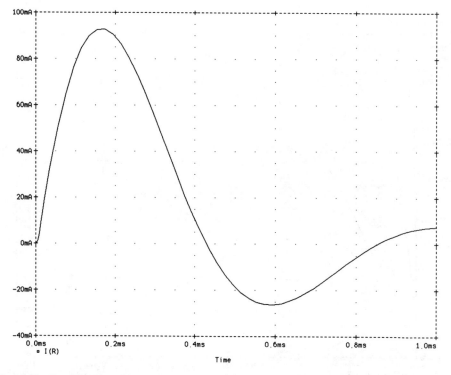

Fig. 6.12 Double-Energy Circuit, Underdamped.

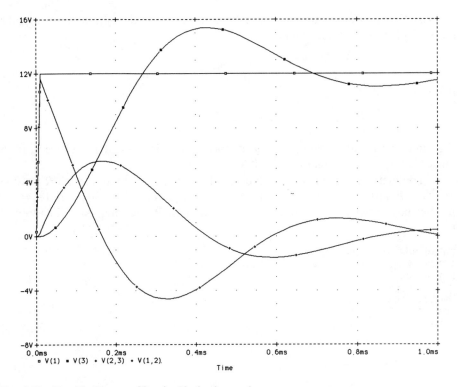

Fig. 6.13 Double-Energy Circuit, Underdamped.

STEP RESPONSE OF AN AMPLIFIER

When a step of voltage is applied to an amplifier, the high-frequency response will determine whether or not the output wave closely resembles the input wave. Consider the amplifier as a low-pass circuit as shown in Fig. 6.14. The output voltage will show exponential rise and fall. Specifically, during the rise the output is given by

$$v_o = V(1 - e^{-t/RC})$$

Fig. 6.14 Step response of an amplifier.

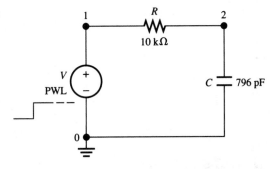

The rise time t_r is a measure of how fast the amplitude of the output can respond to the step of input voltage. Since

$$f_H = \frac{1}{2\pi RC}$$

the rise time is

$$t_r = 2.2RC = \frac{0.35}{f_H}$$

We suggest that to avoid excessive distortion you let $f_H = 1/t_p$, where t_p is the pulse width. This means that $t_r = 0.35t_p$.

To illustrate these features, choose a low-pass circuit model with $R = 10 \text{ k}\Omega$ and $C = 796 \text{ pF}$, giving $f_H = 20$ kHz. From the equations, $t_p = 50 \mu\text{s}$ and $t_r = 17.5 \mu\text{s}$. See how closely these values are matched in the PSpice analysis. The input file is

```
Pulse Response When fH=1/tp
V 1 0 PWL(0,0 0.5us, 1V 50us,1V 50.5us,0)
R 1 2 10k
C 2 0 796pF
.TRAN 0.5us 100us
.PROBE
.END
```

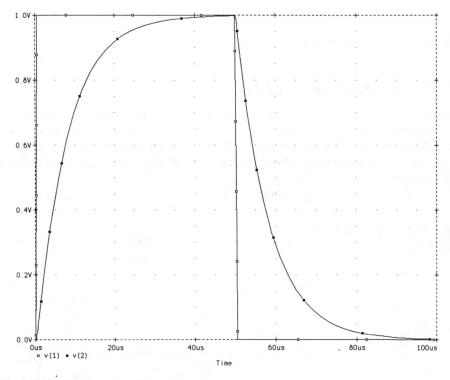

Fig. 6.15 Pulse Response When fH = 1/tp.

Run the analysis, and in Probe plot v(1) and v(2), the input and output voltages. Verify on the output wave that $t_{0.1} = 1$ μs and $t_{0.9} = 18.4$ μs. These represent the times when the output voltage is at one-tenth and nine-tenths of its peak value. The difference in the two times is the rise time. This gives $t_r = 17.4$ μs, in close agreement with our prediction. Refer to Fig. 6.15 for this plot.

What would be the results of using a value of capacitance that is twice as large as the recommended maximum value? Simply run the analysis with the new value, $C = 1.592$ nF. Observe that the output wave does not reach the desired value of 1 V and is generally more distorted.

It is also instructive to see the results when the capacitance is smaller than the maximum recommended value. Run the analysis with $C = 398$ pF. You will see that the square wave is much more faithfully reproduced in the output wave.

LOW-FREQUENCY RESPONSE OF AN AMPLIFIER

When a high-pass circuit such as shown in Fig. 6.16 is used to simulate the low-frequency response of an amplifier, the output is given by

$$v_o = Ve^{-t/RC}$$

When the time constant $\tau = RC$ is too small, the output wave will display an undesirable tilt. Since R is probably fixed as the input resistance to an amplifier stage, the value of C must be made large enough to avoid excessive tilt. For an example, choose $R = 1.59$ kΩ and $C = 10$ μF, and use a 50-Hz square wave for testing. The input file is

```
Tilt of Square Wave for Low-Frequency Response
V 1 0 PWL(0,0 10us,1V 10ms,1V 10.01ms,-1V 20ms,-1V 20.01ms,1V 30ms,1V)
C 1 2 10uF
R 2 0 1.59k
.TRAN 0.15ms 30ms
.PROBE
.END
```

Run the analysis: then plot v(1) and v(2). Find the tilt of the output by comparing the peak on the leading and trailing edges. Verify that these values are 1 V and 0.533 V, giving a tilt of 46.7%. Often a tilt of no more than about 10% is considered desirable. Obviously, a larger value of capacitance is needed. Let $C =$

Fig. 6.16 Low-frequency response of an amplifier.

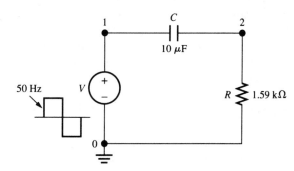

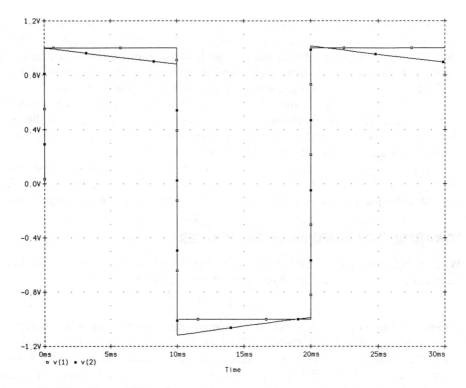

Fig. 6.17   Tilt of Square Wave for Low-Frequency Response.

50 μF and run the analysis again. Verify that the tilt is now 12%. Figure 6.17 shows this plot.

In the laboratory, the response would be observed with an oscilloscope connected to the amplifier output when the input is a square wave of the proper frequency.

CIRCUIT WITH CHARGED CAPACITOR

The circuit in Fig. 6.18 contains capacitance in one branch and inductance in another branch.* A voltage source is present to charge the capacitor; then the voltage source is effectively shortened.

Before a PSpice analysis can be performed, the initial voltages and currents that affect the analysis must be determined. In the description for v_s, it is seen that the applied voltage is constant at 6 V for $t < 0$. In this dc circuit the capacitor is an open circuit, and the inductor is a short circuit. The current from the 6-V source becomes 6 V/3 Ω = 2 A, and the voltage at node I is 4 V. This is the voltage that

---

* This example is from Bobrow (see ''Further Reading,'' p. 5).

Fig. 6.18 Circuit with capacitive branch and inductive branch.

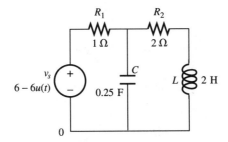

appears across the capacitor at $t = 0$. The current of 2 A passes through R_1, R_2, and L.

At $t = 0$, the applied voltage $v_s = 0$ V, and the circuit becomes Fig. 6.19. This circuit is the subject of the PSpice analysis. The input file becomes

```
Initial Conditions Example
R1 0 1 1
R2 1 2 2
C 1 0 250mF IC=4V
L 2 0 2H IC=2A
.TRAN 0.01ms 4s UIC
.PROBE
.END
```

The input file contains on the C statement an IC value of 4 V, which is the initial capacitor voltage; on the L statement there is an IC of 2 A, which is the initial current through L. Note that an initial condition for a capacitor is limited to a voltage value and that an initial condition for an inductor can only be a current. It is also necessary to append to the *.TRAN* statement a term *UIC*, which means that the transient analysis is to begin with the specified initial values.

Run the analysis and plot both the capacitor voltage and the inductor voltage. Verify that at $t = 0.5$ s, $v_C(0.5 \text{ s}) = -0.860$ V and $v_L(0.5 \text{ s}) = -3.49$ V. The plot is shown in Fig. 6.20.

As an additional exercise, plot both the capacitor current and the inductor current. Observe that $i_C(0) = -6$ A. Since $R_1 = 1$ Ω and $R_2 = 2$ Ω, we expect twice the initial current through R_1 as the current through R_2. This gives 4 A through R_1 and 2 A through R_2. Sketch the circuit and correctly show the directions of the

Fig. 6.19 Circuit at $t = 0$.

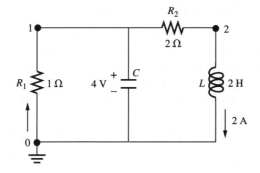

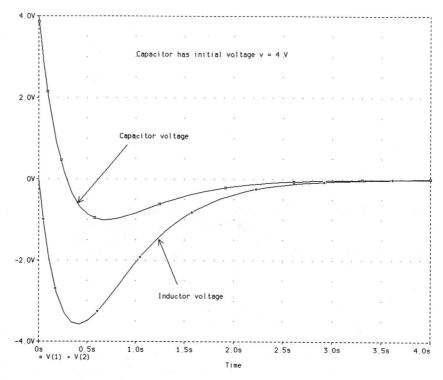

Fig. 6.20 Initial Conditions Example.

various branch currents. While you are observing the current traces, verify that at $t = 0.5$ s, $i_C(0.5$ s$) = -0.4595$ A and $i_L(0.5$ s$) = 1.316$ A. Note that if you have two traces on the same plot, you may choose which one to follow with the cursor by choosing *Cursor*; then with the mouse click on the selected trace marker. For example, click on the small diamond to select the second trace.

Before leaving Probe, plot the currents through both resistors. Verify that at $t = 0$, $i_{R1}(0) = -4$ A and $i_{R2}(0) = 2$A. Observe the directions of the current arrows in Fig. 6.19 in order to understand the current signs (positive and negative).

SWITCH-OPENING CIRCUIT WITH *L* and *C*

Another circuit where the source voltage is removed at $t = 0$ is shown in Fig. 6.21. The initial conditions will be found before attempting the PSpice analysis. There is a dc voltage $V_s = 6$ V applied at $t < 0$. Under this condition, the circuit appears as R_1 and R_2 in parallel. By current division, the current $I_{R1} = 3$ A and the current $I_{R2} = 2$ A. This current also passes through the inductor L. The current through R_2 produces a voltage

$$V(1,2) = R_2 I_{R2} = (3)(2) = 6 \text{ V}$$

This is the initial capacitor voltage.

Fig. 6.21 Circuit with switch
opening at $t = 0$.

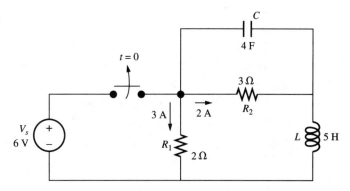

At $t = 0$, the switch is opened, and the circuit for PSpice is as shown in Fig.
6.22. The initial values are shown in the figure. Note the polarity of the initial
capacitor voltage and the direction of the initial inductor current. The input file
thus becomes

```
Switch-Opening Circuit with L,C
R1 1 0 2
R2 1 2 3
C 1 2 4000mF IC=6V
L 2 0 5H IC=2A
.TRAN 0.01ms 16s UIC
.PROBE
.END
```

Run the analysis and verify that when the switch is opened at $t = 0$, $v_C(0) = 6$ V
and $i_L(0) = 2$ A, in keeping with our initial-condition statements. Also verify that
$v_L(0) = -10$ V by plotting $v(2)$, and verify that $i_C(0) = 0$.

How can $v_L(0)$ be predicted using simple circuit analysis just after the switch
is opened? Since the current through the inductor cannot change instantaneously,
the current in R_1 suddenly becomes 2 A (upward) rather than its former value of
3 A (downward). This current of 2 A produces a voltage drop of 4 V with the
polarity shown in Fig. 6.22. Applying KVL to the loop containing R_1, C, and L

Fig. 6.22 Conditions when
switch is opened.

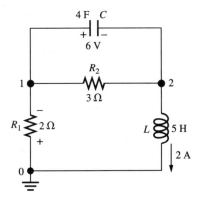

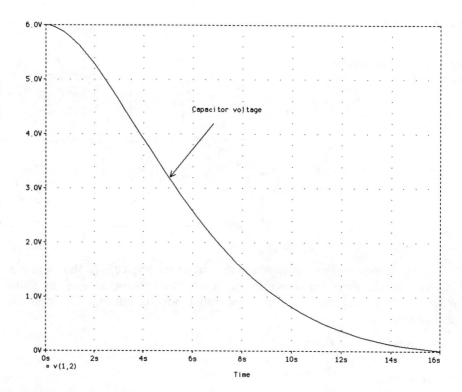

Fig. 6.23 Switch-Opening Circuit with L,C.

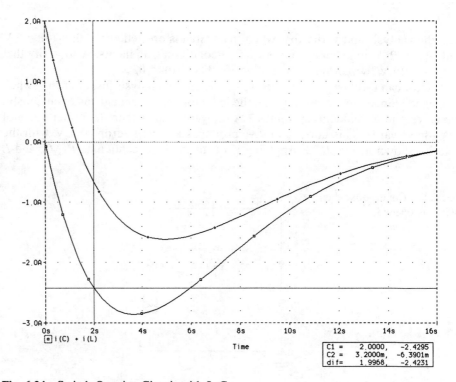

Fig. 6.24 Switch-Opening Circuit with L,C.

240

gives $v_L(0) = -10$ V, confirming the PSpice results. Figure 6.23 shows v(1,2) which is v_C.

While you are still in the Probe analysis, verify currents and voltages at $t = 2$ s, as follows:

$$v_C(2 \text{ s}) = 5.2778 \text{ V}$$
$$v_L(2 \text{ s}) = -3.94 \text{ V}$$
$$i_C(2 \text{ s}) = -2.428 \text{ A}$$
$$i_L(2 \text{ s}) = -0.675 \text{ A}$$

The currents are shown in Fig. 6.24.

CIRCUIT WITH CURRENT SOURCE

The circuit shown in Fig. 6.25 shows a source producing a steady current of 3 A for $t < 0$. Then at $t = 0$, the current becomes 0. Initial conditions for L and C must be determined before the PSpice analysis is undertaken. Prior to $t = 0$, the current through R is 3 A, while the current through the other branch is zero since C appears open for dc conditions. Thus $i_L(0) = 0$. The voltage drop across R is $(2 \text{ }\Omega)(3 \text{ A}) = 6$ V, with the polarity as shown in Fig. 6.26. Since, under dc conditions, there can be no voltage across L, the voltage $v_C(0) = 6$ V. This is enough information to perform the PSpice analysis. The input file is

```
Initial Conditions from Current Source
R 1 0 2
L 1 2 3H
C 2 0 4000mF IC=6V
```

Fig. 6.25 Circuit with current source.

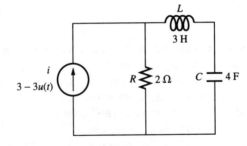

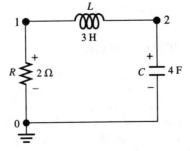

Fig. 6.26 Circuit at $t = 0$.

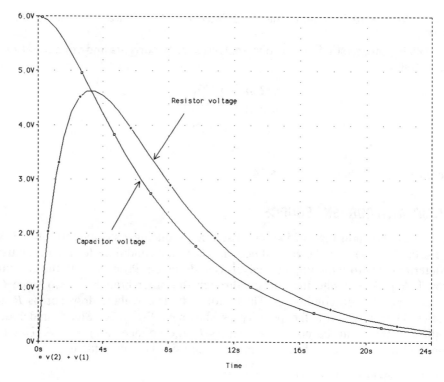

Fig. 6.27 Initial Conditions from Current Source.

```
.TRAN 0.001ms 24s UIC
.PROBE
.END
```

Run the analysis and plot the resistor voltage, the inductor voltage, and the capacitor voltage. Verify the initial conditions for both of these voltages. As an exercise, verify that $v_C(4 \text{ s}) = 4.2169$ V and $v_R(4 \text{ s}) = 4.5456$ V. Without obtaining a trace of v_L, what should be its value at $t = 4$ s? Use KVL to find this value. Resistor and capacitor voltages are shown in Fig. 6.27.

Now plot the current i_C. Note that it has an initial value of zero, since it must have the same current as the inductor. Verify that $i(4 \text{ s}) = 2.2738$ A. This is the current through each element in the CCW direction. Also verify that the maximum current is $i_{max} = 2.3125$ A at $t = 2.4944$ s.

BRIDGE CIRCUIT WITH INITIAL CURRENT

The circuit of Fig. 6.28 has a switch that opens at $t = 0$. Before the switch is opened, the circuit appears as in Fig. 6.29. The inductance has been replaced by a short circuit, indicating that 6 V appears across both R_1 and R_3. This voltage produces a current of 2 A in R_1 and a current of 3 A in R_2 as shown. Since there is no current through the capacitor branch, the current of 3 A must also appear in the inductor. Since the voltage V(1,3) is zero, $v_C = 0$. This information will allow the

Fig. 6.28 Circuit with switch opening at $t = 0$.

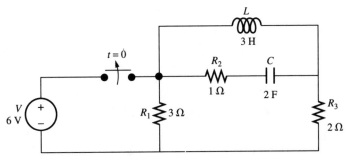

Fig. 6.29 Circuit conditions before switch is opened.

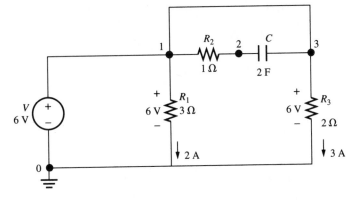

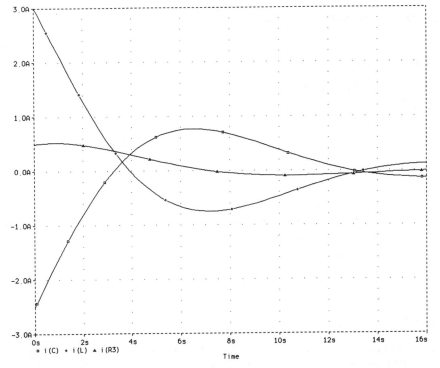

Fig. 6.30 Switch Opening in Bridge Circuit.

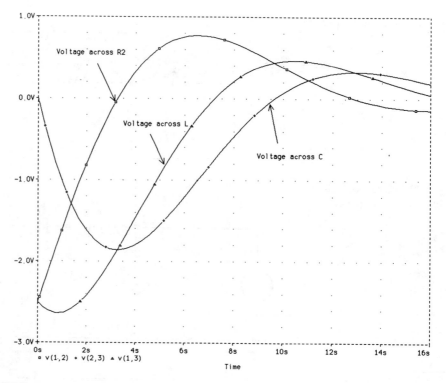

Fig. 6.31 Switch Opening in Bridge Circuit.

PSpice initial conditions to be set, producing the following input file:

```
Switch Opening in Bridge Circuit
R1 0 1 3
R2 1 2 1
R3 3 0 2
L 1 3 3H IC=3A
C 2 3 2000mF
.TRAN 0.001ms 16s UIC
.PROBE
.END
```

Run the analysis and verify the following: $i_C(0) = -2.5$ A, $i_L(0) = 3$ A, $i_{R3}(0) = 0.5$ A, $v_{12}(0) = -2.5$ V, $v_{23}(0) = 0$, and $v_{13}(0) = -2.5$ V. (*Note:* $v_{12}(0)$ means that $v(1,2)$ at $t = 0$.) The current traces are shown in Fig. 6.30, and the voltage traces are shown in Fig. 6.31.

As an exercise, apply KVL to the loop containing R_1, R_2, C, and R_3 to predict i_C at $t = 0$.

A RINGING CIRCUIT

The circuit of Fig. 6.32 might represent a network to be tested with a square-wave input. The input voltage is given as a 1-V source which abruptly changes from 0 to

Fig. 6.32 A ringing circuit.

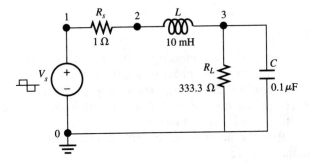

1 V; then at 2 ms there is a 2-V change to −1 V; then at 4 ms there is another abrupt change to 1 V. The problem is to determine how accurately the square wave is reproduced as a voltage across R_L. The input file is

```
Ringing Circuit
Vs 1 0 PWL(0s,0V 0.01ms,1V 2ms,1V 2.01ms,−1V 4ms,−1V 4.01ms,1V)
Rs 1 2 1
L 2 3 10mH
RL 3 0 333.3
C 3 0 0.1uF
```

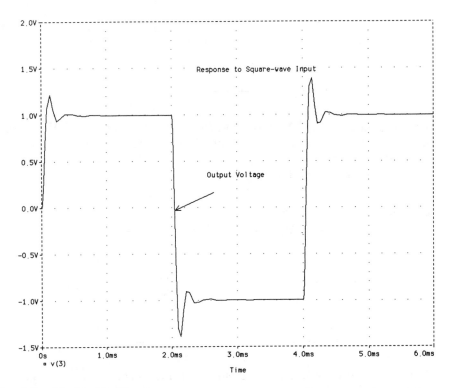

Fig. 6.33 Ringing Circuit.

```
.TRAN 0.05ms 6ms
.PROBE
.END
```

The Probe results of plotting V(3) are shown in Fig. 6.33. In Probe you may also want to display V_s in order to see the deviation in the two traces. Before leaving Probe, remove the voltage traces and plot each of the currents. Of particular interest is the current I(C). The current traces should give you a better understanding of this type of circuit.

Run the analysis again with C decreased by an order of magnitude, and compare the results.

PROBLEMS

6.1 Given the circuit shown in Fig. 6.34, with $V = 10$ V, $R_1 = R = 1$ kΩ, and $C = 200$ μF, obtain a plot of $v_c(t)$ for a time range extending from before the switch is opened to when the capacitor voltage approaches zero. Perform the required PSpice analysis and obtain the Probe graph of v_c.

Fig. 6.34

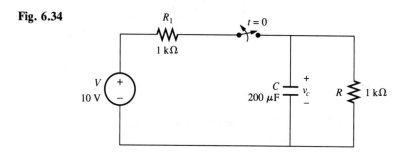

6.2 For the circuit of Fig. 6.35, $V = 10$ V, $R_1 = R = 100$ Ω, and $L = 2$ H. Obtain a plot of $v_L(t)$ for a time range extending from before the switch is opened to when the inductor voltage approaches zero. Perform the required PSpice analysis and obtain the Probe graph of v_L.

Fig. 6.35

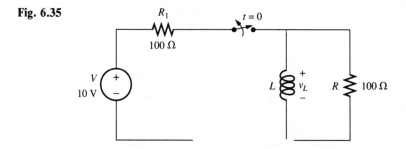

6.3 A double-energy circuit, which is shown in Fig. 6.36, has $V = 20$ V, $R = 100$ Ω, $L = 20$ mH, and $C = 2$ μF. Obtain a plot that shows the current as a function of time, beginning when the switch is closed. Since this circuit has an R value which gives underdamping, plot at least a full cycle of the oscillatory current.

Fig. 6.36

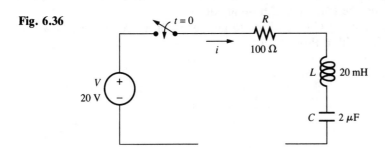

6.4 (a) Increase the value of R in Problem 6.3 to give critical damping, and obtain plots of current and component voltages. Find the maximum positive and negative current values.

(b) Using $R = 250 \ \Omega$, repeat part a. Find the maximum positive and negative values of all component voltages.

6.5 At high frequencies the output capacitance of a voltage amplifier must be taken into account. In Fig. 6.37, $R = 10 \ k\Omega$ and $C = 1 \ nF$ in order to model the output. With a pulse input voltage of 1 V applied for $t_p = 100 \ \mu s$, the output voltage should be a reasonable replica of the input pulse.

(a) Use the method described in the section "Step Response of an Amplifier" to determine the nature of the output voltage. Using Probe, see if the output voltage across C is a reasonable replica of the input pulse.

(b) If you desire a more exact replica of the input voltage, select a new value for t_p and run the analysis again. What are the f_H values for parts a and b?

Fig. 6.37

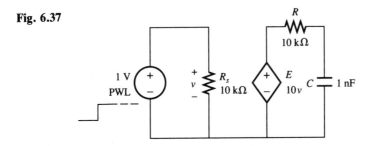

6.6 The discussion on low-frequency response of an amplifier in this chapter stated that often a tilt of no more than 10% is desirable. An approximate formula for tilt is

$$\text{tilt} = \frac{\pi f_L}{f}$$

where $f_L = \frac{1}{2\pi RC}$ and f is the frequency of the square wave. Use an analysis based on the method shown in the example where the test is to be based on a 60-Hz square wave.

(a) Using $R = 1.59 \text{ k}\Omega$ and $C = 10 \ \mu F$, find the tilt on the output by comparing the peak on the leading and trailing edges.

(b) What value of C is required to give a tilt of approximately 10%? Verify your choice with a Probe plot.

Fourier Series and Harmonic Components

One of the powerful features of PSpice is its ability to analyze systems with nonlinear response. For example, a power amplifier that is subjected to large input-signal swings will begin to operate over the nonlinear portion of its characteristics. This will introduce distortion in the output waveshape. In this chapter you will find out how much distortion is present, and you will analyze the harmonic content of the amplifier output.

FUNDAMENTAL AND SECOND-HARMONIC FREQUENCY

We will begin with a simple circuit that will introduce the concepts that are needed for more elaborate circuits. Figure 7.1 shows an input voltage $V_{in,p} = 1$ V. This is a sine wave with $f = 1$ kHz and a peak value of 1 V. ($V_{in} = 1/\sqrt{2}$ rms.) The voltage-dependent voltage source E is used to give an output that is a nonlinear function of the input. In this example, let

$$f(x) = 1 + x + x^2$$

This functional relationship is shown in the E statement in terms of the polynomial coefficients. Recall the polynomial form,

$$f(x) = k_0 + k_{1x} + k_2x^2$$

In our example this means that each of the last three numbers in the E statement will be *one*. We want to perform a Fourier analysis to see which harmonics are present in the output, but first we need to know more about what to expect.

Fig. 7.1 Output voltage is a nonlinear function of input.

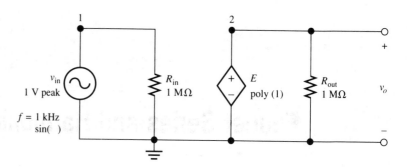

A transient analysis must be performed in order to allow for the Fourier analysis to take place. We will need both a .TRAN and a .FOUR statement. It is customary to run the transient analysis for a full period of the fundamental frequency. In this example, $f = 1$ kHz; therefore $T = 1/f = 1$ ms. The Fourier analysis will produce frequency components through the ninth harmonic. This should be more than adequate for most purposes. If higher harmonics were shown, they would have little meaning due to the accumulation of round-off error in the results.

In order to give a more complete description of the input voltage V_{in}, use the *sin* form for the source. The arguments of sin($a, b, c, . . .$) are a = offset voltage, b = peak value, c = frequency, d = delay, e= damping factor, and f = phase.

The .FOUR statement produces the Fourier analysis, giving the Fourier components of the results of the transient analysis. Arguments for this statement include the frequency and the variables with which you are working. In this example, you will look at V(1) and V(2), which represent the input and output voltage waveforms. The input file is

```
Fourier Analysis; Decomposition of Polynomial
Vin 1 0 sin(0 1 1000); arguments are offset, peak, and frequency
Rin 1 0 1MEG
E 2 0 poly(1) 1,0 1 1 1; last 3 1s are for k0, k1, k2
Rout 2 0 1MEG
.TRAN 1us 1ms
.FOUR 1000 V(1) V(2)
.PROBE
.END
```

Run the analysis (it may be a bit time-consuming); then plot V(1) and V(2). Verify that V(1) is a replica of the input voltage V_{in}. The output voltage should show a dc component and a composite wave peaking at 3 V. From your study of Fourier series, you may observe that this looks like a wave consisting of a fundamental and second harmonic. You may want a printed copy of this graph for future study. Figure 7.2 shows these plots.

We are not through with this analysis. Look at the output file for this circuit, shown in Fig. 7.3, and note the following features. The node voltages show V(1) = 0 V and V(2) = 1 V. This means that the input wave has no offset, but the output has a 1-V offset, since $k_0 = 1$.

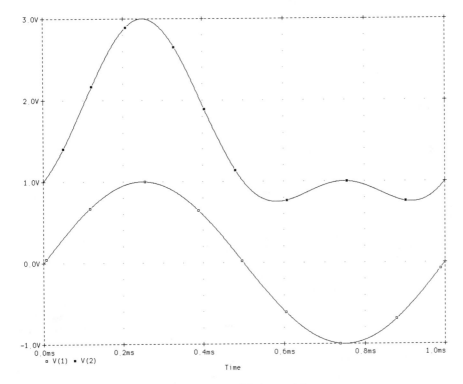

Fig. 7.2 Fourier Analysis; Decomposition of Polynomial.

Figure 7.3 includes a table of Fourier components of V(1), but only a few of the values have real significance. The dc component of 3.5E−10 is close to zero. It is not quite zero due to accumulation of round-off error. Harmonic no. 1 represents the fundamental at $f = 1$ kHz. It shows a Fourier component of 1 and a phase of 2.4E−7 (almost zero). If you think of the components as

$$b_n\sin(nx)$$

then this corresponds to $b_1 = 1$, $n = 1$, where x is the fundamental frequency. The other harmonics can be ignored, since their values are many orders of magnitude smaller than the fundamental. It is from these components that we obtained the Probe trace of V(1). Refer to Fig. 7.3 for these data.

There is another table of Fourier components in Fig. 7.3; these are for V(2). Before looking at the various harmonics, note that there is a dc component in the composite wave of 1.5 V. Why 1.5 V? The value $k_0 = 1$ V is part of it, but the other 0.5 V is associated with b^2. Theory shows that in second-harmonic distortion, $b_0 = b_2$, where b_0 is a dc component introduced into the output. The fundamental frequency is present with $b_1 = 1$ V, and the second harmonic shows $b_2 = 0.5$ V at −90°. The higher harmonics are much smaller in magnitude, and they may be ignored.

```
 Fourier Analysis; Decomposition of Polynomial
Vin 1 0 sin(0 1 1000)
Rin 1 0 1M
E3 2 0 poly(1) 1,0 1 1 1
R3 2 0 1M
.tran 1us 1ms
.four 1000 V(1) V(2)
.probe
.end
```

NODE	VOLTAGE	NODE	VOLTAGE	NODE	VOLTAGE	NODE	VOLTAGE
(1)	0.0000	(2)	1.0000				

FOURIER COMPONENTS OF TRANSIENT RESPONSE V(1)

DC COMPONENT =   3.538473E-10

HARMONIC NO	FREQUENCY (HZ)	FOURIER COMPONENT	NORMALIZED COMPONENT	PHASE (DEG)	NORMALIZED PHASE (DEG)
1	1.000E+03	1.000E+00	1.000E+00	2.403E-07	0.000E+00
2	2.000E+03	6.525E-10	6.526E-10	1.297E+02	1.297E+02
3	3.000E+03	2.530E-09	2.530E-09	1.382E+02	1.382E+02
4	4.000E+03	1.001E-09	1.001E-09	1.554E+02	1.554E+02
5	5.000E+03	2.812E-09	2.812E-09	8.551E+01	8.551E+01
6	6.000E+03	1.019E-09	1.019E-09	1.216E+02	1.216E+02
7	7.000E+03	1.132E-09	1.132E-09	-1.545E+02	-1.545E+02
8	8.000E+03	1.292E-09	1.292E-09	3.312E+01	3.312E+01
9	9.000E+03	2.589E-09	2.589E-09	1.405E+02	1.405E+02

   TOTAL HARMONIC DISTORTION =   5.140876E-07 PERCENT

FOURIER COMPONENTS OF TRANSIENT RESPONSE V(2)

DC COMPONENT =   1.500000E+00

HARMONIC NO	FREQUENCY (HZ)	FOURIER COMPONENT	NORMALIZED COMPONENT	PHASE (DEG)	NORMALIZED PHASE (DEG)
1	1.000E+03	1.000E+00	1.000E+00	-9.470E-07	0.000E+00
2	2.000E+03	5.000E-01	5.000E-01	-9.000E+01	-9.000E+01
3	3.000E+03	2.381E-08	2.381E-08	-9.813E+01	-9.813E+01
4	4.000E+03	2.043E-08	2.043E-08	-1.038E+02	-1.038E+02
5	5.000E+03	2.091E-08	2.091E-08	-8.469E+01	-8.469E+01
6	6.000E+03	2.473E-08	2.473E-08	-9.313E+01	-9.313E+01
7	7.000E+03	2.328E-08	2.328E-08	-9.578E+01	-9.578E+01
8	8.000E+03	2.356E-08	2.356E-08	-9.508E+01	-9.508E+01
9	9.000E+03	1.951E-08	1.951E-08	-9.203E+01	-9.203E+01

   TOTAL HARMONIC DISTORTION =   4.999941E+01 PERCENT

Fig. 7.3

As an exercise in wave synthesis, you may want to sketch the individual waves to see how you might predict the Probe result that you obtained as V(2). Remember to include proper amplitudes and phases for the fundamental and second harmonic as well as the total dc component.

After you have attempted to sketch the composite wave, you will be glad to know that PSpice can do this for you.

DECOMPOSITION AND RECONSTRUCTION OF WAVE

We will create a new input file based on Fig. 7.4, where we introduce two independent current sources. These are added to aid our analysis. They are not intended to modify or interfere with the original circuit in any way.

We used the two current sources so that you can look at the fundamental and second-harmonic frequencies. They supply a 1-Ω resistor in the parallel connection. The new input file is an extension of the previous file and looks like this:

```
Fourier Analysis Decomposition and Reconstruction
Vin 1 0 sin(0 1 1000)
Rin 1 0 1MEG
E 2 0 poly(1), 1,0 1 1 1
Rout 2 0 1MEG
i1 0 3 sin(1 1 1000)
i2 0 3 sin(0.5 0.5 2000 1 1 -90)
r 3 0 1
.tran 1us 1ms
.four 1000 V(1) V(2) V(3)
.probe
.end
```

Before running the analysis, look carefully at the descriptions for i1 and i2. The information used for the reconstruction of the waves comes from your Fourier results of the previous study. Be sure that you understand the various arguments; then run the analysis. In Probe, plot I(i1), I(i2), and I(r). Although these are currents, they may be thought of as numerically equal to voltages, since they are in parallel with $r = 1\ \Omega$. Figure 7.5 shows the results. Now you can see that the first trace represents the fundamental, the second represents the second harmonic, and the third represents the addition of these in the resistor r. Of course, if you prefer, you may plot V(3) instead of I(r). This may be more attractive since the Y-axis will no longer be marked in amperes. Verify that the first two waves add together to give the third wave at various times. In order to show a more compact plot, we used a 1-V offset for the fundamental and a 0.5-V offset for the second harmonic. Actually, the fundamental has a zero offset.

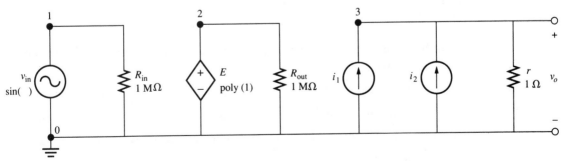

Fig. 7.4 Decomposition and reconstruction of waves.

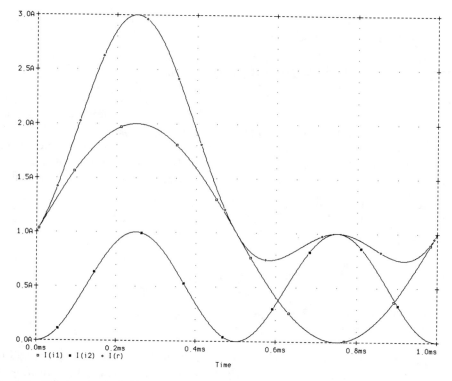

Fig. 7.5 Fourier Analysis Decomposition and Reconstruction.

SECOND-HARMONIC DISTORTION IN A POWER AMPLIFIER

When an amplifier is driven so that it no longer follows the linear portion of its characteristic curve, it will introduce some distortion. A first approximation to this distortion is often made by including the second harmonic, indicating that the transfer function relating i_c to i_b (the collector current and base current) is somewhat parabolic. Usually, the distortion is much less than that indicated by our first, introductory example, which was shown in Fig. 7.1. The polynomial might be more like

$$f(x) = 0.1 + x + 0.2x^2$$

It will be a simple matter to convert the original input file to handle this situation. The E statement will be

```
E 2 0 poly(1) 1,0 0.1 1 0.2; last 3 terms are for k0, k1, k2
```

The entire input file is

```
Fourier Analysis; Second-Harmonic Distortion, Power Amplifier
Vin 1 0 sin(0 1 1000)
Rin 1 0 1M
E 2 0 poly(1) 1,0 0.1 1 0.2
```

```
Rout 2 0 1M
.TRAN 1us 1ms
.FOUR 1000 V(1) V(2)
.PROBE
.END
```

Run the analysis and in Probe plot V(1) and V(2). You will see that both waves look very much like true sine waves. For a more accurate comparison, remove the V(2) trace and plot V(2) − 0.1 instead. This allows both waves to align more closely. When comparing the waves, remember that V(1) is the true sine wave and V(2) is the combination of the fundamental and the second harmonic. In this example, the second-harmonic content is considerably smaller in amplitude than in the previous example. You may want a printed copy of these waves for further study. Figure 7.6 shows the results.

After leaving Probe, look at the output file for this case. The fundamental is exactly the same as in the previous example, but the second harmonic is present in V(2). Note that the overall dc component of the output voltage is 0.2 V, and the second harmonic at $f = 2$ kHz has an amplitude of 0.1 V and an angle of −90°. The other frequency components are much smaller and can be ignored. Finally, look at the total harmonic distortion, which is very close to 10% as expected. The second harmonic distortion is defined as b_2/b_1, where the b values are the coefficients of

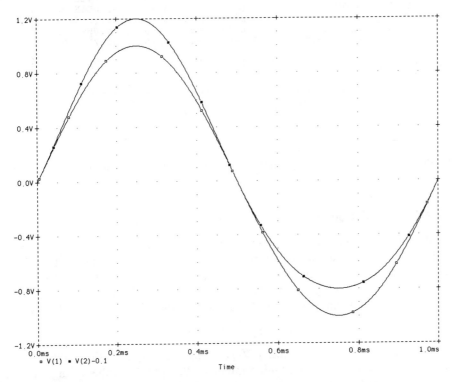

Fig. 7.6 Fourier Analysis; Second-Harmonic Distortion, Power Amplifier

```
Fourier Analysis; Second-Harmonic Distortion, Power Amplifier

****      CIRCUIT DESCRIPTION

Vin 1 0 sin(0 1 1000)
Rin 1 0 1M
E 2 0 poly(1) 1,0 0.1 1 0.2
Rout 2 0 1M
.TRAN 1us 1ms
.FOUR 1000 V(1) V(2)
.PROBE
.END

 NODE    VOLTAGE      NODE    VOLTAGE      NODE    VOLTAGE      NODE    VOLTAGE

(    1)   0.0000  (    2)     .1000

FOURIER COMPONENTS OF TRANSIENT RESPONSE V(2)

 DC COMPONENT =    2.000000E-01

 HARMONIC   FREQUENCY     FOURIER      NORMALIZED     PHASE       NORMALIZED
   NO         (HZ)       COMPONENT     COMPONENT      (DEG)      PHASE (DEG)

    1       1.000E+03    1.000E+00     1.000E+00     6.278E-08    0.000E+00
    2       2.000E+03    1.000E-01     1.000E-01    -9.000E+01   -9.000E+01
    3       3.000E+03    3.368E-09     3.368E-09    -7.744E+01   -7.744E+01
    4       4.000E+03    6.577E-09     6.577E-09    -9.074E+01   -9.074E+01
    5       5.000E+03    3.922E-09     3.922E-09    -1.041E+02   -1.041E+02
    6       6.000E+03    4.060E-09     4.060E-09    -1.054E+02   -1.054E+02
    7       7.000E+03    3.895E-09     3.895E-09    -1.130E+02   -1.130E+02
    8       8.000E+03    2.909E-09     2.909E-09    -6.059E+01   -6.059E+01
    9       9.000E+03    4.354E-09     4.354E-09    -1.238E+02   -1.238E+02

     TOTAL HARMONIC DISTORTION =    9.999881E+00 PERCENT
```

Fig. 7.7

the second harmonic and the fundamental, respectively. Refer to Fig. 7.7 for these data.

INTERMODULATION DISTORTION

We will use the simple circuit shown in Fig. 7.8 to show how two sine waves combine in a nonlinear device using frequencies that are fairly close together. The first frequency $f_1 = 1$ kHz and the second frequency $f_2 = 1.5$ kHz. The nonlinear mixing occurs in the voltage-dependent voltage source e. The polynomial describing the relationship is an extension of that used in the previous example and is

$$f(x) = 1 + x + x^2 + x^3$$

The currents add in $R = 1\ \Omega$, making V(1) numerically equal to the current in R. Thus the input voltage is V(1), which might be thought of as the voltage into a nonlinear mixer. Because the sine waves are of different frequencies, their sum

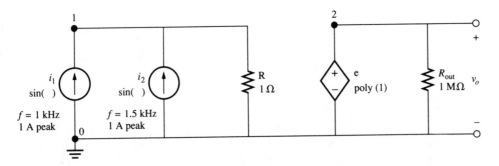

Fig. 7.8 Illustrating intermodulation distortion.

appears as a complex waveform. The input file is

```
Intermodulation Distortion
i1 0 1 sin(0 1 1000)
i2 0 1 sin(0 1 1500)
r 1 0 1
e 2 0 poly(1) 1,0 1 1 1 1
rout 2 0 1M
.tran 50us 50ms 0 50us
.probe
.end
```

Run the analysis and look at V(1). Choose *X*-axis, and set the range from 0 to 10 ms to give a clearer view of the trace. The display shows five cycles of the composite input voltage. Figure 7.9 shows this graph.

To confirm that this is actually the addition of the 1-kHz and 1.5-kHz waves, exit; then choose *X*-axis, Fourier, converting to the time domain. After a brief time, the display in the frequency domain will appear. The range extends to 50 kHz. Change the range to extend only to 4 kHz. Verify that the components are at the proper frequency with the expected amplitudes. Actually at $f = 1$ kHz, the voltage is 0.9928 V, and at $f = 1.5$ kHz, the voltage is 0.9826 V. Remember that in this synthesis, some accumulated error is present. Refer to Fig. 7.10 for the frequency display.

Return to the main menu; then remove the V(1) trace and plot V(2). This represents the output of the mixer, which has been operated on by the polynomial relationship given by $f(x)$. The time domain plot looks somewhat like what you saw at the input V(1). But closer observation will tell you that the two waves differ considerably. There is little hint of the frequency components making up this complex wave, so you need to choose *X*-axis, Fourier. Then expand the display of frequency components by letting the *X*-range from 0 to 5 kHz. Obtain a printed copy of this frequency spectrum for further study. From your studies of modulation-frequency components, you can predict and verify the results. Note that there is a dc component of 2 V along with significant components at 0.5-kHz spacing in the range from 0.5 kHz to 4.5 kHz. See Fig. 7.11 for the frequency spectrum.

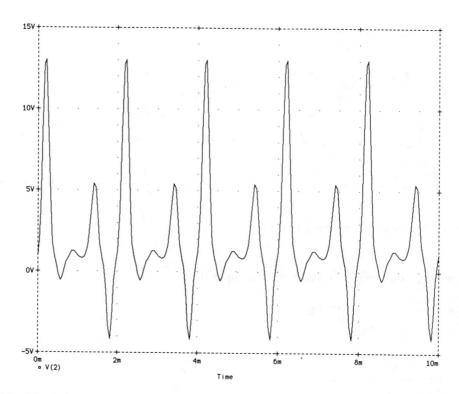

Fig. 7.9 Intermodulation Distortion.

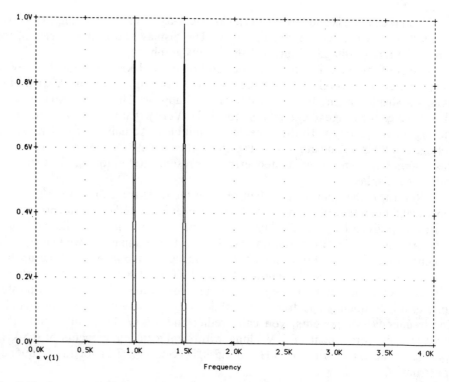

Fig. 7.10 Intermodulation Distortion.

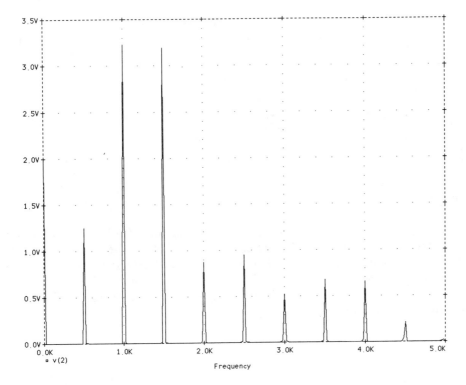

Fig. 7.11 Intermodulation Distortion.

ADDING SINE WAVES

One of the first things you learn in ac steady-state circuit analysis is that when a sine-wave voltage is used in a circuit containing linear elements such as resistors, inductors, and capacitors, the resulting current is also a sine wave of the same frequency. The various voltage drops in the circuit are all sine waves as well, obviously at the same frequency and differing only in amplitude and phase. We will use a simple circuit model to illustrate some of these features. Figure 7.12 shows three voltage sources feeding a circuit containing $R = 1\ \Omega$ along with $R_1 =$

Fig. 7.12 Adding sine waves of the same frequency.

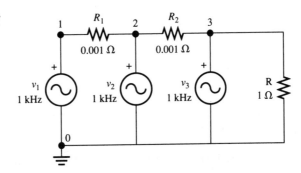

$R_2 = 0.001\ \Omega$. The latter two resistors are required to make the voltage sources practical. With this circuit we can show the addition of sine waves in Probe. The input file is

```
Addition of Sine Waves of the Same Frequency
v1 1 0 sin(0 1 1kHz)
*In the complete expression for sin the terms are
*offset, peak, frequency, delay, damping factor, and phase
v2 2 0 sin(0 1 1kHz 0 0 45); phase=45 degrees
v3 3 0 sin(0 1 1kHz 0 0 90); phase=90 degrees
r1 1 2 0.001
r2 2 3 0.001
R 3 0 1
.tran 0.1ms 2ms
.probe
.end
```

Run the analysis, and in Probe plot v(1), v(2), and v(1) + v(2). The results show the waves v_1 and v_2, with v_2 peaking 45° ahead of v_1. The wave v_3 is clearly the sum of v_1 and v_2, and since each of the waves is at 1 kHz, the resulting wave is a sine wave. Remove these traces and try other combinations, such as $v_1 + v_3$. Try to predict what the peak value of the resultant will be before you look at the graph, which is shown in Fig. 7.13.

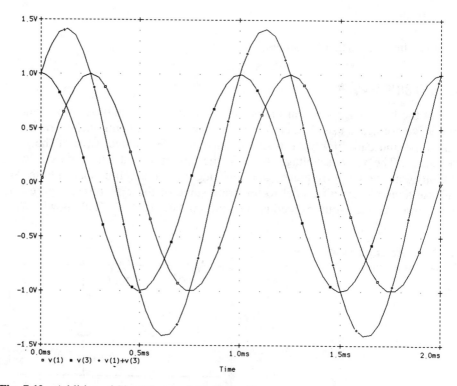

Fig. 7.13 Addition of Sine Waves of the Same Frequency.

ADDING FUNDAMENTAL AND SECOND HARMONIC

The input file based on Fig. 7.12 can easily be modified for a variety of source combinations. Remove v_3 and let v_2 be the second harmonic of v_1. Of course, the resulting wave will have a shape that is no longer a sine wave. In fact, its form will depend on the relative phasing of v_1 and v_2. For our example, we will let the two waves reach their positive peaks together. The input file is

```
Adding Sine Waves; Fundamental and 2nd Harmonic Peaking Together
v1 1 0 sin(0 1 1kHz)
v2 2 0 sin(0 1 2kHz 0 0 −90)
R1 1 2 0.001
R 2 0 1
.tran 0.1ms 1ms
.probe
.end
```

Run the analysis and in Probe plot v(1), v(2), and v(1) + v(2). Since v_1 and v_2 reach their positive peaks together, the resultant wave peaks at 2 V, but when the fundamental reaches its negative peak, the second harmonic has returned to its positive peak, giving a result of zero. The wave of $v_1 + v_2$ is clearly not a sine wave. See Fig. 7.14 for these plots.

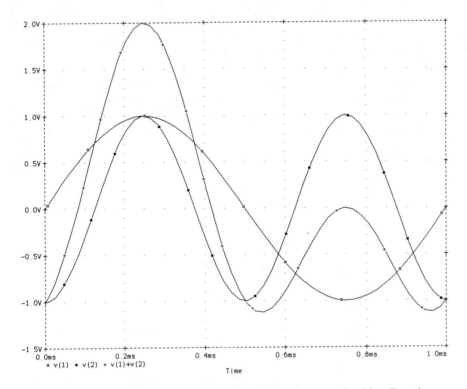

Fig. 7.14 Adding Sine Waves; Fundamental and 2nd Harmonic Peaking Together.

AMPLITUDE MODULATION

An interesting plot of an amplitude-modulated wave may be obtained in PSpice by using a multiplier function on sine waves of widely differing frequency. Figure 7.15 shows the simulator circuit model. The first sinusoidal source is v_1 with a frequency of 1 kHz. The second source is v_2 with a frequency of 20 kHz. The multiplication takes place in the depndent voltage source e. The resistors are required to avoid floating sources. The input file is

```
Multiplier for Modulated Wave
v1 1 0 sin(0 1 1000)
R1 1 0 10k
v2 2 0 sin(0 1 20000)
R2 2 0 10k
e 3 0 poly(2) 1,0 2,0 0 0 0 0 1
R3 2 0 10k
.tran 1us 1ms
.four 1000 v(1) v(2) v(3)
.probe
.end
```

The polynomial statement has for its last five terms *0 0 0 0 1*. Recall that these are coefficient values for $k_0 + k_1 v_1 + k_2 v_2 + k_3 v_1^2 + k_4 v_1 v_2$. All k values are zero except for the last, with $k_4 = 1$. Run the analysis; then in Probe plot v(1) and v(3). This plot deliberately omits the 20-kHz wave, making the results easier to interpret. The composite wave v(3) clearly shows the classical appearance of an amplitude-modulated wave. in this example, both of the input waves v_1 and v_2 have 1-V amplitudes. See these traces in Fig. 7.16.

Before leaving Probe, add the plot of the other input voltage v(2), so that you are now displaying v(1), v(2), and v(3). This now shows the carrier along with the other two waves, giving the complete picture. Print this if you would like a copy of it; then remove the v(2) trace and choose *X*-axis, Fourier. Let the *X*-axis range extend from 0 to 30 kHz. The frequency domain now shows the 1-kHz, 19-kHz, and 21-kHz components. The latter components are the upper- and lower-side-band frequencies produced by this modulation. Observe the amplitude of each wave. Recall the trigonometric identity

$$(\sin \alpha)(\sin \beta) = \tfrac{1}{2}[\cos(\alpha - \beta) - \cos(\alpha + \beta)]$$

Fig. 7.15 Multiplier for modulated wave.

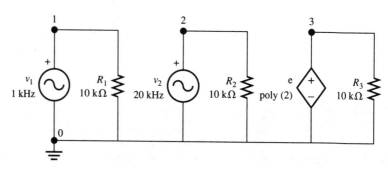

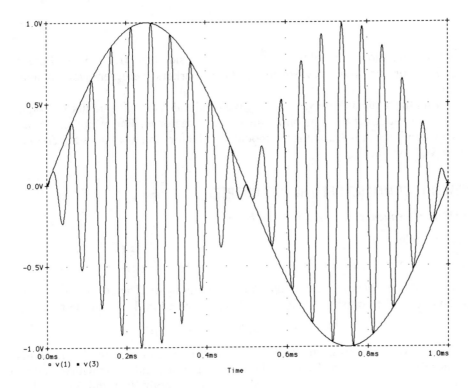

Fig. 7.16 Multiplier for Modulated Wave.

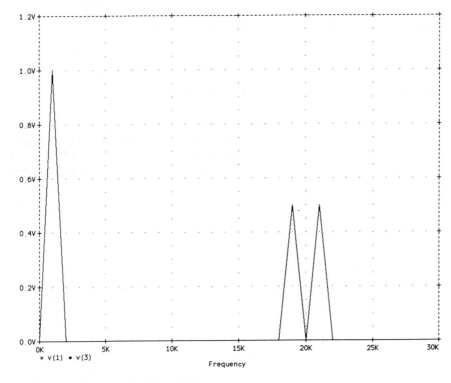

Fig. 7.17 Multiplier for Modulated Wave.

which accounts for the amplitudes of 0.5 V for the side-band frequencies. Refer to Fig. 7.17 for the frequency spectrum. (The markers were removed to avoid clutter.)

Run the analysis with various relative amplitudes for the modulating voltage v_1 to see what effect this has on the modulation factor m. For example, when v_1 has an amplitude of 0.8, what is the modulation factor, and what does the composite wave look like?

NEW DOT COMMAND USED IN THIS CHAPTER

.FOUR *<frequency>* <output variable>*

For example,

```
.FOUR 1kHz V(1) V(2)
```

means perform a decomposition into Fourier components. The decomposition will be done only if a transient analysis has been specified in the input file by using

$$\text{.TRAN } <step\ value> <final\ time>$$

The Fourier analysis gives the dc component, the fundamental, and all harmonics through the ninth. They are shown in magnitude and phase with actual and normalized values. In the example shown above, both V(1) and V(2) will be analyzed into their components.

It is customary to use the *.PROBE* statement in connection with the Fourier analysis; however, *.PRINT* or *.PLOT* may be used as well.

PROBLEMS

7.1   In Fig. 7.18, the polynomial for E is of the form

$$f(x) = x + x^2$$

Using $v_{ipeak} = 1$ V, $f = 1$ kHz, and $V = 1$ V, it is desired that we compare v_o with v_i. Predict what the approximate harmonic content in the output will be; then run a PSpice analysis, which will show the harmonic content of both the

Fig. 7.18

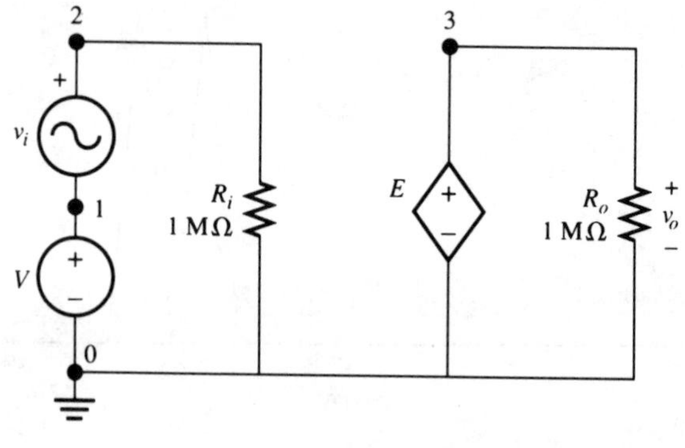

input and output voltages. In the .FOUR statement use voltages V(2,1) and V(3). Look at the output file and determine the harmonic content in V(3).

7.2   In Problem 7.1, use X-axis, Fourier to observe the harmonic content in V(3). Let the X-axis run from 0 to 5 kHz displaying V(2,1) and V(3).

7.3   Run the analysis for Problem 7.1 with

$$f(x) = 2 + 0.1x^2$$

Predict the approximate harmonic content in the output; then plot V(2,1) and V(3) to verify the accuracy of your predictions.

7.4   Figure 7.4 shows a polynomial source. The source E was given as

$$f(x) = 1 + x + x^2$$

Change the source to

$$f(x) = x + x^2$$

and perform the decomposition and reconstruction analysis, modifying i_1 and i_2 as required so that I(r) will be the same shape as V(2).

7.5   In the section "Second-Harmonic Distortion in a Power Amplifier," change the polynomial to

$$f(x) = 0.05 + x + 0.1x^2$$

and run a PSpice analysis as suggested in the text. Plot V(1) and V(2) − 0.05 to compare the ac components of input and output voltage. Predict the dc component of the output voltage, the amplitude and phase of the second harmonic, and the total harmonic distortion. Verify your predictions using Probe and your output file.

7.6   In the section "Intermodulation Distortion," we combined two sine waves of different frequencies. Perform an analysis using $f_1 = 2$ kHz and $f_2 = 2.5$ kHz with the $f(x)$ expression remaining the same. Modify the .TRAN statement as necessary. Follow the steps as given in the text example in order to verify your predictions of frequency components in the output.

7.7   In the section "Adding Sine Waves," Fig. 7.12 showed parallel branches with three voltage sources. The adding of waves was mathematical rather than physical. Modify the circuit so that the voltage sources are all in series; then perform the analysis again. Are the results the same?

7.8   Perform an analysis to add sine waves of the same frequency $f = 1$ kHz, given

$$v_1 = 0.5\underline{/0°}\ \text{V} \qquad v_2 = 1\underline{/45°}\ \text{V} \qquad v_3 = 1.5\underline{/90°}\ \text{V}$$

(a) Find the peak value of $v_1 + v_2$ and the time and the phase angle at which the peak occurs.

(b) Repeat part a for $v_1 + v_3$.

When using the cursor mode and several traces are shown on the same plot, use the [Ctrl] key to select among the traces.

7.9   To illustrate the effect of adding sine waves of slightly different frequencies rather than adding harmonics, perform an analysis as in Problem 7.8 using $v_1 = 1$ V, 1 kHz, $0°$; $v_2 = 1$ V, 1.2 kHz, $0°$; and $v_3 = 1$ V, 1.4 kHz, $0°$.

(a) Plot v_1, v_2, and $v_1 + v_2$. Find the peak value of $v_1 + v_2$.

(b) Plot v_1, v_3, and $v_1 + v_3$. Find the peak value of $v_1 + v_3$.

7.10 Based on the section dealing with amplitude modulation, let $v_1 = 1$ V, 1 kHz and modify v_2 so that the index of modulation will be 0.5. Run the PSpice analysis to illustrate the desired results.

8

Stability and Oscillators

Amplifiers, especially those involving several stages, may be stable or may go into oscillation. When oscillations occur, they are prone to settle at a certain frequency, depending on the combination of components used, including any stray inductance and capacitance. When a portion of the output signal is fed back to the input, there is the possibility that oscillations will be produced.

THE FEEDBACK LOOP

Figure 8.1 shows a block diagram of the basic feedback loop. It includes a mixing junction where the input signal v_i and the feedback signal v_f combine. Actually, the mixing junction gives a phase inversion to the feedback signal as shown by the minus sign in the figure, so that this component appears as a negative input. The difference voltage is

$$v_d = v_i - v_f$$

This voltage is fed into the amplifier, which has a voltage gain A. This creates the output voltage

$$v_o = Av_d = A(v_i - v_f)$$

The output is returned to the summing junction after passing through a feedback network labeled β. The ratio v_f/v_o defines β as the portion of the output voltage that is fed back or returned to the summing junction.

It is assumed that you will study this topic more thoroughly in another textbook. However, some authors use different symbols for the various quantities

Fig. 8.1 The basic feedback loop.

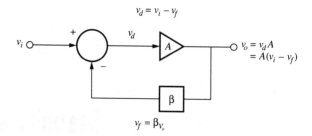

shown here. In this text the loop gain or transmission gain is given by $A\beta$, the product of the gain of the amplifier and the feedback factor. If the effect of the phase inversion is taken into account, the loop gain is—$A\beta$. It is not difficult to show that

$$A_f = \frac{A}{1 + A\beta}$$

This represents the closed-loop gain, which is the gain of the circuit with the feedback network in place.

If the signal is fed back, including its inversion, is identical to the input signal, the external input may be removed, and the amplifier will continue to produce the same output as before. This is expressed by stating that

$$-A\beta = 1$$

is the condition required for oscillation, which is called the Barkhausen criterion. From a practical standpoint, $|A\beta|$ should be slightly greater than unity. Nonlinearity in the circuit components allows the oscillations to continue without building up to larger and larger amplitudes.

In Fig. 8.1, all voltages may be replaced by currents, producing the dual of the previous situation. This means that the input is i_i, the difference is i_d, the output is $i_o = i_d A$, and the feedback signal becomes $i_f = \beta i_o$, making

$$i_d = i_i - i_f$$

The loop gain is still $A\beta$, which becomes $-A\beta$ taking the phase intervention into account. The closed-loop gain is still

$$A_f = \frac{A}{1 + A\beta}$$

THE WIEN-BRIDGE OSCILLATOR WITH INITIAL HELP

As an introduction to the study of oscillators, consider the Wien-bridge circuit of Fig. 8.2. The circuit components have been chosen to give $f_o = 25$ kHz with

$$f_o = \frac{1}{2\pi RC}$$

Fig. 8.2 Wien-bridge oscillator.

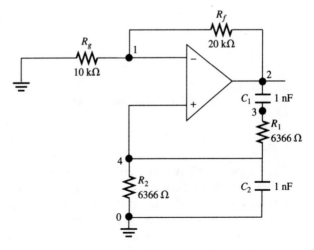

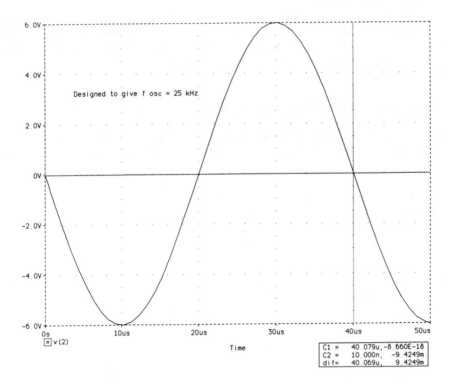

Fig. 8.3 Wien-Bridge Oscillator.

where $R = R_1 = R_2$ and $C = C_1 = C_2$. If C is chosen as 1 nF, then $R = 6366\ \Omega$. For oscillations to be sustained, R_f/R_g must be 2. If R_f is chosen as 20 kΩ, then $R_g = 10$ kΩ. Our problem is to use PSpice to show that oscillations will occur at the desired frequency. This input file is proposed:

```
Wien-Bridge Oscillator
E 2 0 4 1 2E5
Ri 4 1 1E6
Rg 1 0 10k
R1 3 4 6366
R2 4 0 6366
Rf 2 1 20k
C1 2 3 1nF  IC=2V;  initial charge to begin oscillations
C2 4 0 1nF
.PROBE
.TRAN 0.05us 50us UIC
.END
```

The capacitor C_1 is given an initial charge (with specified voltage) in order to create the priming condition required to begin the oscillations. Without this, the PSpice analysis will show that the output voltage is a steady zero value.

The results obtained in Probe are shown in Fig. 8.3. The trace shows the output voltage v(2). Note that the output is a sine wave with a frequency $f = 25$ kHz and an amplitude of 6 V.

Can the actual oscillator be expected to produce a peak output voltage of 6 V? In an attempt to answer this question, run the analysis again, but change the capacitor voltage using IC = 1 V. We will also look at the oscillator from the standpoint of loop gain and phase shift later in this chapter.

THE *LC* OSCILLATOR WITH INITIAL HELP

A tuned-circuit oscillator using two capacitors and one inductor in the feedback loop is called a Colpitts oscillator. The Hartley oscillator is similar, but it uses two

Fig. 8.4 Colpitts oscillator.

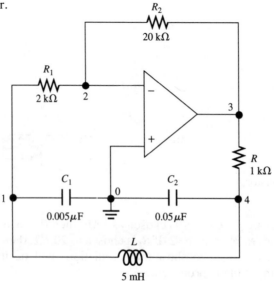

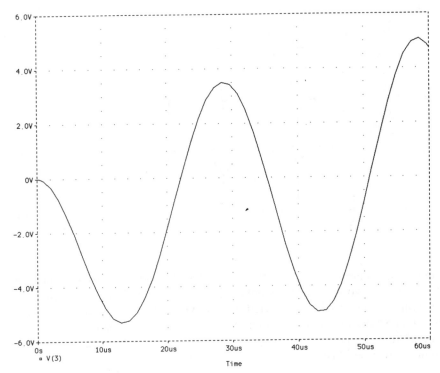

Fig. 8.5 Colpitts Oscillator.

inductors and one capacitor. The Colpitts circuit and its components are shown in Fig. 8.4. The input file is

```
Colpitts Oscillator
E 3 0 0 2 2E5
Ri 0 2 1E6
R1 2 1 10k
R2 2 3 20k
C1 1 0 0.005uF
C2 4 0 0.05uF IC=2V
L 1 4 5mH
.PROBE
.TRAN 0.3us 60us UIC
.END
```

In Probe, when the output voltage V(3) is plotted, the results are as shown in Fig. 8.5. Observe that the oscillations appear to be increasing in amplitude. Since we are using an ideal op amp in the circuit, the onset of nonlinearity does not appear. In the practical circuit the sine wave would begin to show some distortion as well as leveling off. The amplitude of the sine wave would be determined largely by the biasing voltage present in the op amp circuit.

MEASUREMENTS WITH A TEST CIRCUIT

In order to take measurements of the loop gain and phase shift of the feedback loop, we will introduce a test circuit as shown in Fig. 8.6. The circuit consists of

Fig. 8.6 Test circuit for voltage measurement.

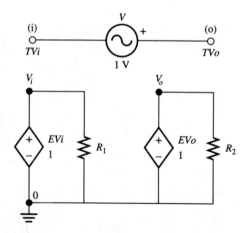

an independent ac voltage $V = 1$ V at the nodes TV_o and TV_i (*o* for *out* and *i* for *in*). There are also two voltage-dependent voltage sources, EV_i and EV_o. The independent voltage source will be inserted at a convenient break point in the feedback path of an oscillator. Each dependent voltage has its accompanying internal resistance so that the nodes V_i and V_o will not appear to be floating. The measuring technique will be illustrated in the following examples.

THE PHASE-SHIFT OSCILLATOR

The classic RC phase-shift oscillator is shown in Fig. 8.7. The output of the op amp is connected to three RC phase-shifting networks. Each network will produce a certain phase shift, and if the total phase shift produced in the three networks is

Fig. 8.7 Phase-shift oscillator.

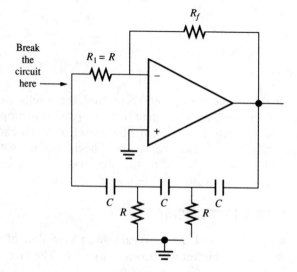

$180°$, oscillations may occur. It is further required that $|A\beta| = 1$. An analysis of the circuit shows that the frequency of oscillation is

$$f_o = \frac{1}{2\pi RC \sqrt{6}}$$

For example, let the desired frequency of oscillation be $f_o = 100$ Hz, and let $C = 0.5\ \mu$F. Then R becomes 1.3 kΩ. The analysis of this circuit also shows that at f_o, $\beta = 1/29$; therefore $|A|$ should equal 29 if oscillations are to be sustained. In practice, $|A|$ is to be made slightly greater than 29 to account for slight variations in components and operating conditions.

This circuit is an inverting op amp where

$$A = \frac{-R_f}{R_1}$$

Since $R_1 = R$, this yields

$$A = -29 = \frac{-R_f}{1.3\ k\Omega}$$

and solving for R_f, we find that $R_f = 37.7$ kΩ. If this value is increased by 5% to allow for the variations previously mentioned, the value of R_f becomes 39.58 kΩ.

In order to look at the loop gain, we will break the feedback loop to allow for measurements with our test circuit. The break will take place at the point marked in Fig. 8.7. The node at this point will thus become two nodes. These two nodes would appear to be floating unless they were shown in a subscript. For this reason, the oscillator is redrawn in Fig. 8.8, where the break involves the two nodes i (for *in*) and o (for *out*). The figure includes labeling of the other nodes in

Fig. 8.8 Phase-shift oscillator with feedback loop broken.

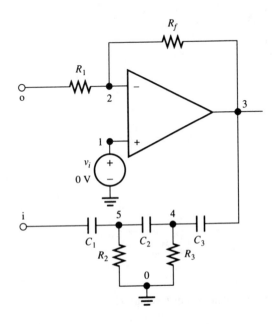

preparation for PSpice input file. It is convenient to put the op amp in a subcircuit as we have done in other examples. The simplified op amp model will have this subcircuit:

```
.subckt iop m p vo; m is inverting, p is noninverting
  rin m p 1E6
  e vo 0 p m 2E5
.ends
```

The next portion of the input file involves the circuit shown in Fig. 8.8, showing the break point that will allow for the loop-gain analysis. The op amp is called from its subcircuit with the X statement. This portion of the input file is

```
.subckt rc i p; i and o are in and out labels for break point
  x 2 1 3 iop; this calls the opamp subcircuit
  vi 1 0 1V
  rf 3 2 39.58k
  rl o 2 1.3k
  r2 5 0 1.3k
  r3 4 0 1.3k
  c1 i 5 0.5uF
  c2 5 4 0.5uF
  c3 4 3 0.5uF
.ends
```

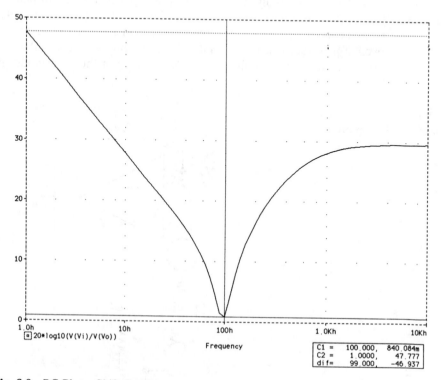

Fig. 8.9 RC Phase-Shift Oscillator.

The last portion of the input file shows the reference to the oscillator subcircuit *rc*; it also contains the statements required to perform the test of the feedback loop. The input statements are

```
* loop-gain test statements
X TVi TVo rc
V TVo TVi AC 1
R1 Vi 0 1E6
EVo Vo 0 TVo 0 1
R2 Vo 0 1E6
.AC DEC 20 1Hz 10kHz
.PROBE
.END
```

Create an input file by combining all three of the segments shown above, and run the analysis. Then plot

$$20*\log 10(V(Vi)/V(Vo))$$

The plot indicates the open-loop gain of the feedback network. Remember that if oscillations are to be sustained, $|A\beta|$ must equal 1. On the logarithmic plot, the value corresponds to zero rather than one. Verify that at $f = 100$ Hz, the plot gives a near-zero response. See Fig. 8.9 for this plot.

Now plot the phase shift with

$$VP(Vi) - VP(Vo)$$

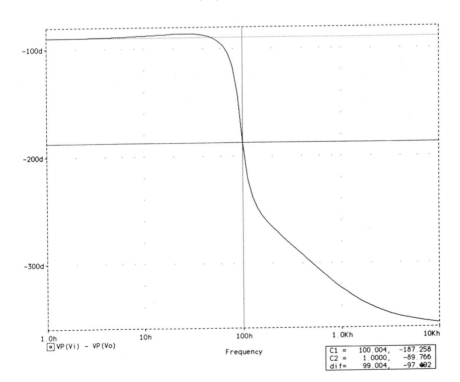

Fig. 8.10 RC Phase-Shift Oscillator.

which will compare the phases of the fed-back signal and the output signal. Confirm that at $f = 100$ Hz, the angle of the trace is $-187°$. Since the open-loop analysis does not include the phase inversion at the input, the total phase shift under closed-loop conditions is actually $367°$. This is close to the desired $360°$, which would mean that the amplifier cannot distinguish between an input signal and a feedback signal, thereby sustaining oscillations. Figure 8.10 shows the phase-shift plot.

The WIEN-BRIDGE OSCILLATOR

For another oscillator example, consider Fig. 8.11, which shows the Wien-bridge oscillator. The bridge involves series elements R_1, C_1 and parallel elements R_2, C_2. An analysis of this circuit shows that

$$f_o = \frac{1}{2\pi RC}$$

We will choose $f_o = 25$ kHz, $C_1 = C_2 = 1$ nF, and $R_g = 10$ kΩ. This leads to $R = R_1 = R_2 = 6366\ \Omega$. In this circuit, $|A\beta| = 1$ is required for oscillations to be sustained. The analysis also shows that at resonance, $\beta = \frac{1}{3}$, calling for an amplifier gain of $A = 3$. Since the gain of the noninverting op amp is

$$A = \frac{R_g + R_f}{R_g}$$

this gives $R_f = 20$ kΩ.

We redraw the circuit in Fig. 8.12 to show the break for the test measurements and the labeling of the nodes. The subcircuit for the oscillator thus becomes

Fig. 8.11 Wien-bridge oscillator.

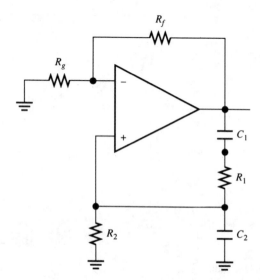

Fig. 8.12 Wien-bridge oscillator with feedback loop open.

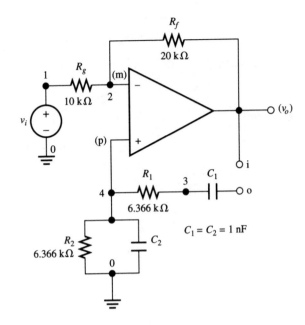

a portion of the input file. The entire input file is

```
Wien-Bridge Oscillator
.subckt wien i o
  x 2 4 i iop
  vi 1 0 0V
  rg 1 2 10k
  rf 2 i 20k
  r1 3 4 6366
  r2 4 0 6366
  c1 o 3 1nF
  c2 4 0 1nF
.ends.
.subckt iop m p vo
  rin m p 1E6
  e vo 0 p m 2E5
.ends
X TVi TVo wien
V TVo TVi AC 1
EVi Vi 0 0 TVi 1
R1 Vi 0 1E6
EVo Vo 0 TVo 0 1
R2 Vo 0 1E6
.AC DEC 40 1kHz 1 MegHz
.PROBE
.END
```

Run the analysis and plot

$$20*\log10(V(Vi)/V(Vo))$$

Verify that at $f = 25.12$ kHz, the gain is at its maximum. This corresponds to a unity gain since this is a dB plot.

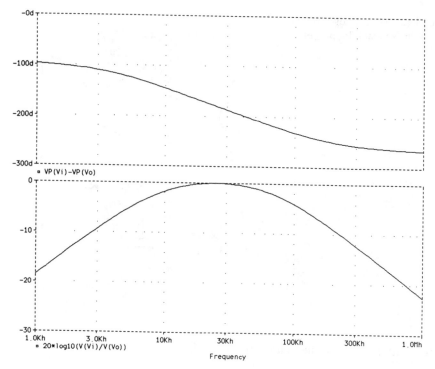

Fig. 8.13   Wien-Bridge Oscillator.

Now select plot control and add another plot that will show the phase shift of the feedback loop. The desired trace is

$$VP(Vi) - VP(Vo)$$

which will tell you whether or not the phase-shift circuit produces the desired phase shift required to sustain oscillations. Verify that the phase-shift plot shows that at $f = 25.3$ kHz, the phase shift is $-180°$. Refer to Fig. 8.13 for these plots.

Another Wien-Bridge Example

For another example, suppose that we are given the component values for the bridge portion of a Wien-bridge oscillator, but we have not solved for the frequency of oscillation. It is our desire to see whether or not oscillations will be sustained and if so at what frequency. The circuit diagram is essentially the same as in the previous example, and Fig. 8.14 shows the component values. The input file is

```
Another Wien-Bridge Example
.subckt wien i o
vi 1a 0 0V
x 1 o 2 iop
r1 1 2 20k
r2 1 1a 20k
```

Fig. 8.14 Another Wien-bridge oscillator with feedback loop broken.

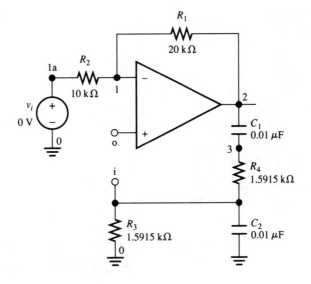

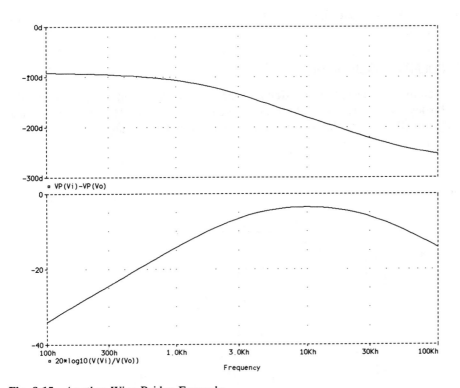

Fig. 8.15 Another Wien-Bridge Example.

```
r3 i 0 1.5915k
r4 3 i 1.5915k
c1 2 3 0.01uF
c2 i 0 0.01uF
.ends
.subckt iop m p vo
rin m p 1E6
e vo 0 p m 2E5
.ends
X TVi TVo wien
V TVo TVi AC 1
EVi Vi 0 0 TVi 1
R1 Vi 0 1E6
EVo Vo 0 TVo 0 1
R2 Vo 0 1E6
.AC DEC 20 100Hz 0.1MegHz
.PROBE
.END
```

Run the analysis, and as in the previous example, plot

$$20*\log10(V(Vi)/V(Vo))$$

Use the cursor mode to show that the *maX* of this plot is at $f = 10$ kHz. In order to verify that this is the frequency at which oscillations will be sustained, plot

$$VP(Vi) - VP(Vo)$$

and show that the phase shift is $-180°$ at a frequency of 10 kHz. Figure 8.15 shows these plots.

THE COLPITTS OSCILLATOR

We often use a general type of oscillator circuit to describe either the Colpitts or the Hartley oscillator. Figure 8.16 shows the general circuit. The impedances are Z_1, Z_2, and Z_3. For oscillations to be produced, it is found that

$$Z_1 + Z_2 + Z_3 = 0$$

Fig. 8.16 Basic circuit for resonant-type oscillator.

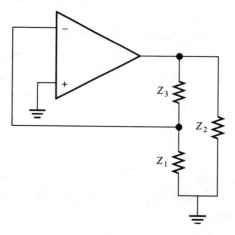

The impedances are usually assumed to be pure reactances, with X_1 and X_2 the same type and X_3 the opposite type. In the Colpitts, X_1 and X_2 are capacitive while X_3 is inductive. Let us choose $C_1 = 0.005 \ \mu F$ and $C_2 = 0.05 \ \mu F$, which gives $L = 5$ mH.

The frequency of oscillation is found to be

$$f_o = \frac{1}{2\pi} \frac{C_1 + C_2}{LC_1 C_2}$$

This gives $f_o = 33.38$ kHz. This is the circuit previously shown in Fig. 8.4. The other components are $R = 1k\Omega$, $R_1 = 10$ kΩ, and $R_2 = 20$ kΩ.

The circuit is broken for test purposes, giving Fig. 8.17. From this the input file becomes

```
Colpitts Oscillator
.subckt colpitts i o
 x 2 1a 3 iop
 vi 1a 0 0V
 r1 o 2 10k
 r2 2 3 20k
 r 3 4 1k
 c1 i 0 0.005uF
 c2 4 0 0.05uF
 L i 4 5mH
.ends
.subckt iop m p vo
 rin m p 1E6
 evo 0 p m 2E5
.ends
X TVi TVo colpitts
V TVo TVi ac 1
EVi Vi 0 0 TVi 1
```

Fig. 8.17 Colpitts oscillator test circuit.

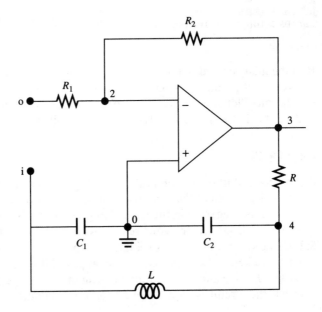

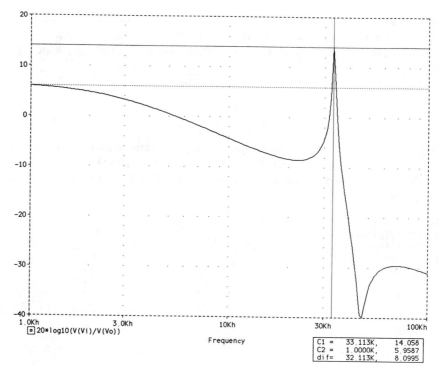

Fig. 8.18 Colpitts Oscillator.

```
R1 Vi 0 1E6
EVo Vo 0 TVo 0 1
R2 Vo 0 1E6
.ac DEC 100 1kHz 100kHz
.PROBE
.END
```

Run the analysis and obtain a plot like the one shown in Fig. 8.18. Note that there
is a resonant point at $f = 33.1$ kHz, which is close to the predicted frequency of
oscillation. Plot the difference in phase between the input and output voltages,
and verify that at $f = 33.5$ kHz, the phase shift is 180°.

PROBLEMS

8.1 A phase-shift oscillator as shown in Fig. 8.7 is to oscillate at $f_o = 1$ kHz. Using
$C = 1 \ \mu F$, select the required components and run an analysis by one of the
methods suggested in the text. Using Probe, verify that the circuit performs as
expected. Document your results with a Probe plot.

8.2 Using the Wien-bridge oscillator as shown in Fig. 8.11, it is desired to have
$f_o = 10$ kHz. Make the required changes in the input file given in the text, and
run a PSpice analysis that includes an initial charge of C_1. Using Probe, verify
that the circuit is capable of sustaining oscillations at the desired frequency.

8.3 Design a Colpitts oscillator to operate at $f_o = 100$ kHz. The circuit shown in Fig. 8.4 may be used as a model. Use the technique of opening the feedback loop to show that oscillations will be sustained at the desired frequency and to demonstrate the phase shift at this frequency.

8.4 For the Colpitts oscillator of Problem 8.3, close the feedback loop and use the stimulus method to show that oscillations occur at $f_o = 100$ kHz. Verify sinusoidal oscillations using Probe.

8.5 Refer to Fig. 8.16 for the general LC oscillator configuration. Design a Hartley oscillator where X_1 and X_2 are inductors and X_3 is a capacitor, such that $f_o = 50$ kHz. Let $L_1 = L_2 = 20$ mH. Assume that there is no mutual coupling between the inductors. Use PSpice/Probe to verify the design.

8.6 An FET phase-shift oscillator equivalent circuit is shown in Fig. 8.19. For oscillations to occur, $|A|$ must be at least 29, requiring an FET with $\mu \geq 29$. Given: $g_m = 5$ mS, $r_d = 500$ kΩ, $C = 0.5$ μF, $R = 1.3$ kΩ, and $R_d = 10$ kΩ. Use the method of opening the feedback loop to determine if oscillations will occur and if so at what frequency.

Fig. 8.19.

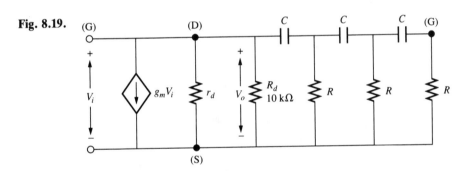

An Introduction to PSpice Devices

In the previous chapters, we have emphasized creating our own models. The models have been those that are customarily used in circuit analysis. These models have been linear and bilateral. Such models should be used whenever possible in order to keep the analysis clear and simple.

Often there is a need for a more complicated model, perhaps one that would require the use of equations that would convey the characteristics of the device. One of the strengths of PSpice is that several such models are available for your use. These models contain features that are usually associated with actual components.

A HALF-WAVE RECTIFIER

To introduce you to the concept of device models, consider the circuit shown in Fig. 9.1. This is the basic circuit for a half-wave rectifier, consisting of an ac-voltage source, a diode, and a resistor. The use of the diode presents a problem. You might model the diode as a closed switch for the positive half cycles of the input voltage and as an open switch for the negative half cycles. If you did this, you would then break the problem into two parts because your models would call for two different representations.

There is no need to do this, however, because PSpice has a built-in model for the diode. In order for you to use this built-in model, you must include a .MODEL statement in your input file. It has the following form:

.MODEL <model name> <model type>

Fig. 9.1 Half-wave rectifier using diode model.

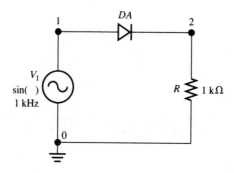

For this example, we will chose *D1* as the model name, and we must use *D* as the model type. This is required if PSpice is to recognize the device as a diode of the built-in model type. In order to see the circuit in action, use this input file:

```
Half-Wave Rectifier Using Built-in Model
v1 1 0 sin(0 12V 1000Hz)
DA 1 2 D1
R 2 0 1k
.MODEL D1 D
.TRAN 0.1ms 1ms
.PROBE
.END
```

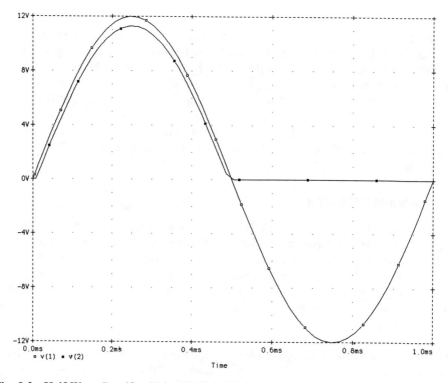

Fig. 9.2 Half-Wave Rectifier Using Built-in Model.

The transient analysis will allow you to look at a full cycle of both the input sine wave at $f = 1$ kHz and the output voltage across the resistor as a function of time. Note the form of the diode statement:

```
DA 1 2 D1
```

The designation *DA* is our choice for this diode name. The name must begin with *D*. It is located between nodes *1* (p) and *2* (n). The last item on the list is *D1*. This must agree with the diode *model name*.

Run the analysis and in Probe plot v(1) and v(2) for a full cycle. This will be for $t = 1$ ms. Observe that the input is a sine wave and that the output has been rectified such that the negative half cycle is missing. The output voltage differs from the input voltage during the positive half cycle by the voltage drop across the diode. Use the cursor mode to determine the diode drop when the input voltage is at its peak. Verify that this voltage drop is 0.72 V. Figure 9.2 shows the input and output waves.

The listing of the model type in the .MODEL statement could be extended to include other parameters. For example, silicon diodes and germanium diodes have different cut-in voltages and different saturation currents. You can specify up to 14 parameters in order to customize a diode. A complete list of these parameters is given in Appendix D under "*D* Diode."

THE BUILT-IN MODEL FOR A DIODE

If you would like to see the characteristic curve for the built-in model of a diode in PSpice, you might consider using the dc sweep. The circuit is shown in Fig. 9.3. This will enable you to look at the point-by-point response, obtaining a characteristic curve in much the same manner as you would in the laboratory. The input file will look like this:

```
Built-in Diode Model for PSpice
V 1 0 10V
R 1 2 100
D1 2 0 DMOD
.DC V -0.5V 10V 0.02V
.MODEL DMOD D
.PROBE
.END
```

Fig. 9.3 Circuit for diode characteristic curve.

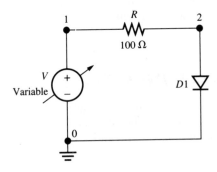

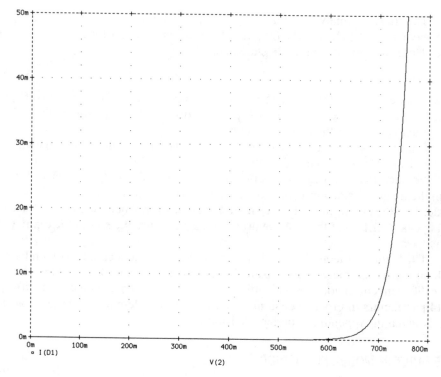

Fig. 9.4   Built-in Diode Model for PSpice.

For this example, we called the diode *D1*, and it is located between nodes *2* and *0*. We chose the model name *DMOD*. The name is specified in the .MODEL statement as the first argument.

In Probe, change the *X*-axis to represent V(2) in the range up to 0.8 V, and plot I(D1) in the range up to 50 mA. This produces a characteristic curve for the built-in diode with no changes in the predefined parameters. See Fig. 9.4 for the diode characteristic curve. You may want to modify the diode model to fit a particular need. For example, the parameter *EG* represents the barrier height. Normally this is 1.1 eV. Changing this to 0.72 eV would be in keeping with a Ge diode. If you change any parameter, you may want to look at the resulting characteristics before proceeding with an analysis of an elaborate circuit using your revised model.

THE FILTERED HALF-WAVE RECTIFIER

In order to smooth the output voltage, place a capacitor across the load resistance, as shown in Fig. 9.5. The capacitor must be large enough to keep the output voltage from dropping significantly during the time when the diode is not conducting.

Fig. 9.5 Half-wave rectifier with capacitor filter.

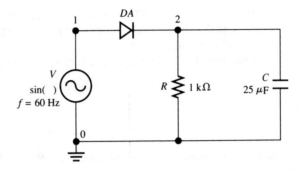

For this classic problem, choose $R = 1$ kΩ, $C = 25$ μF, and $f = 60$ Hz. The frequency is chosen to represent an ordinary voltage supply such as you would expect to find in home or commercial situations. The input file is

```
Half-Wave Rectifier with Capacitor Filter
v 1 0 sin(0 12 60)
DA 1 2 D1
R 2 0 1k
C 2 0 25uF
.MODEL D1 D
.TRAN 0.1ms 33.33ms
.PROBE
.END
```

Run the analysis, and choose a time range from zero to 25 ms. Plot v(1) and v(2). Note that the output voltage follows a pattern like that of the first example until a time just past when the voltages reach their peaks. Then, because the capacitor was charged to the peak voltage, the diode stops conducting. This allows the capacitor to discharge exponentially until the time when the input voltage is large enough to allow conduction to begin again. Refer to Fig 9.6 for the details of this waveshape.

The mathematical treatment of this circuit usually neglects the diode drop. The equation for peak diode current is then

$$I_m = V_m \sqrt{\frac{1}{R_L^2} + \omega^2 C^2}$$

Verify that, using these values, this gives $I_m = 113.7$ mA.

Fig. 9.6 Ideal diode in half-wave, capacitor-filtered circuit.

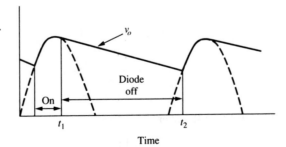

Remove the traces of voltage and plot I(DA), the diode current. Using the cursor mode, check the on-screen current maximum against the calculated value. There should be close agreement.

The equations for determining the cutout angle are

$$\theta = \arctan \omega C R_L$$

and

$$\omega t_1 = \pi - \theta = 180° - \theta$$

Using these values leads to $\theta = 83.94°$ and $\omega t_1 = 96.06°$. Remove the trace of diode current and plot v(2) again. Verify that the cutout occurs at the predicted location. The cut-in point can easily be found from the graph of v(2). It is the point where the voltage begins to rise after its exponential decay. You should be able to demonstrate that this occurs at $v(2)_{min} = 6.484$ V when $t = 18.37$ ms. Subtract the time for one cycle from this and show that this corresponds to 36.8°.

The maximum value of output voltage is simply the maximum value of the input voltage minus the diode drop. This is found to be $v(2)_{max} = 11.28$ V. The ripple voltage is $V_r = 11.28 - 6.484 = 4.796$ V. See Fig. 9.7 for the traces of v(1) and v(2).

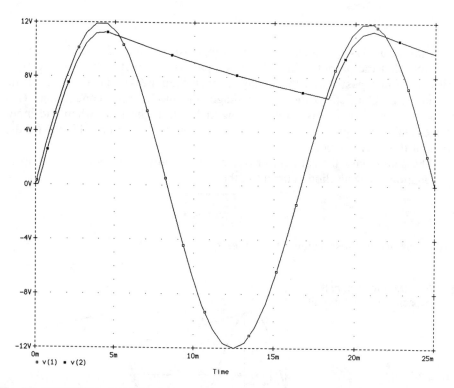

Fig. 9.7 Half-Wave Rectifier.

Now it is possible to observe the effects of changing the value of C in order to lessen the ripple voltage. Change the value of capacitance to $C = 50 \ \mu\text{F}$ and run the analysis again. Plot v(1) and v(2). Verify that the ripple is reduced to $V_r = 2.806$ V.

THE FULL-WAVE RECTIFIER

In order to better utilize the input voltage, use a full-wave rectifier, such as that shown in Fig. 9.8. The voltages v_1 and v_2 are taken from a center-tapped transformer with the ground connection at the center tap. The output voltage is v_o across R. For this connection, the input file becomes

```
Full-Wave Rectifier
v1 1 0 sin(0 12 60Hz)
v2 0 3 sin(0 12 60Hz)
R 2 0 1k
D1 1 2 DA
D2 3 2 DA
.MODEL DA D
.TRAN 0.1ms 25ms
.PROBE
.END
```

Run the analysis and in Probe plot v(1), v(3), the two input voltages, and v(2), the output voltage. Here you see full-wave rectification with conduction over the entire cycle. These traces are shown in Fig. 9.9. Remove the voltage traces; then plot i(R), the output current. Verify that the current reaches a peak i(R)$_{\text{max}}$ = 11.24 mA each half cycle. Does this agree with your calculated value? See Fig. 9.10 for the load-current plot.

Fig. 9.8 Full-wave rectifier, PSpice-model circuit.

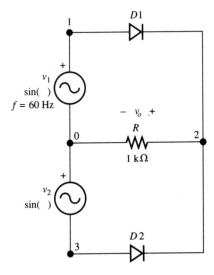

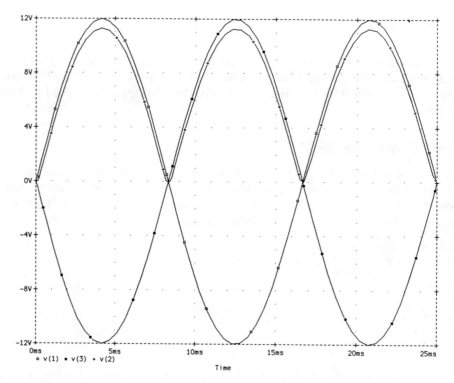

Fig. 9.9 Full-Wave Rectifier.

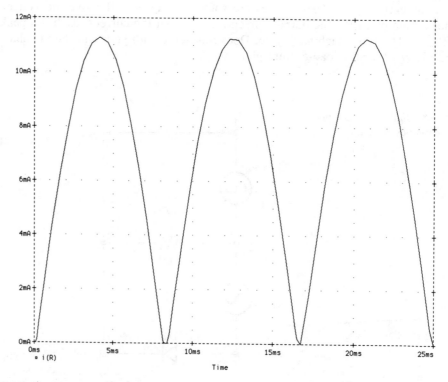

Fig. 9.10 Full-Wave Rectifier.

FULL-WAVE RECTIFIER WITH FILTER

Place a 25-μF capacitor across R and add the statement

```
C 2 0 25uF
```

to the input file of the example used in the previous section.

After you run the analysis, plot the output voltage v(2) along with the input voltages v(1) and v(3). Observe the effects of the capacitor on the ripple voltage. Using the cursor mode, verify that v(2)$_{max}$ = 11.25 V and v(2)$_{min}$ = 8.79 V, giving V_r = 2.46 V. These traces are shown in Fig. 9.11.

SIMPLE DIODE CLIPPER

A clipper is used to transmit part of an input voltage of an arbitrary waveform to its output terminals. When a diode is biased by placing it in series with a dc voltage, the clipping action takes place. Figure 9.12 shows one such circuit. The input file is

```
Diode Clipping Circuit
vi 1 0 sin(0 12V 60Hz)
DA 2 3 D1
R 1 2 1k
```

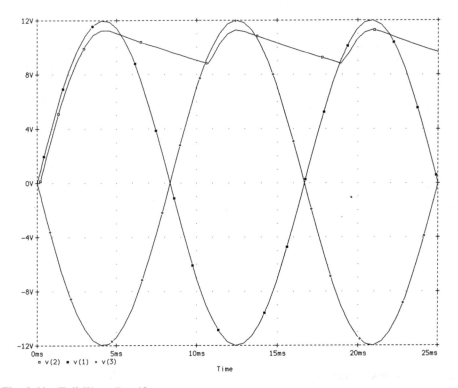

Fig. 9.11 Full-Wave Rectifier.

```
VR 3 0 8V
.MODEL D1 D
.TRAN 0.1ms 25ms
.PROBE
.END
```

Run the analysis and plot the input voltage v(1) and the output voltage v(2). Can you predict what will be the clipping level? Why is it not at 8 V? These traces are shown in Fig. 9.13.

Fig. 9.12 A simple diode clipper.

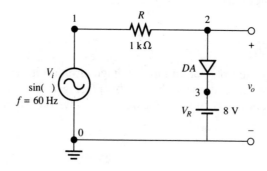

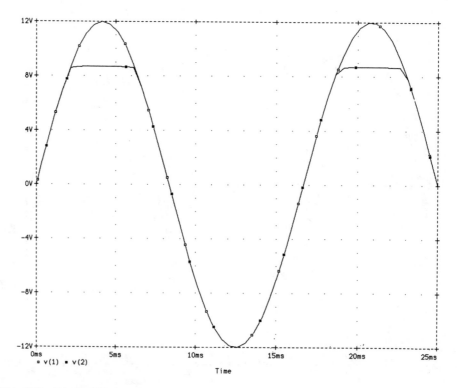

Fig. 9.13 Diode Clipping Circuit.

A DOUBLE-ENDED CLIPPER

Double-ended clipping is used to convert a sine wave into a square wave. A simple back-to-back connection of two avalanche diodes is shown in Fig. 9.14. The avalanche diodes are selected to have a Zener (breakdown) voltage of 2.4 V. The built-in diode model can easily be converted to an avalanche diode through the use of the parameter *BV* for breakdown voltage. This is shown in the input file:

```
Double-Ended Clipper Using Avalanche Diodes
vi 1 0 sin(0 24V 60Hz)
DA 3 2 D1
DB 3 0 D1
R 1 2 1k
.MODEL D1 D(BV=2.4V)
.TRAN 0.1ms 25ms
.PROBE
.END
```

Run the analysis; then plot the input voltage v(1) and the output voltage v(2). Note that the output is clipped on both ends due to the action of the back-to-back diodes. Does the clipping occur such that the output voltage swings between ±2.4 V? Verify that the output voltage peaks at 3.628 V. You should also look at v(2) alone, where the squared output is more evident. Refer to Fig. 9.15 for these traces.

The transfer characteristic is often shown for this circuit. You can look at this curve by changing the *X*-axis to represent v(1), then plotting v(2). This plot shows the output voltage over the complete range of input-voltage swings. Note that this plot shows some overtracing. This is due to the transient analysis being based on the sine wave input. You can avoid this problem by using a dc sweep. Revise the input file to become

```
Double-Ended Clipper Using Avalanche Diodes
VI 1 0 24V
DA 3 2 D1
DB 3 0 D1
R 1 2 1k
.MODEL D1 D(BV=2.4V)
.DC VI -24 24 0.1
.PROBE
.END
```

Fig. 9.14   Double-ended clipper using avalanche diodes.

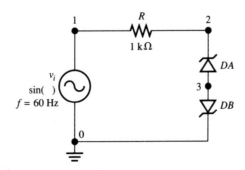

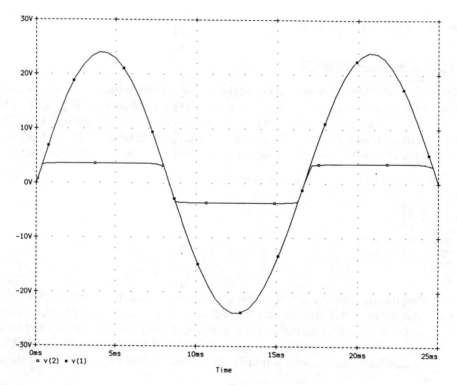

Fig. 9.15 Double-Ended Clipper Using Avalanche Diodes.

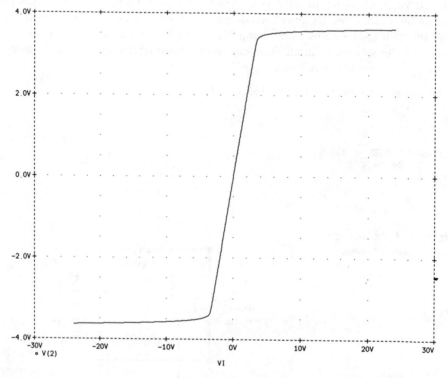

Fig. 9.16 Double-Ended Clipper Using Avalanche Diodes.

Run the analysis and obtain the improved transfer characteristic. The input voltage *VI* is shown on the *X*-axis. Plot V(2) on the *Y*-axis. Refer to Fig. 9.16 for this trace.

VARIABLE LOAD RESISTOR FOR MAXIMUM POWER

We have considered the maximum power theorem for both dc and ac circuits. In both cases the load was set and the analysis made. If we desired to change the load, we made the change in the input file and ran the analysis again. There is a way, however, to let the load resistance change without running another analysis. We will outline this method next.

The circuit of Fig. 9.17 shows a 12-V dc source with $R_i = 5 \Omega$ connected to a variable R_L. In order to produce the variable R_L, it is necessary to use a .MODEL statement for the resistor. It will look like this:

```
.MODEL RL RES
```

where *RL* is the chosen model name and *RES* is the required model type. By using the model, you will be able to include a .DC sweep statement showing a range of values for *RL*. This statement will be

```
.DC RES RL(R) 0.1 10 0.1
```

where *RES* is the required sweep-variable name and RL(R) uses our choice for the model name; the (R) is the required resistor-device name. The entire input file is

```
Maximum Power with Variable Load Resistor
V 1 0 12V
RI 1 2 5
RLOAD 2 0 RL 1
.MODEL RL RES
.DC RES RL(R) 0.1 10 0.1
.PROBE
.END
```

Note the statement for *RLOAD*. The last parameter given is a scale factor of 1. This is a required value, and the analysis will not run without it. Its purpose is to allow for various scale factors (multipliers) if required. This might be useful if there are several resistors, each based on the same model.

Fig. 9.17 Maximum power with variable load resistor.

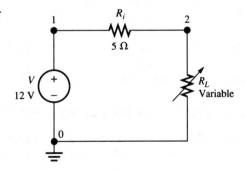

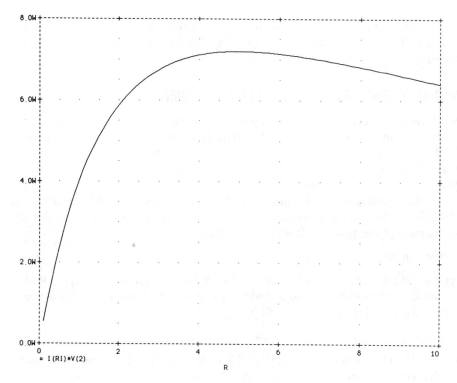

Fig. 9.18   Maximum Power with Variable Load Resistor.

Run the analysis and plot

$$I(RI)*V(2)$$

giving the power delivered to the load resistor. Verify that the peak occurs when $R = 5\ \Omega$, representing $RLOAD = 5\ \Omega$. Use the cursor to show that $P_{max} = 7.2$ W. Figure 9.18 shows this plot.

BUILT-IN MODEL FOR THE BIPOLAR-JUNCTION TRANSISTOR

We have waited a long time to introduce the built-in model for the BJT. Although one of the main strengths of PSpice lies in the wide range and versatility of its built-in models, at the same time these complex models can strike fear into the heart of the beginner. For example, the BJT device Q has 40 parameters that the user may specify. If you take a look under ''Q Bipolar Transistor'' in Appendix D, you will see just how comprehensive these parameters are. Many of them will probably be unfamiliar to you, and they are beyond the scope of our discussion.

OUTPUT CHARACTERISTICS OF THE COMMON-EMITTER TRANSISTOR

To introduce the BJT model, we will use the common-emitter biasing circuit of Fig. 9.19. If you were to do a laboratory investigation of the output characteristics

Fig. 9.19 BJT in *CE* connection to show output characteristics.

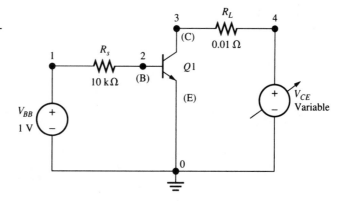

of a BJT, you might use this circuit. One characteristic curve would be obtained by keeping the input current I_B constant while you varied the voltage V_{CE}. Most students are familiar with this exercise. We will look at this from the SPICE point of view.

The transistor will be called *Q1* with the model name *BJT*. We choose both of these, and the required statement becomes

```
Q1 3 2 0 BJT
```

The nodes are in the order of *collector, base,* and *emitter,* respectively. The model statement will be

```
.MODEL BJT NPN
```

where *BJT* is chosen to agree with our *Q1* designation and *NPN* is the required model type for an *npn* transistor. The entire input file will be

```
BJT PSpice Model Characteristics
VBB 1 0 1V
RS 1 2 10k
RL 3 4 0.01
Q1 3 2 0 BJT; 3=collector, 2=base, 1=emitter
VCE 4 0 5V
.MODEL BJT NPN
.DC VCE 0 15V 0.1V
.PROBE
.END
```

Run the analysis and plot −I(RL). The minus sign is correct with respect to the R_L statement shown in the file. Use the cursor mode to find I_{Cmax}. This should be $I_{Cmax} = 2.07$ mA. This characteristic curve is shown in Fig. 9.20. Remove this trace and plot I(RS) to obtain the input current I_B. Verify that this has a maximum value of $I_B = 20.7$ μA. It is obvious from these two values that $h_{FE} = 100$, in agreement with the model parameter *BF*. You may specify other values for *BF* as necessary for a particular transistor model. See Appendix D under ''*Q* Bipolar Transistor'' for a list of all of the transistor parameters.

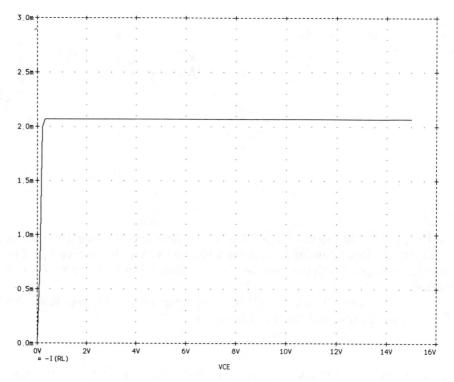

Fig. 9.20   BJT PSpice Model Characteristics.

INPUT CHARACTERISTICS OF THE COMMON-EMITTER TRANSISTOR

The input characteristics may be obtained from an input file that contains reference to the built-in model as follows:

```
BJT Input Characteristics
IBB 0 1 100uA
Rs 1 0 1000k
RL 2 3 1k
Q1 2 1 0 BJT
VCC 3 0 12V
.MODEL BJT NPN
.DC IBB 0 100uA 1uA
.PROBE
.END
```

Reference to Fig. 9.21 shows that this *npn*-transistor model has a value for V_{BE} in the active region of about 0.8 V. Since this is about 0.1 V too high for a BJT, the standard model will produce results that differ from those obtained when you created your own model for bias analysis.

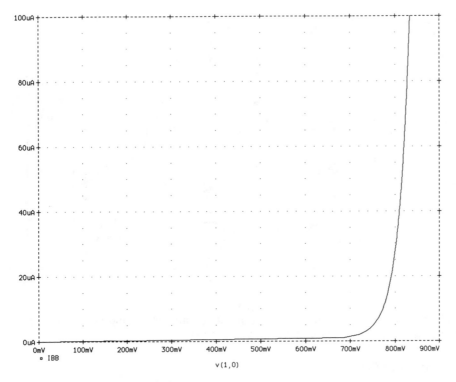

Fig. 9.21   BJT Input Characteristics.

OUTPUT CHARACTERISTICS OF THE JFET

A family of curves representing the output characteristics of the JFET are of value in determining the drain current saturation value, the pinch-off voltage, and so forth. The construction of these curves is left as an exercise for you and is the subject of Problem 9.8.

OTHER ACTIVE SEMICONDUCTOR DEVICES

Appendices A, B, and D contain listings of other active devices including "*B* GaAsFET" and "*M* MOSFET." From an academic point of view you are encouraged to use your own models for transistors and other devices. This will allow you to decide which model is most appropriate for which situation. In more advanced applications, the built-in models will allow you to set parameters that could not otherwise be taken into account.

THE DIFFERENTIAL AMPLIFIER

The differential amplifier is used as the first stage of the operational amplifier. In its elementary form it looks like Fig. 9.22. For an analysis we will use the built-in

Fig. 9.22 The differential amplifier (difference mode).

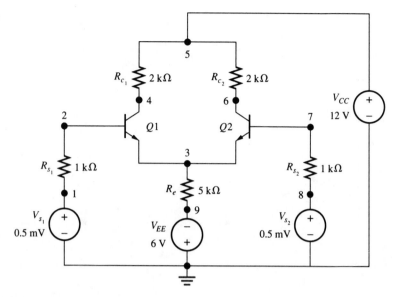

model for the *npn* transistor, giving a matched pair for *Q1* and *Q2*. Also, we will choose $R_{s1} = R_{s2} = 1$ kΩ and $R_{c1} = R_{c2} = 2$ kΩ.

Difference-Mode Gain

The difference-mode gain of the differential amplifier is found by setting $V_{s1} = -V_{s2} = V_s/2$. The gain is approximated as

$$A_d = \frac{v_o}{v_s} = \frac{h_{fe}R_c}{2(R_s + h_{ie})}$$

For our analysis, choose $V_s = 1$ mV, giving $V_{s1} = 0.5$ mV and $f = 1$ kHz. Based on the built-in model, calculate the expected value for A_d. The input file will look at bias voltages and ac operation. In order to obtain the ac voltages in the output file, use the .PRINT statement. This leads to the following input file:

```
Model for Differential Amplifier
VS1 1 0 AC 0.5mV ; Vs1 = -Vs2 = Vs/2
VS2 0 8 AC 0.5mV ; This gives difference-mode operation
RS1 2 1 1k
RS2 7 8 1k
RE 3 9 5k
RC1 4 5 2k
RC2 5 6 2k
VCC 5 0 12V
VEE 0 9 6V
Q1 4 2 3 BJT
Q2 6 7 3 BJT
.AC LIN 1 1000Hz 1000Hz
.MODEL BJT NPN
.PRINT AC V(1) V(2) V(3) V(4) V(5) V(6) V(7) V(8)
.TF V(4) VS1
.END
```

Run the analysis and obtain a printed copy of the output file. If you remove the unnecessary information, you should be able to reduce this file to a single page of less than 60 lines. Verify that $A_d = V_o/V_{s1} = V(4)/V(1) = 33.4$. Note how the input voltages V_{s1} and V_{s2} are referenced in the input file with respect to their nodes. Figure 9.23 shows this output file.

Common-Mode Gain

For common-mode operation it is necessary to set $V_{s1} = V_{s2} = V_s$. The common-mode gain may be approximated by

$$A_c = \frac{-h_{fe}R_c}{R_s + h_{ie} + (1 + h_{fe})2R_e}$$

Using the known values for the transistor, predict what this gain will be. The input file will be modified so that the input voltages are shown as

```
VS1 1 0 AC 1mV; Vs1 = Vs2
VS2 8 0 AC 1mV; This gives common-mode operation
```

After making these changes, run the analysis and obtain a printed copy of the output file as before. Verify that $A_c = V(4)/V(1) = 0.197$. This output file is shown in Fig. 9.24.

Transfer Characteristics of the Differential Amplifier

An important aspect of differential-amplifier operation is found by investigating its transfer-characteristic curve. The use of the built-in model for the transistor will make this task easy. Since we are interested in small values of differential input voltage, we will use a dc sweep for the input voltage range of -0.5 V to 0.5 V. For this analysis, V_{s2} will be given a fixed value of 1 mV; the sweep will apply to V_{s1}. Thus the input file becomes

```
Transfer Characteristics of Differential Amplifier
VS1 1 0 1mV ; this input will vary from −0.5 V to 0.5 V
VS2 8 0 1mV ; this input will remain fixed
RS1 2 1 1k
RS2 7 8 1k
RE 3 9 5k
RC1 4 5 2k
RC2 5 6 2k
VCC 5 0 12V
VEE 0 9 6V
Q1 4 2 3 BJT
Q2 6 7 3 BJT
.MODEL BJT NPN
.OP
.DC VS1 -0.5 0.5 0.01
.PROBE
.END
```

Run the analysis and in Probe plot $-I(RC1)$. Compare the results with the transfer characteristics given in a text dealing with this subject. Note that the linear portion of this transfer curve is somewhat limited. Can you approximate the

```
Model for Differential Amplifier

****        CIRCUIT DESCRIPTION

VS1 1 0 AC 0.5mV ; Vs1 = -Vs2 = Vs/2
VS2 0 8 AC 0.5mV ; This gives difference-mode operation
RS1 2 1 1k
RS2 7 8 1k
RE 3 9 5k
RC1 4 5 2k
RC2 5 6 2k
VCC 5 0 12V
VEE 0 9 6V
Q1 4 2 3 BJT
Q2 6 7 3 BJT
.AC LIN 1 1000Hz 1000Hz
.MODEL BJT NPN
.PRINT AC V(1) V(2) V(3) V(4) V(5) V(6) V(7) V(8)
.TF V(4) VS1
.END

    ****        BJT MODEL PARAMETERS

            BJT
            NPN
    IS   100.000000E-18
    BF   100
    NF    1
    BR    1
    NR    1

NODE    VOLTAGE    NODE    VOLTAGE    NODE    VOLTAGE    NODE    VOLTAGE
(    1)    0.0000  (    2)   -.0052   (    3)   -.7624   (    4)   10.9630
(    5)   12.0000  (    6)  10.9630   (    7)   -.0052   (    8)    0.0000
(    9)   -6.0000

    VOLTAGE SOURCE CURRENTS
    NAME          CURRENT
    VS1          -5.186E-06
    VS2           5.186E-06
    VCC          -1.037E-03
    VEE          -1.048E-03

    TOTAL POWER DISSIPATION    1.87E-02   WATTS

****        SMALL-SIGNAL CHARACTERISTICS
    V(4)/VS1 = -1.681E+01
    INPUT RESISTANCE AT VS1 =  1.190E+04
    OUTPUT RESISTANCE AT V(4) =  2.000E+03

FREQ         V(1)         V(2)         V(3)         V(4)         V(5)
1.000E+03   5.000E-04   4.165E-04   9.123E-19   1.670E-02   1.000E-20

FREQ         V(6)         V(7)         V(8)
1.000E+03   1.670E-02   4.165E-04   5.000E-04
```

Fig. 9.23

```
Model for Differential Amplifier

****      CIRCUIT DESCRIPTION

VS1 1 0 AC 1mV ; Vs1 = Vs2
VS2 8 0 AC 1mV ; This gives common mode operation
RS1 2 1 1k
RS2 7 8 1k
RE 3 9 5k
RC1 4 5 2k
RC2 5 6 2k
VCC 5 0 12V
VEE 0 9 6V
Q1 4 2 3 BJT
Q2 6 7 3 BJT
.AC LIN 1 1000Hz 1000Hz
.MODEL BJT NPN
.PRINT AC V(1) V(2) V(3) V(4) V(5) V(6) V(7) V(8)
.TF V(4) VS1
.END

   ****      BJT MODEL PARAMETERS

            BJT
            NPN
       IS   100.000000E-18
       BF   100
       NF   1
       BR   1
       NR   1

 NODE    VOLTAGE      NODE    VOLTAGE      NODE    VOLTAGE      NODE    VOLTAGE
(    1)    0.0000   (    2)    -.0052   (    3)    -.7624   (    4)   10.9630
(    5)   12.0000   (    6)   10.9630   (    7)    -.0052   (    8)    0.0000
(    9)   -6.0000

     VOLTAGE SOURCE CURRENTS
     NAME        CURRENT
     VS1         -5.186E-06
     VS2         -5.186E-06
     VCC         -1.037E-03
     VEE         -1.048E-03

     TOTAL POWER DISSIPATION   1.87E-02   WATTS

  ****      SMALL-SIGNAL CHARACTERISTICS
       V(4)/VS1 = -1.681E+01
       INPUT RESISTANCE AT VS1 =  1.190E+04
       OUTPUT RESISTANCE AT V(4) =  2.000E+03

  FREQ         V(1)         V(2)         V(3)         V(4)         V(5)
  1.000E+03   1.000E-03   9.990E-04   9.941E-04   1.969E-04   1.000E-20

  FREQ         V(6)         V(7)         V(8)
  1.000E+03   1.969E-04   9.990E-04   1.000E-03
```

Fig. 9.24

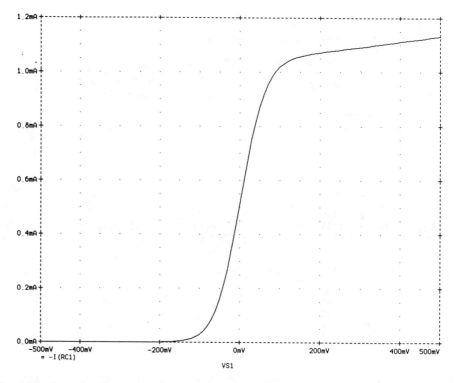

Fig. 9.25 Transfer Characteristics of Differential Amplifier.

linear range for V_{s1} and I_{RC1}? The X-axis is usually normalized to the volt-equivalent of temperature and is given as

$$\frac{V_{B1} - V_{B2}}{V_T}$$

Recall that $V_T = 26$ mV at room temperature and the linear portion of the curve is limited to about ± 26 mV. Refer to Fig. 9.25 for the transfer characteristic curve.

The differential amplifier is a good limiter, and if the input voltage exceeds ± 100 mV at room temperature, the output becomes saturated. Verify these statements from your printed copy of the transfer curve. Can you produce the mirror image of this transfer curve? This can be obtained by keeping V_{s1} fixed while varying V_{s2}.

LOGIC GATES

The evaluation version of PSpice contains a sampling of the TTL devices that are available in the production version of the software. These include the *7402* as a two-input NOR gate. the *7404* inverter, the *7405* open-collector inverter, the *7414* Schmidt trigger, the *7474* D-type flip-flop, the *74107 JK*-type flip-flop, and the *74393* 4-bit binary counter.

THE 7402 NOR GATE

A circuit involving a single NOR gate is shown in Fig. 9.26. The two inputs, *A* and *B*, are shown as pulse trains of various widths and amplitudes of 1 V. The NOR gate is referenced with a subroutine call where the nodes *1, 2,* and *3* refer to inputs *A* and *B* and output *Y*, respectively. The call lists the device as *7402*. The entire input file is shown in the output listing of Fig. 9.28, which is an abbreviated

Fig. 9.26 NOR gate with two inputs.

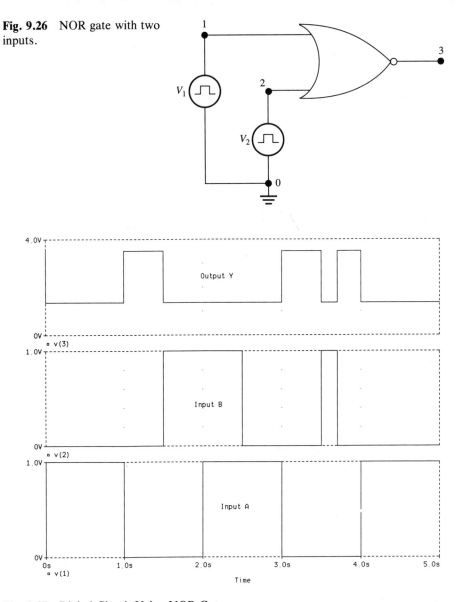

Fig. 9.27 Digital Circuit Using NOR Gate.

version of the actual output file consisting of several pages. The listing of the
model parameters has been omitted to save space. Note that analog-to-digital
interface statements in the form of subroutine calls are automatically generated
from the *7402* subroutine. These are for each of the three nodes of the NOR gate.
Statements for the digital power supply are also automatically produced. Note the
listing of $G_DPWR = 5$ V along with the other node voltages. The Probe plot of
Fig. 9.27 shows the two input waves along with the output wave. This is a simple
timing diagram for our gate circuit. The A/D interfaces are also shown in Fig.
9.28.

```
 Digital Circuit Using NOR gate
VCC 4 0 5V
X 1 2 3 7402
V1 1 0 PWL(0s 0V 0.1ms 1V 1s 1V 1.0001s 0V
+2s 0V 2.0001s 1V 3s 1V 3.0001s 0V 4s 0V 4.0001s 1V 5s 1V)
V2 2 0 PWL(0s 0V 1.5s 0V 1.50001s 1V 2.5s 1V 2.50001s 0V 3.5s 0V 3.50001s 1V
+3.7s 1V 3.70001s 0V 5s 0V)
R 4 3 100k
.lib eval.lib
.tran 0.01ms 5s
.probe
.end

**** Generated AtoD and DtoA Interfaces ****
* Analog/Digital interface for node 1
* Moving X.U1:IN1 from analog node 1 to new digital node 1$AtoD
X$1_AtoD1 1 1$AtoD $G_DPWR $G_DGND AtoD_STD
+       PARAMS: CAPACITANCE=   0
*
* Analog/Digital interface for node 2
* Moving X.U1:IN2 from analog node 2 to new digital node 2$AtoD
X$2_AtoD1 2 2$AtoD $G_DPWR $G_DGND AtoD_STD
+       PARAMS: CAPACITANCE=   0
*
* Analog/Digital interface for node 3
*
* Moving X.U1:OUT1 from analog node 3 to new digital node 3$DtoA
X$3_DtoA1 3$DtoA 3 $G_DPWR $G_DGND DtoA_STD
+       PARAMS: DRVH= 96.4   DRVL= 104   CAPACITANCE=   0
*
* Analog/Digital interface power supply subckt
X$DIGIFPWR 0 DIGIFPWR

.END   ;(end of AtoD and DtoA interfaces)

 ****      Digital Input MODEL PARAMETERS

 ****      Digital Output MODEL PARAMETERS

 ****      INITIAL TRANSIENT SOLUTION      TEMPERATURE =   27.000 DEG C

NODE   VOLTAGE      NODE   VOLTAGE      NODE   VOLTAGE      NODE   VOLTAGE
(   1)    0.0000 (   2)    0.0000 (   3)    3.5028 (   4)    5.0000
($G_DGND) 9.972E-12                 ($G_DPWR)    5.0000
(X$DIGIFPWR.REF) 9.972E-12
 DGTL NODE : STATE  DGTL NODE : STATE  DGTL NODE : STATE  DGTL NODE : STATE
   1$AtoD) : 0      ( 2$AtoD) : 0      ( 3$DtoA) : 1
```

Fig. 9.28

As an exercise, change the timing patterns for the two inputs and run the analysis again. From your knowledge of the operaton of the NOR gate, it is easy to verify the results. When the author attempted to increase the voltage levels of the input waves, the transient analysis was interrupted before successful completion.

Note that each logic device is modeled by a subcircuit. The subcircuit name is the part name. Look at the ASCII file *EVAL.LIB* under

```
.subckt 7402 A B Y
```

to see how a typical digital entry is shown in the library as a subroutine. Other devices, such as NAND gates, are modeled in a similar fashion. The production version of PSpice contains hundreds of digital devices. For further information, see *The Design Center Analysis Reference Manual*, pp. 259–92 published by the MicroSim Corporation, version 5.1, January 1992.

NEW PSPICE STATEMENTS USED IN THIS CHAPTER

D[*name*] *<+node> <−node> <model name>* [*area*]

For example,

```
DA 1 2 D1
```

means that a certain diode *DA* is used in the circuit between nodes *1* (p) and *2* (n). The model for the diode is to be described in a .MODEL statement that bears the model name *D1*. There might be several diodes, *DA*, *DB*, and *DC*, for example, based on the same model.

J[*name*] *<drain node> <gate node> <source node> <model name>* [*area*]

For example,

```
J 5 4 2 JFET
```

means that a certain junction FET is used in the circuit connected among nodes *5* (drain), *4* (gate), and *2* (source). The model for the FET is to be described in a .MODEL statement that has the model name *JFET*. The circuit may contain several FETs, *J*, *J1*, and *J2*, for example, based on the same model.

Q[*name*] *<collector node> <base node> <emitter node>* [*substrate node*]
                    <model name> [*area*]

For example,

```
Q1 3 2 0 BJT
```

means that a certain bipolar transistor *Q1* is connected among nodes *3* (collector), *2* (base), and *0* (emitter). The model for the bipolar transistor is to be described in a .MODEL statement that bears the model name *BJT*. There might be several transistors, *Q1*, *Q2*, and *Q3*, for example, which are based on the same model.

R[*name*] *<+node> <−node>* [*model name*] *<value>*

For example,

```
RLOAD 2 0 RL 1
```

means that a resistor *RLOAD* is connected between nodes *2* and *0*. Also, this resistor is modeled as *RL* in a .MODEL statement. The last value (shown after *RL* as *1*) represents a scale factor of unity. It or some other scale factor must be included. A scale factor of 2 would indicate a doubling of the value described in the model.

One of the reasons for using a resistor based on a model is to allow for a .DC statement where *RES* is chosen as the sweep variable. Refer to the dot statements for more information.

NEW DOT COMMANDS

.MODEL <*model name*> <*type*>

For example,

```
.MODEL D1 D
```

is used to define a diode model. The *D* is used to show that the model is in the PSpice library of devices. The list of library models includes the following:

- CAP (capacitor)
- IND (inductor)
- RES (resistor)
- D (diode)
- NPN (*npn*-type BJT)
- PNP (*pnp*-type BJT)
- NFJ (*n*-channel JFET)
- PFJ (*p*-channel JFET)
- NMOS (*n*-channel MOSFET)
- PMOS (*p*-channel MOSFET)
- GASFET (*n*-channel GaAs MOSFET)
- ISWITCH (current-controlled switch)
- VSWITCH (voltage-controlled switch)
- CORE (nonlinear, magnetic-core transformer)

A more complete form of the model statement is

.MODEL <*model name*> <*type*> [(<parameter name> = <*value*>)]*

For example,

```
.MODEL D1 D(IS=1E-12 N=1.2 VJ=0.9 BV=10)
```

means that some of the 14 possible diode parameters have been selected to have values other than their default ones. The asterisk positioned after the brackets means that the enclosed item may be repeated. Appendix D contains a list of each library device with its various model parameters and their default values.

PROBLEMS

9.1 A half-wave rectifier as shown in Fig. 9.1 is to have the following parameters: $IS = 1E-9$ A, $VJ = 0.8$ V, $IBV = 1E-6$ A, $EG = 0.72$ eV. Run an analysis like the one described in the text, and compare the results with those previously obtained. What differences are observed?

9.2 A diode circuit containing a biasing voltage in series with an ac source is shown in Fig. 9.29. Use the built-in diode model in your analysis. Given: $V = 0.8$ V and $R = 1$ kΩ. Describe the ac voltage as a sin( ) function with a 0.2-V peak at $f = 1$ kHz.

(a) Perform an analysis to plot voltages v(2,1) and v(3) in Probe. Does the diode conduct for a full cycle? Verify and explain your answer.

(b) Run the analysis again with $V = 0.6$ V. Explain the results.

(c) Run the analysis again with $V = 0.4$ V. Explain the results.

Fig. 9.29

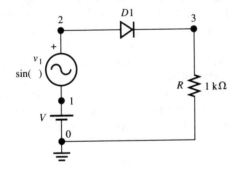

9.3 The diode characteristic curve shown in Fig. 9.4 is for the built-in diode model. Produce a diode characteristic for the diode of Problem 9.1. Describe any differences in the two curves.

9.4 A full-wave rectifier with a capacitor filter using $C = 25$ μF is discussed in the text. Use an analysis like the one suggested to determine the ripple and the average value of output voltage when $C = 10$ μF.

9.5 The circuit shown in Fig. 9.30 is similar to Fig. 9.12 except that the diode is

Fig. 9.30

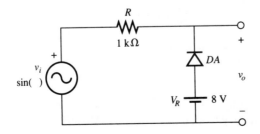

reversed. When $v_i = 12$ V peak, $f = 60$ Hz, and $V_R = 8$ V, run an analysis to show the input and output voltages. Predict the output voltage waveshape and compare with the Probe results.

9.6   The circuit shown in Fig. 9.31 uses the same values as those of Problem 9.5. The output is taken across R and V_R in series. Predict the output voltage waveshape; then run the PSpice analysis for comparison.

Fig. 9.31

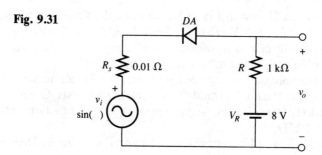

9.7   Using the circuit shown in Fig. 9.32 with the built-in model for the BJT, find the Q-point voltages and currents. Let $h_{FE} = 80$ (BF=80 in the .MODEL statement). Note that the circuit is the same as Fig. 3.1. Compare the results obtained here with those given in Chapter 3. Note that there are noticeable differences in the two sets of values. In Fig. 3.2, VA was chosen as 0.7 V, representing a realistic value for V_{BE} in the active region. Using the built-in model, your output file should show $V_{BE} = 0.806$ V. This difference is responsible for the change in base current and so forth between the two sets of answers. This problem points out one of the reasons why your own models (such as those developed in Chapter 3) are often preferred over the built-in models.

Fig. 9.32

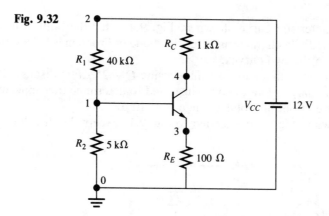

9.8   The circuit shown in Fig. 9.33 is to be used to obtain the drain characteristics of the FET. Let V_{DS} vary from 0 to 18 V in 0.2-V increments. Use the built-in

Fig. 9.33

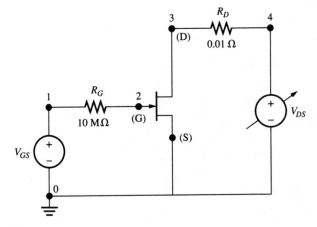

model for the JFET, with the designation *NJF* (for the *n*-channel junction FET).

(a) On the first analysis, use $V_{GS} = 0$ V. Find the maximum drain current and the pinch-off voltage.

(b) Change the value of V_{GS} to -1 V and rerun the analysis.

9.9 Figure 3.7 shows a *CE h*-parameter model for a BJT. If the *h* parameters are not used and the built-in model is chosen, the circuit must be modified to show V_{CC} and a dc biasing voltage for the base. Typically, the full circuit might look like Fig. 9.34. Use the built-in model for the BJT with $h_{FE} = 50$ (BF=50), and create an input file that will find the low-frequency voltage gain. Use $f = 5$ kHz for a single-frequency analysis. Compare the results with those obtained using the *h*-parameter model. Which method is preferred for this type of problem? Why?

Fig. 9.34

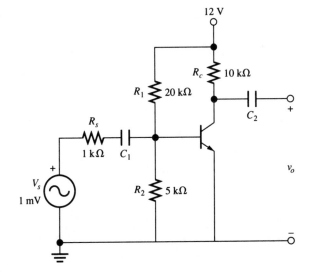

The BJT and Its Model

THE BIPOLAR-JUNCTION TRANSISTOR (BJT)

This chapter emphasizes the use of the library model of the BJT as compared to the *h*-parameter model or other simplified models. PSpice has a generic, built-in model for the BJT which uses the parameters shown in Appendix D under the topic "*Q* Bipolar Transistor." It will prove helpful to look at both the input and the output characteristics of this model. This will provide an introduction to the use of such models, which will be desirable if we are to use the models in various circuits.

Output Characteristics

A test circuit is shown in Fig. 10.1. The circuit shows a variable voltage supply V_{CC} and a variable current supply I_B. The transistor is called Q_1. When the built-in model for a BJT is used, the designation for the device must begin with Q. The input file becomes

```
BJT Output Characteristics
VCC 4 0 10V
IB 0 1 25uA
RB 1 2 0.01
RC 4 3 0.01
Q1 3 2 0 BJT; the designation BJT is our choice
.MODEL BJT NPN(BF=80)
.dc VCC 0 10V 0.05V IB 5uA 25uA 5uA
.PROBE
.END
```

Fig. 10.1 Test circuit for BJT
characteristics.

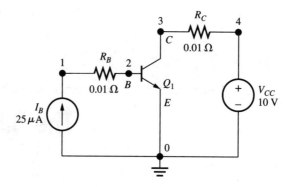

The .MODEL statement shows our choice *BJT* for the model name and *NPN*
(required) for the model type. The default value for *BF*, which represents the
forward beta (h_{FE}), is *100*; it has been changed to *80* as an item of choice. Other
parameters may be changed as desired; otherwise, default values will be used.

The *.dc* statement contains an outer loop to sweep V_{CC} and an inner loop to
sweep I_B. *Caution:* If you attempt to let the value of I_B begin at 0 μA, the dc sweep
will not run successfully.

Run the analysis and in Probe plot I(RC). This graph is shown in Fig. 10.2. It
is helpful to label the various traces with their input current values. The *Y*-axis

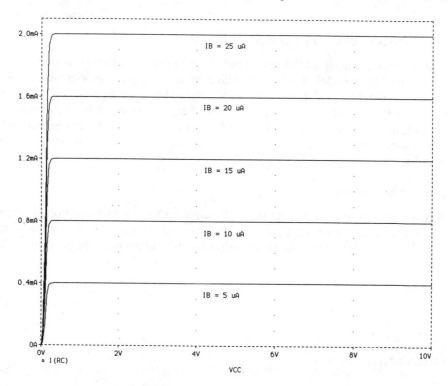

Fig. 10.2 BJT Output Characteristics.

Fig. 10.3 Test circuit for BJT input characteristics.

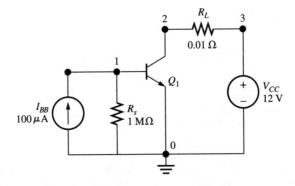

range was changed to read 0 2.1*m*. (If you type in 0 2.1*mA*, the *A* will not appear on the plot.)

Note that when $I_B = 25$ μA and V_{CC} (actually V_{CE}) is above a fraction of a volt, the collector current I_C, shown on the graph as I(RC), is 2.0 mA. This is in keeping with the value $h_{FE} = 80$.

Input Characteristics

To obtain the input characteristics, an alternate circuit, shown in Fig. 10.3, may be used. The current source I_{BB} is made practical by being placed in parallel with

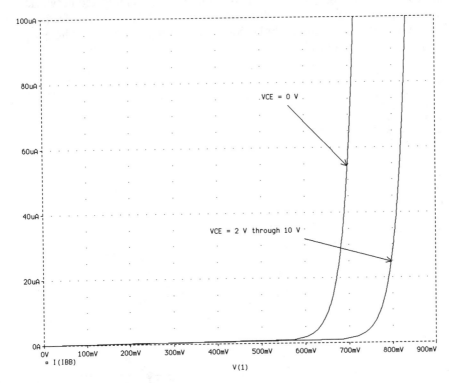

Fig. 10.4 BJT Input Characteristics.

the resistor R_s. The input file is

```
BJT Input Characteristics
IBB 0 1 100uA
Rs 1 0 1000k
RL 2 3 0.01
Q1 2 1 0 BJT
VCC 3 0 10V
.MODEL BJT NPN(BF=80)
.dc IBB 0 100uA 1uA VCC 0V 10V 2V
.PROBE
.END
```

When the analysis is completed, change the X-axis to show V(1), and plot I(IBB). You will obtain a plot that shows only two distinct traces. The first, nearest to the origin, is for $V_{CE} = 0$ V. The other is for all other values of V_{CE}. These are shown in Fig. 10.4. You may wish to run the analysis with the zero value for V_{CC} omitted. The results will confirm that the first trace is no longer shown. Note that when the built-in model for Q is used, V_{BE} will be approximately 0.8 V for typical values of base current.

A COMMON-EMITTER BJT AMPLIFIER

A simple common-emitter circuit is shown in Fig. 10.5. The input loop might be the result of applying Thevenin's theorem to a more complicated network. There is no coupling capacitor. We will assume that our analysis is for a frequency of 5 kHz, where the capacitor might be considered as a short circuit. The given value of h_{FE} is 50. The input file is

```
CE Amplifier, BJT Model
VCC 5 0 18V
VBB 3 2 0.8V
RS 1 2 1k
RL 4 5 10k
Q1 4 3 0 BJT NPN(BF=50)
.TF V(4) VS
.OP
vs 1 0 ac 1mV
.AC lin 1 5 kHz 5kHz
.PRINT ac I(RS) I(RL) V(3) V(4)
.END
```

Fig. 10.5 Common-emitter BJT amplifier.

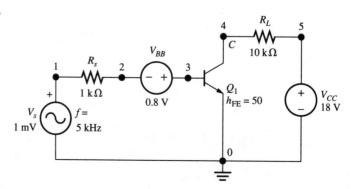

A frequency of 5 kHz is shown in the .AC statement. The .PRINT ac statement will allow us to find the specified currents and voltages. The output file from the PSpice analysis is shown in Fig. 10.6. Various formulas might be used to predict the voltage gain V(4)/V(3); for example,

$$A_V = -h_{fe} \frac{R_L}{h_{ie}} = \frac{-50(10 \text{ k}\Omega)}{1.1 \text{ k}\Omega} = -455$$

using an assumed value $h_{ie} = 1.1$ kΩ. Then using voltage division between R_s and h_{ie}, the voltage gain from the source V(4)/VS is found to be -238. From the PSpice output file under small-signal characteristics, V(4)/VS is given as -233. The results are in close agreement.

The input resistance at VS given by PSpice is 2.144 kΩ. Subtracting R_s (1 kΩ) from this gives $h_{ie} = 1.144$ kΩ, which is close to the assumed value. The output resistance is given as 10 kΩ. In the actual case, this would be the parallel equivalent of R_L and h_{oe}. But if we assume that $h_{oe} \geq R_L$, then the value is approximately R_L. Note that if a capacitor had been included in the input loop, this method of finding the small-signal characteristics would not have given useful results.

The last lines of output contain the ac analysis. The frequency is 5 kHz, the base current is 0.46665 μA, and the collector current is 23.32 μA. In order to check these values by conventional circuit analysis, the ac base current is to be found. This is

$$I_b = \frac{V_s}{R_s + h_{ie}} = \frac{1 \text{ mV}}{1 \text{ k}\Omega + 1.14 \text{ k}\Omega} = 0.476 \text{ } \mu\text{A}$$

The ac collector current is

$$I_c = h_{fe}I_b = 50(0.476 \text{ } \mu\text{A}) = 23.37 \text{ } \mu\text{A}$$

These values are in close agreement with the PSpice results.

Returning to the dc analysis, it is our desire to calculate the dc base current, given by

$$I_B = \frac{V_{BB} - V_{BE}}{R_s} = \frac{0.8 \text{ V} - 0.7774 \text{ V}}{1 \text{ k}\Omega} = 22.6 \text{ } \mu\text{A}$$

This is a presumptuous calculation, since V_{BE} has been taken from the PSpice results. Now it becomes obvious that this circuit is more useful for illustration than for practical purposes, since slight changes in V_{BB} or V_{BE} will cause large changes in I_B. Next, the dc collector current is found as $h_{FE}I_B$, giving a value of 1.13 mA, and the collector voltage is

$$V_C = V_{CC} - R_L I_C = 18 \text{ V} - (10 \text{ k}\Omega)(1.13 \text{ mA}) = 6.7 \text{ V}$$

The total power dissipation shown by PSpice is the product of the magnitudes of the dc source voltages and currents. Verify that this is 20.4 mW.

A final item of interest in Fig. 10.6 is the listing called BIPOLAR-JUNC-TION TRANSISTORS. Our selected name *Q1*, along with the selected model name *BJT*, is shown, followed by a list of 16 quiescent-point values (not all are

```
  CE Amplifier, BJT Model
VCC 5 0 18V
VBB 3 2 0.8V
RS 1 2 1k
RL 4 5 10k
Q1 4 3 0 BJT
.MODEL BJT NPN(BF=50)
.TF V(4) VS
.OP
vs 1 0 ac 1mV
.AC lin 1 5kHz 5kHz
.PRINT ac I(RS) I(RL) V(3) V(4)
.END


****        BJT MODEL PARAMETERS
                 BJT
                 NPN
            IS  100.000000E-18
            BF   50
            NF    1
            BR    1
            NR    1


  NODE    VOLTAGE      NODE    VOLTAGE      NODE    VOLTAGE      NODE    VOLTAGE
(   1)     0.0000    (   2)     -.0226    (   3)     .7774    (    4)    6.6929
(   5)    18.0000


     VOLTAGE SOURCE CURRENTS
     NAME         CURRENT
     VCC          -1.131E-03
     VBB          -2.261E-05
     VS           -2.261E-05


     TOTAL POWER DISSIPATION   2.04E-02   WATTS

**** BIPOLAR JUNCTION TRANSISTORS
NAME          Q1
MODEL         BJT
IB            2.26E-05
IC            1.13E-03
VBE           7.77E-01
VBC           -5.92E+00
VCE           6.69E+00
BETADC        5.00E+01
BETAAC        5.00E+01

  ****        SMALL-SIGNAL CHARACTERISTICS
       V(4)/VS = -2.332E+02

       INPUT RESISTANCE AT VS =  2.144E+03

       OUTPUT RESISTANCE AT V(4) =  1.000E+04

  ****        AC ANALYSIS                         TEMPERATURE =   27.000 DEG C
    FREQ         I(RS)        I(RL)        V(3)         V(4)
    5.000E+03    4.665E-07    2.332E-05    5.335E-04    2.332E-01
```

Fig. 10.6

shown in Fig. 10.6). These values apply to the actual bias conditions of the circuit. They will change when the *Q*-point currents and voltages change. For example, if the transistor should go into saturation, the *BETADC* value would be much lower. After you have studied the results, move to a more practical circuit and do a more comprehensive analysis.

BIASING CASE STUDY

A circuit with a more stable operating point than the previous one is shown in Fig. 10.7. This is called a self-bias or emitter-bias circuit. The input file is

```
Biasing Case Study
VCC 2 0 12V
R1 2 1 40k
R2 1 0 3.3k
RC 2 3 4.7k
RE 4 0 220
Q1 3 1 4 Q2N2222A
.DC VCC 12V 12V 12V
.PRINT DC I(RC) I(R1) I(R2) I(RE)
.OP
.LIB EVAL.LIB; this calls in the library file EVAL.LIB
.END
```

There is no .MODEL statement in this input file. Instead, the transistor is identified as *Q2N2222A*. This is the designation for one of the transistors that comes with the evaluation version of PSpice from MicroSim Corporation. Other BJTs in this version are shown in Appendix E, along with various parts in the evaluation library. The line beginning with *.LIB* is required to make use of the resources in the library.

Earlier releases of the evaluation version used *NOM.LIB* for all the devices or as a master library which called in the other libraries such as *DNOM.LIB* for diodes and *QNOM.LIB* for BJTs. The production version, on the other hand, allows the use of over 3500 analog devices. All of the examples in this book are

Fig. 10.7 Self-bias circuit.

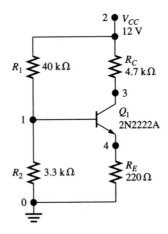

```
 Biasing Case Study
VCC 2 0 12V
R1 2 1 40k
R2 1 0 3.3k
RC 2 3 4.7k
RE 4 0 220
Q1 3 1 4 Q2N2222A
.DC VCC 12V 12V 12V
.PRINT DC I(RC) I(R1) I(R2) I(RE)
.OP
.LIB EVAL.LIB
.END


    ****        BJT MODEL PARAMETERS
                 Q2N2222A
                  NPN
            IS   14.340000E-15
            BF   255.9
            NF   1
            VAF  74.03


    ****      DC TRANSFER CURVES              TEMPERATURE =   27.000 DEG C
     VCC         I(RC)        I(R1)       I(R2)        I(RE)
     1.200E+01   1.114E-03   2.777E-04   2.707E-04   1.121E-03


    ****      SMALL SIGNAL BIAS SOLUTION      TEMPERATURE =   27.000 DEG C
   NODE   VOLTAGE      NODE   VOLTAGE      NODE   VOLTAGE      NODE   VOLTAGE
   (   1)    .8933   (   2)   12.0000   (   3)    6.7651   (   4)     .2466


       VOLTAGE SOURCE CURRENTS
       NAME         CURRENT
       VCC          -1.391E-03


    TOTAL POWER DISSIPATION   1.67E-02  WATTS

    ****      OPERATING POINT INFORMATION     TEMPERATURE =   27.000 DEG C

**** BIPOLAR JUNCTION TRANSISTORS
NAME          Q1
MODEL         Q2N2222A
IB            6.96E-06
IC            1.11E-03
VBE           6.47E-01
VBC           -5.87E+00
VCE           6.52E+00
BETADC        1.60E+02
GM            4.29E-02
RPI           4.12E+03
RX            1.00E+01
RO            7.17E+04
CBE           5.40E-11
CBC           3.47E-12
CBX           0.00E+00
CJS           0.00E+00
BETAAC        1.77E+02
FT            1.19E+08
```

Fig. 10.8

based on using the evaluation version of the library. The production version contains various libraries, one for each type of device. For example, there is *DIODE.LIB* for diodes and *DIGITAL.LIB* for digital devices.

Since the libraries are ASCII files, containing text rather than programming code, it is helpful to bring the contents of a library file into a text editor so that the parameters of a device may be found. The January, 1992, evaluation version of PSpice contains a file *EVAL.LIB* of 52.5 kbytes of information. The designation for *Q2N2222A* consists of three lines showing the various parameters and their assigned values and two lines of comments showing, for example, the *source* (National), the *pid* (19), and the *case* (TO18).

Run the analysis and check the operating-point voltages and currents. Verify that $V_{CE} = 6.5185$ V and that the collector current $I_C = 1.114$ mA. Note that although *Q1* has a maximum forward beta h_{FE} of 255.9, under the operating-point information BETADC is given as 160, in keeping with a base current $I_B = 6.96\ \mu A$. The output file is shown in Fig. 10.8. (Not all model parameters are shown.)

The AC Analysis

In order to illustrate how this circuit behaves as a common-emitter amplifier, several components will be added. Refer to Fig. 10.9 for the new circuit. An ac source voltage of 10 mV (peak value), a source resistance R_s of 50 Ω, and capacitors C_b and C_b have been added. The input file becomes

```
Biasing Case Study
VCC 2 0 12V
Vs 1a 0 ac 10mV
Rs 1a 1b 50
Cb 1b 1 15uF
Ce 4 0 15uF
R1 2 1 40k
R2 1 0 3.3k
RC 2 3 4.7k
RE 4 0 220
Q1 3 1 4 Q2N2222A
.DC VCC 12V 12V 12V
```

Fig. 10.9 Common-emitter amplifier.

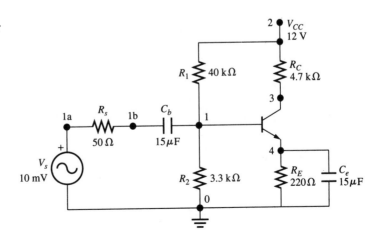

```
 Biasing Case Study
VCC 2 0 12V
Vs 1a 0 ac 10mV
Rs 1a 1b 50
Cb 1b 1 15uF
Ce 4 0 15uF
R1 2 1 40k
R2 1 0 3.3k
RC 2 3 4.7k
RE 4 0 220
Q1 3 1 4 Q2N2222A
.DC VCC 12V 12V 12V
.PRINT DC I(RC) I(R1) I(R2) I(RE)
.OP
.opt nopage nomod
.ac LIN 1 5kHz 5kHz
.PRINT ac i(RC) i(RE) i(RS)
.PRINT ac v(1) v(1b) v(3) v(4)
.LIB EVAL.LIB
.END


 ****     DC TRANSFER CURVES                    TEMPERATURE =   27.000 DEG C
  VCC        I(RC)         I(R1)        I(R2)       I(RE)
   1.200E+01   1.114E-03    2.777E-04   2.707E-04   1.121E-03


 ****     SMALL SIGNAL BIAS SOLUTION           TEMPERATURE =   27.000 DEG C
 NODE    VOLTAGE      NODE    VOLTAGE      NODE    VOLTAGE      NODE    VOLTAGE
 (   1)    .8933    (    2)   12.0000    (    3)    6.7651    (    4)     .2466
 (  1a)    0.0000   (   1b)    0.0000


    VOLTAGE SOURCE CURRENTS
    NAME          CURRENT
    VCC          -1.391E-03
    Vs            0.000E+00


    TOTAL POWER DISSIPATION   1.67E-02   WATTS

 ****     OPERATING POINT INFORMATION          TEMPERATURE =   27.000 DEG C

 **** BIPOLAR JUNCTION TRANSISTORS
 NAME        Q1
 MODEL       Q2N2222A
 IB          6.96E-06
 IC          1.11E-03
 VBE         6.47E-01
 VBC        -5.87E+00
 VCE         6.52E+00
 BETADC      1.60E+02
 BETAAC      1.77E+02


 ****     AC ANALYSIS                          TEMPERATURE =   27.000 DEG C

  FREQ       I(RC)         I(RE)        I(RS)
   5.000E+03   3.888E-04    3.772E-06   5.523E-06
  FREQ       V(1)          V(1b)        V(3)         V(4)
   5.000E+03   9.724E-03    9.725E-03   1.827E+00   8.299E-04
```

Fig. 10.10

```
.PRINT DC I(RC) I(R1) I(R2) I(RE)
.OP
.opt nopage nomod; suppress banner and model parameters
.ac LIN 1 5kHz 5kHz; a sweep is necessary for ac analysis
.PRINT ac i(RC) i(RE) i(RS)
.PRINT ac v(1) v(1b) v(3) v(4)
.LIB EVAL.LIB
.END
```

In this input file, V_s is identified as an ac voltage, and an ac sweep is called for. Without the *.ac LIN* statement there will be no ac analysis information in the output file.

Run the analysis and verify that the bias voltage and current values have not changed from what they were in the previous output file. In fact, all of the operating-point information is the same as before. Refer to Fig. 10.10 for the output file.

In addition to the previous results, we have asked for several ac currents and voltages. Verify that v(3)/v(1) = 188 and v(3)/vs = 182.7. The ac output current is 0.3888 mA, and the ac input current is 5.523 μA, giving a curent gain of 70.4 from the source.

As an exercise, place a small resistor $R_B = 0.01\ \Omega$ in series with the base of the transistor. Call for the current through R_B in the *.PRINT ac* statement; then run the analysis and find the current gain from base to collector. It will not be the

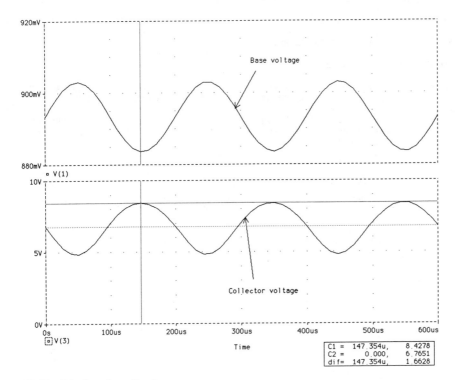

Fig. 10.11 Biasing Case Study.

```
Biasing Case Study

VCC 2 0 12V
Vs 1a 0 sin(0 10mV 5kHz)
Rs 1a 1b 50
Cb 1b 1 15uF
Ce 4 0 15uF
R1 2 1 40k
R2 1 0 3.3k
RC 2 3 4.7k
RE 4 0 220
Q1 3 1 4 Q2N2222A
.opt nopage nomod
.TRAN 0.02ms 0.6ms
.PROBE
.FOUR 5kHz V(3)
.LIB EVAL.LIB
.END

 ****      INITIAL TRANSIENT SOLUTION      TEMPERATURE =   27.000 DEG C
  NODE   VOLTAGE      NODE   VOLTAGE      NODE   VOLTAGE      NODE   VOLTAGE
 (   1)     .8933  (    2)   12.0000  (    3)    6.7651  (    4)      .2466
 (  1a)    0.0000  (   1b)    0.0000

    VOLTAGE SOURCE CURRENTS
    NAME          CURRENT
    VCC         -1.391E-03
    Vs           0.000E+00

    TOTAL POWER DISSIPATION   1.67E-02   WATTS

 ****      FOURIER ANALYSIS                TEMPERATURE =   27.000 DEG C

FOURIER COMPONENTS OF TRANSIENT RESPONSE V(3)

 DC COMPONENT =   6.754234E+00

 HARMONIC    FREQUENCY      FOURIER     NORMALIZED      PHASE       NORMALIZED
    NO          (HZ)       COMPONENT    COMPONENT       (DEG)      PHASE (DEG)

     1       5.000E+03     1.781E+00    1.000E+00     -1.752E+02    0.000E+00
     2       1.000E+04     1.342E-01    7.534E-02      1.018E+02    2.770E+02
     3       1.500E+04     4.550E-03    2.554E-03     -8.778E+00    1.664E+02
     4       2.000E+04     2.816E-03    1.581E-03     -1.121E+02    6.305E+01
     5       2.500E+04     2.697E-03    1.514E-03     -1.203E+02    5.487E+01
     6       3.000E+04     2.654E-03    1.490E-03     -1.269E+02    4.828E+01
     7       3.500E+04     2.570E-03    1.443E-03     -1.331E+02    4.212E+01
     8       4.000E+04     2.497E-03    1.402E-03     -1.381E+02    3.712E+01
     9       4.500E+04     2.436E-03    1.367E-03     -1.447E+02    3.054E+01

    TOTAL HARMONIC DISTORTION =   7.546790E+00  PERCENT
```

Fig. 10.12

same as the current found as I_c/h_{fe}, where h_{fe} is *BETAAC*. Can you give an explanation for this?

It is instructive to use Probe to look at the ac voltages at various points in the circuit. Change the input file to contain the following:

```
Biasing Case Study
VCC 2 0 12V
Vs 1a 0 sin(0 10mV 5kHz); arguments are offset, peak, and
frequency
Rs 1a 1b 50
Cb 1b 1 15uF
Ce 4 0 15uF
R1 2 1 40k
R2 1 0 3.3k
RC 2 3 4.7k
Q1 3 1 4 Q2N2222A
.opt nopage nomod
.TRAN 0.02ms 0.6ms
.PROBE
.FOUR 5kHz V(3)
.LIB EVAL.LIB
.END
```

The voltage source is now shown not simply as an *ac* source but rather as a *sin( )* source. The arguments are *offset, amplitude,* and *frequency*. The wave-shapes may be displayed by including the *.PROBE* statement. Run the analysis; then plot V(3) and V(1) as shown in Fig. 10.11 on page 325. In this figure, the cursor has been used to find the maximum value of collector voltage. Note that the collector voltage is 180° out of phase with the base voltage. Use the cursor to find *maX* and *Min* values. Verify that the base voltage has a peak-to-peak value of 19.4 mV, while the collector has a corresponding value of 3.62 V. This gives a voltage gain $A_V = 187$, in agreement with the previous ac analysis.

Finally, the output file shown in Fig. 10.12 contains the results of the Fourier analysis of the output voltage V(3). The dc component is 6.75 V, which is not quite the same as the bias voltage at the collector due to the harmonic content of the output voltage. The fundamental frequency (5 kHz) shows a value of 1.781 V. This is the peak value, with the peak-to-peak value being 3.562 V. The plot of collector voltage shows a peak-to-peak value of 3.634 V. The two are not expected to be the same, again due to harmonic content. The second harmonic of the output voltage is 0.1342 V, which is an order of magnitude less than the fundamental. Higher harmonics are almost negligible, giving a total harmonic distortion of about 7.5%.

CE AMPLIFIER WITH UNBYPASSED EMITTER RESISTOR

When a common-emitter amplifier uses an unbypassed emitter resistor, the voltage gain of the circuit is reduced, but the frequency response is improved. The circuit, with its current-series feedback, is shown in Fig. 10.13. In this analysis, the built-in model for the BJT will be used, and the value of h_{FE} will be *80*. This is

Fig. 10.13 CE amplifier with unbypassed emitter resistor.

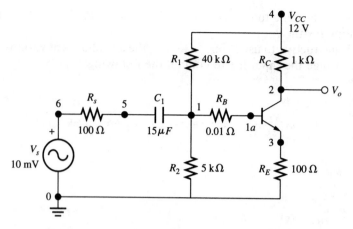

the input file:

```
Analysis of CE Amplifier with Unbypassed RE
VCC 4 0 12V
R1 4 1 40k
R2 1 0 5k
RC 4 2 1k
RE 3 0 100
Rs 6 5 100
Rb 1 1a 0.01
C1 5 1 15uF
Q1 2 1a 3 BJT
.MODEL BJT NPN(BF=80)
.OP
vs 6 0 ac 10mV
.ac LIN 1 5kHz 5kHz
.PRINT ac i(RB) i(RC) i(RS) v(1) v(2) v(3)
.END
```

The dc analysis of this circuit was given in the "PSpice Overview" chapter at the beginning of the book and will not be repeated here.

Turning our attention to the ac analysis, using conventional circuit analysis, the voltage gain (from base to collector) is approximated as

$$A_V = \frac{-R_C}{R_E} = \frac{-1000 \ \Omega}{100 \ \Omega} = -10$$

But for small values of R_C, this could introduce up to about a 10% error (on the high side). A more accurate equation is

$$A_V = \frac{-R_C}{R_i} = \frac{-R_C}{h_{ie} + (1 + h_{fe})R_E} = \frac{-1000 \ \Omega}{1100 \ \Omega + (1 + 80)100 \ \Omega} = -8.7$$

Since this is the voltage gain from base to collector, voltage division is used to find the voltage gain from the source to the collector:

$$A_{Vs} = \frac{A_V R_p}{R_p + R_s} = \frac{-8.7(3000\ \Omega)}{3000\ \Omega + 100\ \Omega} = -8.4$$

where R_p is the parallel combination of R_1, R_2, and R_i.

Now examine the output file, Fig. 10.14, to see how the PSpice analysis compares to the conventional results. PSpice shows an overall voltage gain v(2)/vs = -8.878. The two values differ by a little more than 5%.

```
Analysis of CE Amplifier with Unbypassed RE

VCC 4 0 12V
R1 4 1 40k
R2 1 0 5k
RC 4 2 1k
RE 3 0 100
RS 6 5 100
Rb 1 1a 0.01
C1 5 1 15uF
Q1 2 1a 3 BJT
.MODEL BJT NPN(BF=80)
.OP
vs 6 0 ac 10mV
.ac LIN 1 5kHz 5kHz
.PRINT ac i(RB) i(RC) i(RS) v(1) v(2) v(3)
.END

    ****      BJT MODEL PARAMETERS
                 BJT
                 NPN
            IS  100.000000E-18
            BF   80


 NODE   VOLTAGE     NODE   VOLTAGE    NODE   VOLTAGE    NODE    VOLTAGE
(   1)    1.1464  (    2)    8.6345  (    3)    .3408  (    4)   12.0000
(   5)    0.0000  (    6)    0.0000  (   1a)   1.1464

       VOLTAGE SOURCE CURRENTS
       NAME        CURRENT
       VCC        -3.637E-03
       vs          0.000E+00

       TOTAL POWER DISSIPATION   4.36E-02   WATTS

**** BIPOLAR JUNCTION TRANSISTORS
NAME         Q1
MODEL        BJT
IB           4.21E-05
IC           3.37E-03
VBE          8.06E-01
VBC         -7.49E+00
VCE          8.29E+00
BETADC       8.00E+01
BETAAC       8.00E+01

    ****    AC ANALYSIS                     TEMPERATURE =   27.000 DEG C
    FREQ       I(RB)       I(RC)       I(RS)       V(1)        V(2)
    5.000E+03   1.110E-06   8.878E-05   3.286E-06   9.671E-03   8.878E-02
    FREQ       V(3)
    5.000E+03   8.989E-03
```

Fig. 10.14

Fig. 10.15 Finding the input resistance.

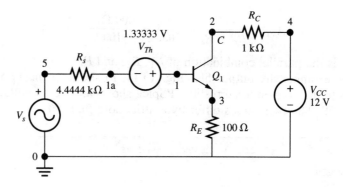

```
    Find Input Resistance

VCC 4 0 12V
VTh 1 1a 1.33333V
RC 4 2 1k
RE 3 0 100
Rs 5 1a 4.4444k
Q1 2 1 3 BJT
.MODEL BJT NPN(BF=80)
.TF v(2) vs
.OP
vs 5 0 ac 10mV
.ac LIN 1 5kHz 5kHz
.PRINT ac i(Rs) i(RC) v(1) v(2) v(3)
.END

    NODE   VOLTAGE      NODE   VOLTAGE      NODE   VOLTAGE      NODE   VOLTAGE

    (   1)   1.1464  (    2)    8.6345  (    3)    .3408  (     4)   12.0000
    (   5)   0.0000  (   1a)   -.1870

       VOLTAGE SOURCE CURRENTS
       NAME           CURRENT

       VCC            -3.366E-03
       VTh            -4.207E-05
       vs             -4.207E-05

       TOTAL POWER DISSIPATION   4.04E-02   WATTS

    ****      SMALL-SIGNAL CHARACTERISTICS

         V(2)/vs = -6.079E+00

         INPUT RESISTANCE AT vs =  1.316E+04

         OUTPUT RESISTANCE AT V(2) =  1.000E+03

    FREQ        I(Rs)        I(RC)        V(1)         V(2)         V(3)

     5.000E+03   7.599E-07   6.079E-05   6.623E-03   6.079E-02   6.155E-03
```

Fig. 10.16

Finding the Input Resistance

It is desirable to find the input resistance as seen by the ac source. If we simply use a statement such as

`.TF v(4) vs`

the results will be disappointing. You might try this and see what happens. Removing the capacitor C_1 from the circuit will not work either, since this will upset the bias conditions.

Another approach is shown in Fig. 10.15, where the Thevenin voltage and Thevenin resistance are shown in the input loop along with V_s. This maintains the proper bias voltages and currents and will allow for the transfer function to be used.

Study the output file of Fig. 10.16 (which also contains the input listing) to see that the bias voltages have not changed. This listing shows the input resistance as 13.16 kΩ. At the base of the transistor, $R_i = 13.16 \text{ k}\Omega - 4.444 \text{ k}\Omega = 8.7 \text{ k}\Omega$. The calculated value of 9.2 kΩ differs from this by slightly more than 5%. We conclude that the methods are in adequate agreement.

USING OUR OWN MODEL WITH THE *h* PARAMETERS

Finding voltage gains, current gains, and input resistance using the PSpice model will now be compared to the use of our own model for the *CE* amplifier based on

Fig. 10.17 CE amplifier for ac analysis.

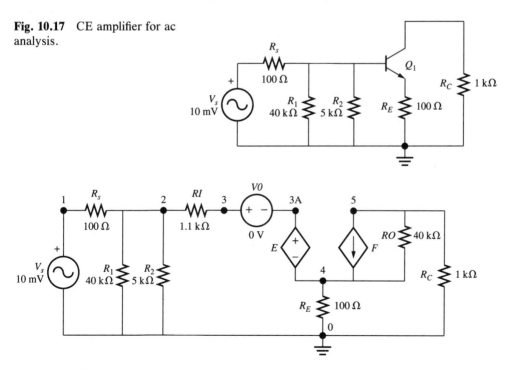

Fig. 10.18 CE amplifier using *h*-parameter model.

```
CE Amplifier with Re

vs 1 0 10mV
VO 3 3A 0
E 3A 4 5 4 2.5E-4
F 5 4 VO 80
Rs 1 2 100
R1 2 0 40k
R2 2 0 5k
RI 2 3 1.1k
RO 5 4 40k
RC 0 5 1k
RE 4 0 100
.TF V(5) Vs
.OP
.dc vs 10mV 10mV 10mV
.PRINT dc i(RC) i(Rs) i(RI) i(RE)
.END


    vs          I(RC)          I(Rs)          I(RI)          I(RE)

   1.000E-02   8.402E-05   3.256E-06   1.079E-06   8.510E-05


   NODE    VOLTAGE      NODE    VOLTAGE      NODE    VOLTAGE      NODE    VOLTAGE

(    1)     .0100  (     2)      .0097  (    3)       .0085  (     4)       .0085
(    5)    -.0840  (    3A)      .0085


      VOLTAGE SOURCE CURRENTS
      NAME           CURRENT

      vs           -3.256E-06
      VO            1.079E-06


    TOTAL POWER DISSIPATION   3.26E-08  WATTS

**** VOLTAGE-CONTROLLED VOLTAGE SOURCES

NAME           E
V-SOURCE    -2.313E-05
I-SOURCE     1.079E-06

**** CURRENT-CONTROLLED CURRENT SOURCES

NAME           F
I-SOURCE     8.634E-05

  ****      SMALL-SIGNAL CHARACTERISTICS

      V(5)/vs = -8.402E+00

      INPUT RESISTANCE AT vs =  3.071E+03

      OUTPUT RESISTANCE AT V(5) =  9.987E+02
```

Fig. 10.19

the h parameters. The method was introduced in Chapter 3, "Transistor Circuits."

The h-Parameter Analysis

The circuit of Fig. 10.13 is considered from the ac point of view. The supply V_{CC} becomes a ground, C_1 is assumed to be a short circuit, and R_1 is in parallel with R_2. The circuit for analysis becomes Fig. 10.17 on page 331, and using the h parameters this becomes Fig. 10.18 on page 331.

Recall that the maximum amount of information can be obtained by pretending that we are doing a dc analysis rather than an ac analysis. This allows us to find the desired small-signal characteristics, including voltage gain and input resistance. The input file will not be repeated here, but it is shown in the output listing of Fig. 10.19. The results compare favorably with those obtained by using the built-in model for the BJT.

PHASE RELATIONS IN THE *CE* AMPLIFIER

When a *CE* amplifier uses an emitter resistor R_E for bias stability, it is bypassed with a capacitor C_E so that the input signal will see the emitter as a ground point in the circuit. If we look at the ac waveshapes at the collector and the emitter, it is of interest to compare what the gain will be with and without C_E. This will also allow us to examine a potential problem when we use the *.TRAN* statement to look at ac steady-state values.

The Amplifier without the Emitter Capacitor

Refer to Fig. 10.13 for the circuit which is shown without C_E. The input file for the analysis is

```
Phase Relations in CE Amplifier
VCC 4 0 12V
R1 4 1 40k
R2 1 0 5k
RC 4 2 1k
RE 3 0 100
Rs 6 5 100
RB 1 1A 0.01
C1 5 1 15uF
Q1 2 1A 3 BJT
.MODEL BJT NPN (BF=80)
vs 6 0 sin(0 10mV 5kHz)
.TRAN 0.02ms 0.2ms
.PROBE
.END
```

Run the analysis and in Probe plot the collector voltage v(2), the source voltage v(6), and the emitter voltage v(3). Note that the source voltage and the emitter voltage are in phase, while the collector voltage is 180° out of phase. Verify that the ac peak value of v(2) is 88.75 mV and the ac peak value of v(3) is 9 mV, compared to the ac peak value of v(6), which is 10 mV. The transient analysis has successfully been used to allow us to look at steady-state values, and the results have been what we would expect when compared to those obtained by conventional analysis methods. Compare your plots to those shown in Fig. 10.20.

The Amplifier with the Emitter Capacitor

The amplifier is designed, however, to use a capacitor C_E across R_E. Let us insert the required line in the input file:

```
CE 3 0 10uF
```

and run the analysis again. In Probe plot the emitter voltage alone (to fill the entire screen), and notice that the wave appears distorted. If we plot several cycles of this voltage, we will see that the wave appears to be going through a transient phase before it begins to settle down to what we would have expected to see in the beginning. In the laboratory, the oscilloscope would show the wave correctly, so why does Probe show it differently? It is because we are using a transient analysis in a circuit with reactive elements. Therefore we must be careful and look for potential problems such as this.

Plot v(2) and verify that v(2) = 0.929 V ac peak and that v(3) max = 3.5 mV ac peak. Note that the collector voltage also appears slightly distorted, with an

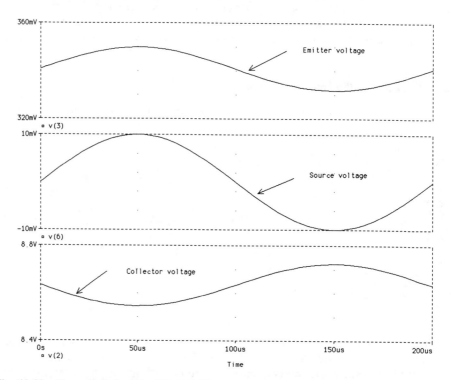

Fig. 10.20 Phase Relations in CE Amplifier.

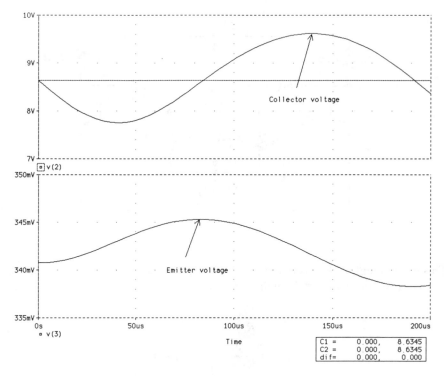

Fig. 10.21 Phase Relations in CE Amplifier with Emitter-Bypass Capacitor.

axis-crossing value of 8.6345 V as shown in Fig. 10.21, a $maX = 9.614$ V, and a $Min = 7.756$ V.

Verify that at $f = 5$ kHz, the capacitor is not an ideal short circuit. Compute the impedance of the parallel combination of R_E and C_E. It is $Z = 3.18\underline{/-88°}\ \Omega$.

As an exercise, plot the current through C_E and the current through R_E. The emitter current can be plotted as $-IE(Q1)$ for comparison. Note the phase relations among the various currents and between the emitter voltage and the source voltage.

THE BJT FLIP-FLOP

A BJT flip-flop using *npn* transistors is shown in Fig. 10.22. For proper operation, this bistable multivibrator (or binary) should allow one transistor to operate well below cutoff while the other transistor is in saturation. Let us assume in the beginning that Q_1 is off and Q_2 is on. Using conventional analysis techniques,

$$V_1 = \frac{V_{BB}R_2}{R_2 + R_3} = -1.57\ V$$

Fig. 10.22 A BJT flip-flop.

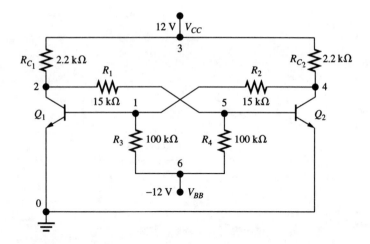

This is sufficient back bias to cut off Q_1. To determine the collector current for the other transistor, I_{RC2} and I_{R2} will be found:

$$I_{RC2} = \frac{V_{CC}}{R_{C2}} = 5.45 \text{ mA}$$

$$I_{R2} = \frac{V_{CC}}{R_2 + R_3} = 0.104 \text{ mA}$$

assuming that $V_4 = 0$.

The collector current in Q_2, which will be called I_{C2}, is the difference of these:

$$I_{C2} = I_{RC2} - I_{R2} = 5.35 \text{ mA}$$

It should be noted that the minimum base current I_{B2} required to saturate Q_2 is

$$I_{B2min} = \frac{I_{C2}}{h_{FE}} = \frac{5.35 \text{ mA}}{30} = 0.18 \text{ mA}$$

The base current is found by calculating its two components:

$$I_{R1} = \frac{V_{CC}}{R_{C1} + R_1} = 0.70 \text{ mA}$$

$$I_{R4} = \frac{0 - V_{BB}}{R_4} = 0.12 \text{ mA}$$

assuming that $V_5 = 0$.

The base current in Q_2, which is called I_{B2}, is the difference of these:

$$I_{B2} = I_{R1} - I_{R4} = 0.58 \text{ mA}$$

a value well above the minimum required for saturation.

Since the circuit is symmetrical, had we assumed that Q_1 was on and Q_2 was off, the analysis would have produced similar results with the roles of the transistors swapped.

```
  BJT Flip-flop
VCC 3 0 12V
VBB 6 0 -12V
RC1 3 2 2.2k
RC2 3 4 2.2k
R1 2 5 15k
R2 4 1 15k
R3 1 6 100k
R4 5 6 100k
Q1 2 1 0 QN
Q2 4 5 0 QN
.MODEL QN NPN(IS=1E-9 BF=30 BR=1 TF=0.2ns TR=5ns)
.NODESET V(4)=0.15V
.OP
.DC VCC 12V 12V 12V
.PRINT DC I(RC1) I(RC2) I(R1) I(R2)
.END

    ****      BJT MODEL PARAMETERS
                  QN
                  NPN
              IS    1.000000E-09
              BF    30
              BR    1
              TF    200.000000E-12
              TR    5.000000E-09


  VCC           I(RC1)       I(RC2)       I(R1)        I(R2)
   1.200E+01   6.742E-04    5.421E-03    6.742E-04    1.050E-04

  NODE    VOLTAGE     NODE    VOLTAGE    NODE    VOLTAGE    NODE    VOLTAGE
(    1)   -1.5012   (    2)   10.5170  (    3)   12.0000  (    4)     .0736
(    5)     .4037   (    6)  -12.0000

     VOLTAGE SOURCE CURRENTS
     NAME          CURRENT
     VCC         -6.095E-03
     VBB          2.290E-04

     TOTAL POWER DISSIPATION    7.59E-02   WATTS

**** BIPOLAR JUNCTION TRANSISTORS
NAME         Q1           Q2
MODEL        QN           QN
IB          -1.05E-09    5.50E-04
IC           1.02E-09    5.32E-03
VBE         -1.50E+00    4.04E-01
VBC         -1.20E+01    3.30E-01
VCE          1.05E+01    7.36E-02
BETADC      -9.78E-01    9.66E+00
BETAAC       0.00E+00    2.83E+01
FT           0.00E+00    3.06E+08
```

Fig. 10.23

The PSpice Analysis

In order to run a PSpice analysis, we will assume that Q_1 is off, as we did in the conventional analysis. To allow for this in the input file, a *.NODESET* statement is required. The input file then becomes

```
BJT Flip-flop
VCC 3 0 12V
VBB 6 0 −12V
RC1 3 2 2.2k
RC2 3 4 2.2k
R1 2 5 15k
R2 4 1 15k
```

```
 BJT Flip-flop
VCC 3 0 12V
VBB 6 0 -12V
RC1 3 2 2.2k
RC2 3 4 2.2k
R1 2 5 15k
R2 4 1 15k
R3 1 6 100k
R4 5 6 100k
Q1 2 1 0 QN
Q2 4 5 0 QN
.MODEL QN NPN(IS=1E-9 BF=30 BR=1 TF=0.2ns TR=5ns)
.NODESET V(2)=0.15V
.OP
.DC VCC 12V 12V 12V
.PRINT DC I(RC1) I(RC2) I(R1) I(R2)
.END

    ****       BJT MODEL PARAMETERS
                  QN
                  NPN
            IS    1.000000E-09
            BF    30
            BR    1
            TF    200.000000E-12
            TR    5.000000E-09

  VCC        I(RC1)       I(RC2)       I(R1)        I(R2)
   1.200E+01  5.421E-03    6.742E-04    1.050E-04    6.742E-04

  NODE   VOLTAGE    NODE   VOLTAGE    NODE   VOLTAGE    NODE   VOLTAGE
 (   1)    .4037  (    2)    .0736  (    3)  12.0000  (    4)  10.5170
 (   5)  -1.5012  (    6)  -12.0000

   VOLTAGE SOURCE CURRENTS
   NAME          CURRENT
   VCC          -6.095E-03
   VBB           2.290E-04

   TOTAL POWER DISSIPATION   7.59E-02   WATTS

**** BIPOLAR JUNCTION TRANSISTORS
NAME       Q1          Q2
MODEL      QN          QN
IB         5.50E-04    -1.05E-09
IC         5.32E-03    1.02E-09
VBE        4.04E-01    -1.50E+00
VBC        3.30E-01    -1.20E+01
VCE        7.36E-02    1.05E+01
BETADC     9.66E+00    -9.78E-01
BETAAC     2.83E+01    0.00E+00
FT         3.06E+08    0.00E+00
```

Fig. 10.24

```
R3  1  6  100k
R4  5  6  100k
Q1  2  1  0  QN
Q2  4  5  0  QN
.MODEL QN NPN(IS=1E-9 BF=30 BR=1 TF=0.2ns TR=5ns)
.NODESET V(4)=0.15V;  guess for Q2 on (in saturation)
.OP
.DC VCC 12V 12V 12V
.PRINT DC I(RC1) I(RC2) I(R1) I(R2)
.END
```

The .NODESET value for V(4) = 0.15 V is a preliminary guess which is used in the beginning of the PSpice solution. When the iteration process of the solution is completed, this value is likely to change.

Run the PSpice analysis and observe that the node voltages and quiescent currents are in close agreement with those obtained by conventional circuit analysis. Also note that under the heading BIPOLAR-JUNCTION TRANSISTORS the operating-condition values of voltages, currents, and betas are close to those that were expected. The results are shown in Fig. 10.23.

It is interesting to run the analysis with the opposite initial conditions, setting Q_1 on and Q_2 off. This is brought about by using an initial guess for V(2) = 0.15 V instead of setting V(4) at that value. The results show that the roles of the two BJTs have reversed, and the various voltages and currents assume their counterpart values. Refer to Fig. 10.24 for the output file.

THE ASTABLE MULTIVIBRATOR

The collector-coupled astable multivibrator is a free-running circuit (*astable* means "not stable"). This circuit can be difficult to handle using PSpice, and unless some trial-and-error is used, the iteration process may fail to converge. The circuit of Fig. 10.25 shows two BJTs in a symmetrical configuration. A transistor is chosen with $h_{FE} = 80$. We may predict the period of oscillation as

$$T = 0.693(R_1C_1 + R_2C_2) = 1.386RC$$

assuming that $R_1 = R_2 = R$ and $C_1 = C_2 = C$. Using off-the-shelf component values for the resistor and capacitor values, let $R = 38$ kΩ and $C = 100$ pF. In our

Fig. 10.25 Collector-coupled astable multivibrator.

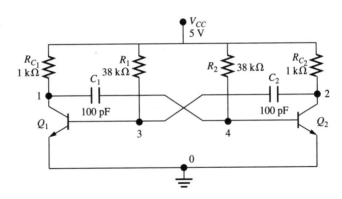

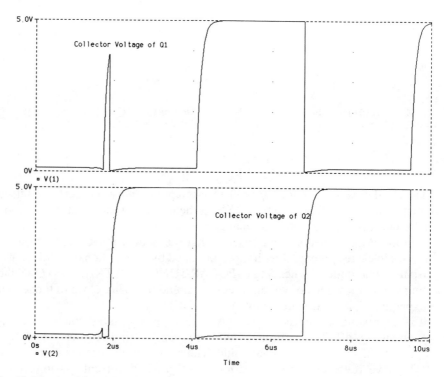

Fig. 10.26 Astable Multivibrator.

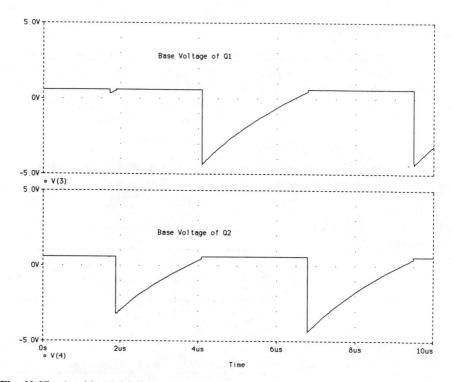

Fig. 10.27 Astable Multivibrator.

```
 Astable Multivibrator
VCC 5 0 5V
RC1 5 1 1k
RC2 5 2 1k
R1 5 3 38k
R2 5 4 38k
C1 1 4 100pF
C2 2 3 100pF
Q1 1 3 0 QN
Q2 2 4 0 QN
.MODEL QN NPN (IS=1E-12 BF=80 BR=1 TF=0.2ns TR=5ns)
.NODESET V(1)=0 V(3)=0
.OP
.PRINT DC I(RC1) I(RC2) I(R1) I(R2)
.TRAN 0.1us 10us
.PROBE
.END


   ****      BJT MODEL PARAMETERS
              QN
              NPN
         IS   1.000000E-12
         BF   80
         BR   1
         TF   200.000000E-12
         TR   5.000000E-09


   ****    SMALL SIGNAL BIAS SOLUTION      TEMPERATURE =  27.000 DEG C
  NODE   VOLTAGE     NODE   VOLTAGE     NODE   VOLTAGE     NODE    VOLTAGE
 (  1)    .1171    (   2)     .1171    (   3)     .5776    (   4)      .5776
 (  5)   5.0000


     VOLTAGE SOURCE CURRENTS
     NAME           CURRENT
     VCC           -9.999E-03


     TOTAL POWER DISSIPATION   5.00E-02  WATTS

   ****     OPERATING POINT INFORMATION      TEMPERATURE =  27.000 DEG C
 **** BIPOLAR JUNCTION TRANSISTORS
 NAME         Q1        Q2
 MODEL        QN        QN
 IB          1.16E-04  1.16E-04
 IC          4.88E-03  4.88E-03
 VBE         5.78E-01  5.78E-01
 VBC         4.61E-01  4.61E-01
 VCE         1.17E-01  1.17E-01
 BETADC      4.20E+01  4.20E+01
 BETAAC      7.91E+01  7.91E+01


   ****    INITIAL TRANSIENT SOLUTION      TEMPERATURE =  27.000 DEG C
  NODE   VOLTAGE     NODE   VOLTAGE     NODE   VOLTAGE     NODE    VOLTAGE
 (  1)    .1171    (   2)     .1171    (   3)     .5776    (   4)      .5776
 (  5)   5.0000


     VOLTAGE SOURCE CURRENTS
     NAME           CURRENT
     VCC           -9.999E-03


     TOTAL POWER DISSIPATION   5.00E-02  WATTS
```

Fig. 10.28

example, this gives a period $T = 5.267$ μs and a frequency $f = 190$ kHz. The input file is

```
Astable Multivibrator
VCC 5 0 5V
RC1 5 1 1k
RC2 5 2 1k
```

Fig. 10.29   Emitter-coupled multivibrator.

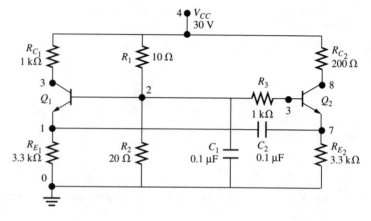

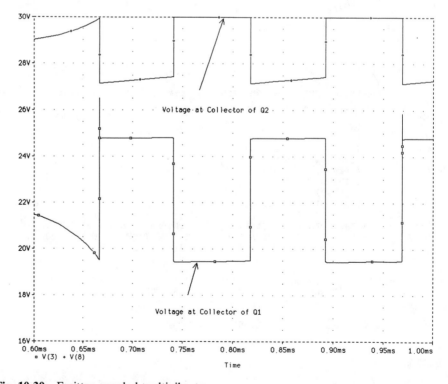

Fig. 10.30   Emitter-coupled multivibrator.

```
    Emitter-coupled multivibrator
   VCC 4 0 30V
   R1 4 2 10
   R2 2 0 20
   R3 2 3 1k
   RC1 4 3 1k
   RC2 4 8 200
   RE1 1 0 3.3k
   RE2 7 0 3.3k
   C1 2 0 0.1uF
   C2 1 7 0.1uF
   Q1 3 2 1 QN
   Q2 8 3 7 QN
   .MODEL QN NPN(IS=1E-12 BF=30 BR=1 TF=0.2ns TR=5ns)
   .NODESET V(3)=25V
   .OP
   .PRINT DC I(RC1) I(RC2) I(RE1) I(RE2)
   .TRAN 0.5us 1ms
   .PROBE
   .END


    ****      BJT MODEL PARAMETERS
                  QN
                  NPN
          IS    1.000000E-12
          BF    30
          BR    1


    ****      SMALL SIGNAL BIAS SOLUTION       TEMPERATURE =   27.000 DEG C
     NODE    VOLTAGE      NODE    VOLTAGE     NODE    VOLTAGE     NODE    VOLTAGE
    (   1)   19.4310    (    2)   20.0120   (   3)   22.0520   (    4)   30.0000
    (   7)   21.4680    (    8)   28.7410


        VOLTAGE SOURCE CURRENTS
        NAME        CURRENT
        VCC        -1.013E+00


        TOTAL POWER DISSIPATION    3.04E+01   WATTS

    ****      OPERATING POINT INFORMATION     TEMPERATURE =   27.000 DEG C
    **** BIPOLAR JUNCTION TRANSISTORS
    NAME         Q1            Q2
    MODEL        QN            QN
    IB           1.90E-04      2.10E-04
    IC           5.70E-03      6.30E-03
    VBE          5.81E-01      5.84E-01
    VBC         -2.04E+00     -6.69E+00
    VCE          2.62E+00      7.27E+00
    BETADC       3.00E+01      3.00E+01
    BETAAC       3.00E+01      3.00E+01


    ****      INITIAL TRANSIENT SOLUTION      TEMPERATURE =   27.000 DEG C
     NODE    VOLTAGE      NODE    VOLTAGE     NODE    VOLTAGE     NODE    VOLTAGE
    (   1)   19.4310    (    2)   20.0120   (   3)   22.0520   (    4)   30.0000
    (   7)   21.4680    (    8)   28.7410


        VOLTAGE SOURCE CURRENTS
        NAME        CURRENT
        VCC        -1.013E+00
```

Fig. 10.31

```
R1 5 3 38k
R2 5 4 38k
C1 1 4 100pF
C2 2 3 100pF
Q1 1 3 0 QN
Q2 2 4 0 QN
.MODEL QN NPN(IS=1E-12 BF=80 BR=1 TF=0.2ns TR=5ms)
.NODESET V(1)=0 V(3)=0
.OP
PRINT DC I(RC1) I(RC2) I(R1) I(R2)
.TRAN 0.1us 10us
.PROBE
.END
```

The *.TRAN* statement deserves discussion. Since the period is known to be slightly over 5 μs, the analysis should run for perhaps 8 μs to 12 μs, in order to give the waveshapes enough time to settle to their expected forms. When a time of 12 μs was used, the iterations did not converge, and the plots were incorrect. Attempts to use other step times were also unsuccessful, although some combinations might be found that will produce proper results. The interval of 10 μs and the step of 0.1 μs did produce good results, however. Collector-voltage waveshapes are shown in Fig. 10.26 on page 340, and base-voltage waveshapes are shown in Fig. 10.27 on page 340. Using the cursor, you should verify that $T = 5.4$ μs, giving $f = 185$ kHz. These values are in close agreement with our predictions. Small-signal bias voltages and the initial transient voltages are shown in the output file of Fig. 10.28 on page 341.

AN EMITTER-COUPLED BJT MULTIVIBRATOR

Another multivibrator is shown in Fig. 10.29 on page 342. This is an emitter-coupled multivibrator, using off-the-shelf components. The details of the analysis are found in Millman and Taub, *Pulse, Digital, and Switching Waveforms*. For their analysis, it is assumed that Q_1 saturates and Q_2 does not. We set the initial voltage at the collector of Q_1 to a value of 25 V. You may want to try several different values of nodeset voltage and compare the results. The Millman and Taub analysis estimates the period of oscillation to be $T = 145.6$ μs.

As an exercise, create your own input file for this circuit. Verify that the Probe results show $T = 151.4$ μs. The traces of collector voltages are shown in Fig. 10.30 on page 342. Note that the X-axis has been plotted for the time interval 0.6 ms to 1.0 ms. This is done by changing the X-axis range to 0.6m 1.0m (omit the s after the m). The output file is shown in Fig. 10.31 on page 343.

PROBLEMS

10.1 The test circuits of Figs. 10.1 and 10.3 are designed to use a typical *npn* transistor. Using data for the 2N3251 *pnp* transistor ($h_{FE} = 180$), produce sets of output and input characteristics. Design an input file that will create the traces in Probe. Provide identifying labels for each of the curves.

10.2 (a) The circuit of Fig. 10.32 has $h_{FE} = 100$. Find the Q point using PSpice; then compare the results with your calculations assuming $V_{BE} = 0.7$ V.
(b) In the PSpice analysis assume that $h_{FE} = 50$ and find the Q point.

Fig. 10.32

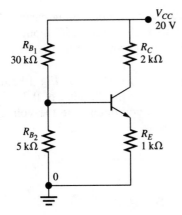

10.3 Determine the Q point for the circuit shown in Fig. 10.33 assuming that $h_{FE} = 60$ and $V_{BE} = 0.7$ V. Verify your calculations with a PSpice analysis using the built-in transistor model.

Fig. 10.33

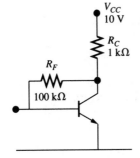

10.4 The circuit of Fig. 10.34 is to be used as a CE amplifier. It is desired that the Q point will allow maximum swing in collector current without undue distortion. The transistor has $h_{FE} = 50$.

Fig. 10.34

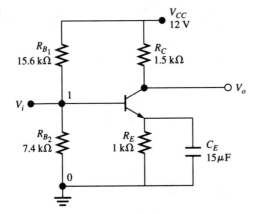

(a) Find the quiescent collector current and voltage using PSpice.

(b) Run a PSpice/Probe analysis with a sinusoidal input v_i, and determine the practical limit of the input-voltage swing. What is the collector-current swing under this condition?

10.5   A *CE* amplifier with an unbypassed R_E is shown in Fig. 10.35. The transistor has $h_{FE} = 100$. The input signal has a peak value of 0.2 V. Using PSpice/Probe, show the output waveform and determine the voltage gain.

Fig. 10.35

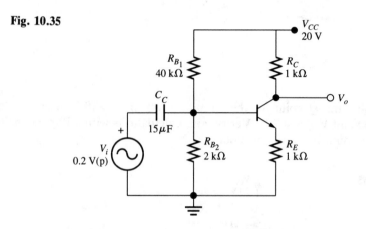

10.6   An emitter-coupled circuit is shown in Fig. 10.36. Use a PSpice analysis to find the quiescent collector currents and voltages for Q_1 and Q_2. Each transistor has $h_{FE} = 100$.

Fig. 10.36

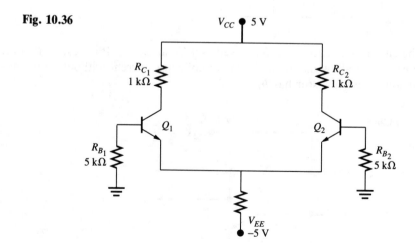

The Field-Effect Transistor

The built-in models for FETs in PSpice are designated with name beginning with *J* (for junction FETs), *M* (for MOSFETs), and *B* (for GaAsFETs). Before any of the devices are used, it is desirable to obtain sets of characteristic curves, so that the operating-point voltages and currents are properly predicted.

OUTPUT CHARACTERISTICS FOR THE JFET

The evaluation version comes with two *n*-channel JFETs in the library *EVAL.LIB*. The *J2N3819* has been chosen to obtain a useful set of output characteristics. The circuit of Fig. 11.1 yields this input file:

```
Output Characteristics for JFET J2N3819
VGS 1 0 0V
VDD 2 0 12V
JFET 2 1 0 J2N3819
.DC VDD 0 12V 0.8V VGS 0 −4V 1V
.PROBE
.LIB EVAL.LIB
.END
```

The nested loop for *.DC* will allow for up to five traces, with V_{GS} taking on integer values between 0 and −4 V. Run the analysis and observe that only four traces are shown. Refer to Fig. 11.2. Since the top trace is for $V_{GS} = 0$ V, the other traces are for −1 V, −2 V, and −3 V. There is no trace for $V_{GS} = -4$ V, since this

Fig. 11.1 Circuit for JFET characteristics.

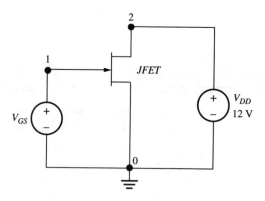

is below the pinch-off value. It is now obvious that pinch off occurs at a value of -3 V. Now we are in a position to design a biasing circuit based on this JFET, since we know what values of V_{GS} are useful and what the resulting I_D values will be.

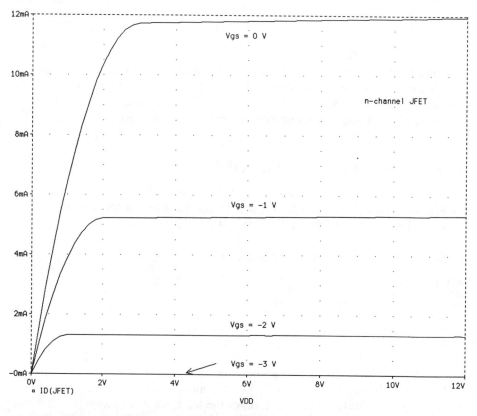

Fig. 11.2 Output Characteristics for JFET J2N3819.

INPUT CHARACTERISTICS FOR THE JFET

For the input characteristics, the nested loop for .*DC* will use V_{GS} in the outer loop, giving this as the *X*-axis variable. Values of V_{DD} will range from 2 V to 10 V in 4-V steps, giving three traces. If you watch the generation of these on the screen in Probe, you will see that the first trace, for $V_{DD} = 2$ V, is below the other two, as expected. Labeling the curves for future use is helpful. Here is the input file:

```
Input Characteristics for JFET
VGS 1 0 0V
VDD 2 0 10V
JFET 2 1 0  J2N3819
.DC VGS -3 0 0.05V VDD 2V 10V 4V
.PROBE
.LIB EVAL.LIB
.END
```

The characteristics are shown in Fig. 11.3, along with appropriate labels.

Since the evaluation library contains only *n*-channel JFETs, if a *p*-channel JFET is needed, you may wish to either insert a .*MODEL* statement in the input

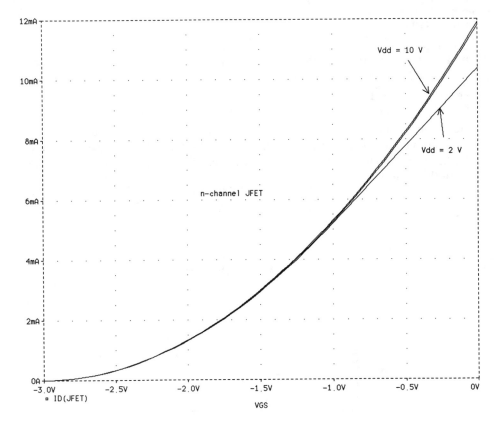

Fig. 11.3 Input Characteristics for JFET.

file or modify the library *EVAL.LIB* to include one or more such devices. The entry for the *J2N3819* is as follows:

```
.model J2N3819 NJF (Beta=1.304m Betatce=-.5 Rd=1 Rs=1 Lambda=2.25m
+Vto=-3
+Vtotc=-2.5m Is=33.57f Isr=322.4f N=1 Nr=2 Xti=3 Alpha=311.7
+Vk=243.6 Cgd=1.6p M=.3622 Pb=1 Fc=.5 Cgs=2.414p Kf=9.882E-18
+Af=1)
*National pid=50 case=TO92
*88-08-01 rmn BVmin=25
```

Note that $V_{to} = -3$ V is given as the threshold voltage. When a *p*-channel JFET is used, its model should be *PJF* (rather than *NJF*), and a positive value of V_{to} must be used.

JFET BIASING CIRCUIT

A source self-bias circuit is shown in Fig. 11.4. The built-in model for the *n*-channel JFET is used with several of the default parameters changed as shown in the input file.

```
n-channel JFET bias circuit
VDD 4 0 24V
RG 1 0 0.5MEG
RS 2 0 770
RD 4 3 8.8k
JFET 3 1 2 JM
.MODEL JM NJF(RD=10 RS=10 VTO=-3 BETA=1.65mS)
.DC VDD 24V 24V 24V
.OP
.PRINT DC I(RD) I(RS) I(RG)
.END
```

The output is shown in Fig. 11.5. In order to see if the PSpice results agree with conventional circuit analysis, the value of I_{DSS} should be known. Run an analysis like the one shown in Fig. 11.2, and verify that for this JFET, $I_{DSS} = 1.78$ mA. Using this, we find

$$g_{mo} = \frac{-2I_{DSS}}{V_P} = \frac{-2(1.78 \text{ mA})}{-3 \text{ V}} = 1.87 \text{ mS}$$

Fig. 11.4   JFET self-bias circuit.

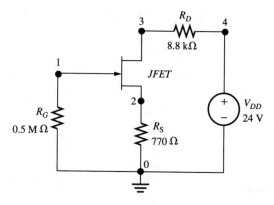

```
    n-channel JFET bias circuit

VDD 4 0 18V
RG 1 0 0.5MEG
RS 2 0 770
RD 4 3 8.8k
JFET 3 1 2 JM
.MODEL JM NJF(RD=10 RS=10 VTO=-3V BETA=0.2m)
.DC VDD 18V 18V 18V
.OP
.PRINT DC I(RD) I(RS) I(RG)
.END

    ****      Junction FET MODEL PARAMETERS

            JM
            NJF
     VTO    -3
    BETA    200.000000E-06
     RD     10
     RS     10

    ****    DC TRANSFER CURVES          TEMPERATURE =   27.000 DEG C

   VDD        I(RD)       I(RS)       I(RG)

   1.800E+01   9.915E-04   9.915E-04   1.006E-11

    ****    SMALL SIGNAL BIAS SOLUTION      TEMPERATURE =   27.000 DEG C

 NODE   VOLTAGE     NODE   VOLTAGE     NODE   VOLTAGE     NODE   VOLTAGE

 (   1) 5.029E-06  (   2)     .7635  (   3)    9.2744  (   4)    18.0000

    VOLTAGE SOURCE CURRENTS
    NAME          CURRENT

    VDD           -9.915E-04

    TOTAL POWER DISSIPATION   1.78E-02   WATTS

    ****     OPERATING POINT INFORMATION     TEMPERATURE =   27.000 DEG C

**** JFETS

NAME          JFET
MODEL         JM
ID            9.92E-04
VGS           -7.63E-01
VDS           8.51E+00
GM            8.91E-04
GDS           0.00E+00
CGS           0.00E+00
CGD           0.00E+00
```

Fig. 11.5

Next, we use the value of I_{DS} from PSpice to find V_{GS}. In the following equation, I_{DS} is the drain current under saturation conditions:

$$I_{DS} = I_{DSS}\left(1 - \frac{V_{GS}}{V_P}\right)^2 = 0.992 \text{ mA} = 1.78\text{mA}\left(1 + \frac{V_{GS}}{3}\right)^2$$

Rearranging and solving for V_{GS} gives $V_{GS} = -0.78$ V. Then

$$g_m = g_{mo}\left(1 - \frac{V_{GS}}{V_P}\right) = 1.187 \text{ mS}\left(1 - \frac{0.78}{3}\right) = 0.88 \text{ mS}$$

The values for V_{GS} and g_m are in close agreement with those shown in Fig. 11.5.

THE JFET AMPLIFIER

The biasing circuit of the example shown above can be made into a voltage amplifier with the addition of two capacitors and an ac voltage source, as shown in Fig. 11.6. The input file is designed to show the ac analysis at $f = 5$ kHz.

```
n-channel JFET Amplifier circuit
VDD 4 0 18V
vi 1a 0 ac 1mV
Cb 1a 1 15uF
Cs 2 0 15uF
RG 1 0 0.5MEG
RS 2 0 770
RD 4 3 8.8k
JFET 3 1 2 JM
.MODEL JM NJF (RD=10 RS=10 VTO=-3V BETA=0.2m)
.DC VDD 18V 18V 18V
.OP
.PRINT DC I(RD) I(RS) I(RG)
.ac lin 1 5kHz 5kHz
.PRINT ac i(RD) v(3) v(1) v(2)
.END
```

The output file is shown in Fig. 11.7. Verify from your analysis that the ac drain voltage V(3) is 7.77 mV, giving a voltage gain of 7.77. This is in agreement

Fig. 11.6 JFET amplifier.

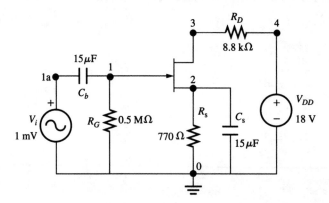

```
    n-channel JFET Amplifier circuit

VDD 4 0 18V
vi 1a 0 ac 1mV
Cb 1a 1 15uF
Cs 2 0 15uF
RG 1 0 0.5MEG
RS 2 0 770
RD 4 3 8.8k
JFET 3 1 2 JM
.MODEL JM NJF(RD=10 RS=10 VTO=-3V BETA=0.2m)
.DC VDD 18V 18V 18V
.OP
.PRINT DC I(RD) I(RS) I(RG)
.ac lin 1 5kHz 5kHz
.PRINT ac i(RD) v(3) v(1) v(2)
.END

    VDD           I(RD)         I(RS)          I(RG)

   1.800E+01   9.915E-04    9.915E-04    1.006E-11

  NODE    VOLTAGE     NODE    VOLTAGE     NODE    VOLTAGE     NODE    VOLTAGE

(    1) 5.029E-06  (    2)     .7635  (    3)    9.2744  (    4)    18.0000
(   1a)    0.0000

     VOLTAGE SOURCE CURRENTS
     NAME              CURRENT

     VDD             -9.915E-04
     vi               0.000E+00

     TOTAL POWER DISSIPATION    1.78E-02   WATTS

**** JFETS


NAME           JFET
MODEL          JM
ID             9.92E-04
VGS            -7.63E-01
VDS            8.51E+00
GM             8.91E-04
GDS            0.00E+00
CGS            0.00E+00
CGD            0.00E+00

  ****     AC ANALYSIS                   TEMPERATURE =    27.000 DEG C

   FREQ        I(RD)        V(3)        V(1)        V(2)

   5.000E+03   8.828E-07   7.768E-03   1.000E-03   1.873E-06
```

Fig. 11.7

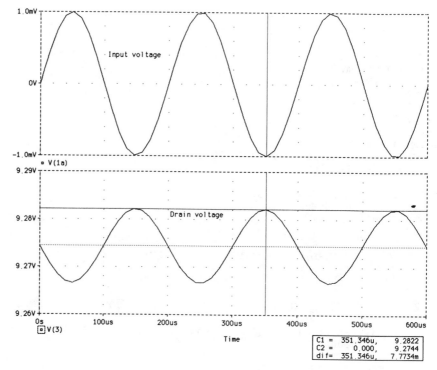

Fig. 11.8 h-channel JFET Amplifier circuit.

with the approximate equation

$$A_V = g_m R_D = (0.891 \text{ mS})(8.8 \text{ k}\Omega) = 7.8$$

JFET Waveshapes

The input file can be modified slightly to produce the waveshapes at the drain and input (source) as shown in Fig. 11.8. Note that the cursor has been positioned to show V(3) *maX*, which is the peak of the ac drain voltage. The input voltage is shown as *sin(0 1mV 5kHz)*, giving it a 1-mV peak value. Run the analysis and show that the drain voltage has a peak value of 9.2822 mV and a valley value of 9.2666 mV. This gives a peak-to-peak value of 15.6 mV, which is 8.8 mV peak. This agrees with the ac analysis previously shown. The output file of Fig. 11.9 shows the modification of the input file and the Fourier components of the output voltage. Note that the dc value is the quiescent value of 9.2744 V, and that there is negligible distortion.

THE POWER MOSFET

For a case study involving a MOSFET, THE *EVAL.LIB* model of such a device will be chosen. This is the *IRF150*, which is an *n*-type power MOSFET. In order

```
   n-channel JFET Amplifier circuit

VDD 4 0 18V
vi 1a 0 sin(0 1mV 5kHz)
Cb 1a 1 15uF
Cs 2 0 15uF
RG 1 0 0.5MEG
RS 2 0 770
RD 4 3 8.8k
JFET 3 1 2 JM
.MODEL JM NJF(RD=10 RS=10 VTO=-3V BETA=0.2m)
.opt nopage nomod
.TRAN 0.02ms 0.6ms
.PROBE
.FOUR 5kHz V(3)
.END

 ****     INITIAL TRANSIENT SOLUTION      TEMPERATURE =   27.000 DEG C

 NODE    VOLTAGE     NODE    VOLTAGE     NODE    VOLTAGE     NODE    VOLTAGE

 (   1) 5.029E-06  (    2)    .7635   (    3)   9.2744   (    4)   18.0000

 (  1a)    0.0000

     VOLTAGE SOURCE CURRENTS
     NAME          CURRENT
     VDD         -9.915E-04
     vi           0.000E+00

   TOTAL POWER DISSIPATION   1.78E-02  WATTS

 ****     FOURIER ANALYSIS              TEMPERATURE =   27.000 DEG C

FOURIER COMPONENTS OF TRANSIENT RESPONSE V(3)

 DC COMPONENT =   9.274381E+00

 HARMONIC    FREQUENCY      FOURIER     NORMALIZED     PHASE      NORMALIZED
    NO         (HZ)        COMPONENT    COMPONENT      (DEG)     PHASE (DEG)

     1       5.000E+03     7.679E-03    1.000E+00    -1.797E+02    0.000E+00
     2       1.000E+04     2.155E-05    2.806E-03    -1.014E+02    7.829E+01
     3       1.500E+04     2.311E-05    3.009E-03    -1.076E+02    7.208E+01
     4       2.000E+04     2.231E-05    2.905E-03    -1.139E+02    6.578E+01
     5       2.500E+04     2.154E-05    2.805E-03    -1.189E+02    6.079E+01
     6       3.000E+04     2.067E-05    2.692E-03    -1.247E+02    5.507E+01
     7       3.500E+04     1.949E-05    2.538E-03    -1.300E+02    4.974E+01
     8       4.000E+04     1.848E-05    2.406E-03    -1.352E+02    4.449E+01
     9       4.500E+04     1.723E-05    2.244E-03    -1.399E+02    3.983E+01

     TOTAL HARMONIC DISTORTION =   7.599231E-01 PERCENT
```

Fig. 11.9

to become familiar with its characteristics, we will look at its output and input family of curves.

The Output Characteristics

In order to obtain the output characteristics, the circuit shown in Fig. 11.10 will be used. The input file is

```
n-channel MOSFET Output Characteristics
VDD 2 0 12V
VGS 1 0 0V
MFET 2 1 0 0 IRF150; drain, gate, source, and substrate
.DC VDD 0 12V 0.8V VGS 0 8V 1V
.LIB EVAL.LIB
.PROBE
.END
```

The figure shows that the source and substrate are connected together, as required. The output characteristics are shown in Fig. 11.11. As an example of the large drain currents involved, it is seen that when V_{GS} = 5 V, the saturation current is greater than 7 A. The library entry for the *IRF150* shows V_{to} = 2.831 V as the zero-bias threshold voltage. This is a positive voltage for an *n*-channel device.

The Input Characteristics

For the input characteristics, several values of V_{DD} will be used as shown in the following file:

```
Input Characteristic for MOSFET
VGS 1 0 0V
VDD 2 0 10V
MOS 2 1 0 0 IRF150
.DC VGS 0 8V 0.1V VDD 2V 10V 4V
.PROBE
.LIB EVAL.LIB
.END
```

The resulting plot is shown in Fig. 11.12. It can be seen that the threshold value of V_{GS} is slightly below 3 V. Also note that the family of curves for V_{DD} = 6 V or more will be indistinguishable.

Fig. 11.10 MOSFET circuit for characteristics.

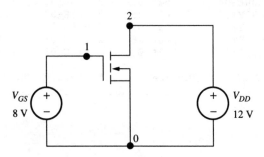

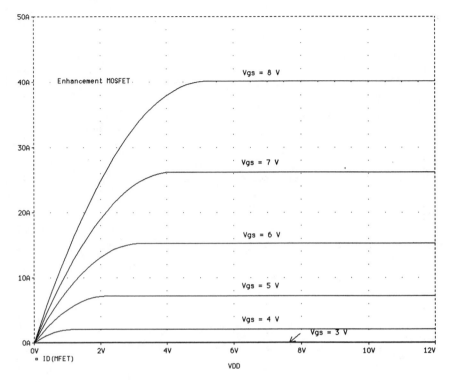

Fig. 11.11   n-channel MOSFET Output Characteristics.

The MOSFET Amplifier

A power amplifier using the *IRF150* is shown in Fig. 11.13. Since large source and drain currents will be involved, the values of R_d and R_s are 2 Ω and 0.5 Ω, respectively. A voltage divider is provided using R_1 and R_2 to give a value of $V_{GS} = 4.7$ V. The input file is

```
n-channel Power MOSFET Amplifier
VDD 4 0 18V
vi 1 0 ac 0.5V
R1 4 2 330k
R2 2 0 220k
Rd 4 3 2
Rs 5 0 0.5
Cb 1 2 15uF
Cs 5 0 15uF
MFET 3 2 5 5 IRF150
.DC VDD 12V 12V 12V
.OP
.PRINT DC I(RD) I(R1) I(R2) I(Rs)
.ac lin 1 5kHz 5kHz
.PRINT ac i(Rd) v(2) v(3)
.LIB EVAL.LIB
.END
```

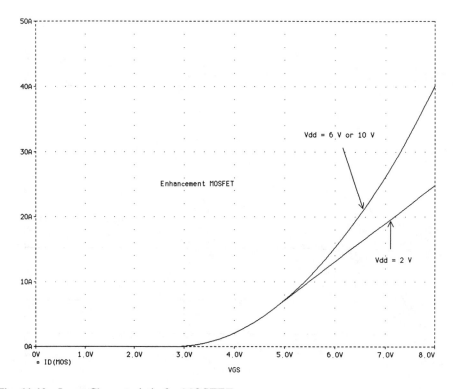

Fig. 11.12   Input Characteristic for MOSFET.

The output file is shown in Fig. 11.14. Both dc and ac results are shown. From the dc portion of the results, the quiescent drain current (and source current) is $I_D = 1.781$ A, giving a drain voltage $V(3) = 7.827$ V and a source voltage $V(5) = 2.543$ V.

Fig. 11.13   Power MOSFET amplifier.

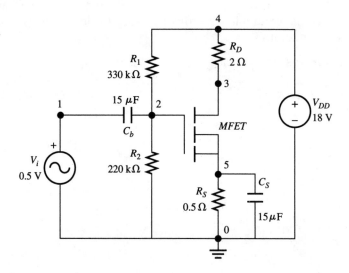

```
     n-channel Power MOSFET Amplifier
VDD 4 0 18V
vi 1 0 ac 0.5V
R1 4 2 330k
R2 2 0 220k
Rd 4 3 2
Rs 5 0 0.5
Cb 1 2 15uF
Cs 5 0 15uF
MFET 3 2 5 5 IRF150
.DC VDD 12V 12V 12V
.OP
.PRINT DC I(RD) I(R1) I(R2) I(Rs)
.ac lin 1 5kHz 5kHz
.PRINT ac i(Rd) v(2) v(3)
.LIB EVAL.LIB
.END


    ****      MOSFET MODEL PARAMETERS
                  IRF150
                  NMOS
         LEVEL    3
           VTO    2.831
           PHI    .6
            RD    1.031000E-03
            RS    1.624000E-03
            RG    13.89


   VDD          I(RD)        I(R1)        I(R2)        I(Rs)
   1.200E+01    1.781E+00    2.182E-05    2.182E-05    1.781E+00
  NODE   VOLTAGE      NODE    VOLTAGE     NODE   VOLTAGE      NODE    VOLTAGE
(    1)   0.0000   (    2)    7.2000   (    3)   7.8271   (    4)   18.0000
(    5)   2.5432


    VOLTAGE SOURCE CURRENTS
    NAME         CURRENT
    VDD          -5.086E+00
    vi            0.000E+00


    TOTAL POWER DISSIPATION   9.16E+01   WATTS

**** MOSFETS
NAME         MFET
MODEL        IRF150
ID           5.09E+00
VGS          4.66E+00
VDS          5.28E+00
VBS          0.00E+00
VTH          2.83E+00
GM           5.60E+00

   ****     AC ANALYSIS                 TEMPERATURE =   27.000 DEG C

   FREQ         I(Rd)        V(2)        V(3)

  5.000E+03    7.536E-01    4.999E-01    1.507E+00
```

Fig. 11.14

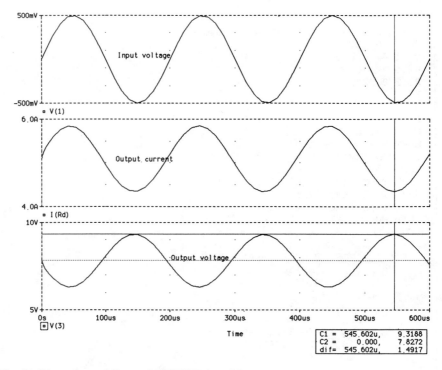

Fig. 11.15   n-channel Power MOSFET Amplifier.

The ac analysis shows an input voltage $v_i = 0.5$ V and an output voltage at the drain v(3) = 1.5 V, giving a voltage gain of 3. The ac output current has a value $i_D = 0.7536$ A. All of these are peak values.

The Waveshapes

The input file may be modified slightly to allow for plots of the output voltage, the output current, and the input voltage. Figure 11.15 shows these waveshapes. The cursor shows the *maX* value of the output voltage V(3) = 9.3188 V along with the quiescent value of 7.8272 V, giving a peak value of 1.4916 V. This is in close agreement with the ac voltage shown in the previous analysis, affirming the voltage gain of *three*. The modification of the output file along with the Fourier components of the output voltage are shown in Fig. 11.16. Note that there is a small second harmonic, giving a total distortion of just above one-half of one percent.

THE GALLIUM ARSENIDE FET

PSpice includes a built-in model for the *n*-channel GaAsFET. The device name begins with *B*. No entries are included in the evaluation library; however, you may specify model parameters or simply use the default values, which are shown

```
   n-channel Power MOSFET Amplifier

VDD 4 0 18V
vi 1 0 sin(0 0.5V 5kHz)
R1 4 2 330k
R2 2 0 220k
Rd 4 3 2
Rs 5 0 0.5
Cb 1 2 15uF
Cs 5 0 15uF
MFET 3 2 5 5 IRF150
.opt nopage nomod
.TRAN 0.02ms 0.6ms
.PROBE
.FOUR 5kHz v(3)
.LIB EVAL.LIB
.END
```

```
 ****      INITIAL TRANSIENT SOLUTION          TEMPERATURE =   27.000 DEG C

 NODE    VOLTAGE     NODE    VOLTAGE     NODE    VOLTAGE     NODE    VOLTAGE
(    1)   0.0000  (    2)    7.2000  (    3)    7.8271  (    4)   18.0000
(    5)   2.5432
```

```
     VOLTAGE SOURCE CURRENTS
     NAME          CURRENT

     VDD           -5.086E+00
     vi             0.000E+00
```

```
     TOTAL POWER DISSIPATION   9.16E+01  WATTS
```

```
 ****      FOURIER ANALYSIS                    TEMPERATURE =   27.000 DEG C

FOURIER COMPONENTS OF TRANSIENT RESPONSE V(3)

 DC COMPONENT =   7.819596E+00

 HARMONIC   FREQUENCY    FOURIER    NORMALIZED    PHASE      NORMALIZED
   NO          (HZ)     COMPONENT   COMPONENT     (DEG)     PHASE (DEG)

     1       5.000E+03   1.490E+00   1.000E+00   -1.703E+02   0.000E+00
     2       1.000E+04   7.828E-03   5.254E-03    1.286E+02   2.989E+02
     3       1.500E+04   2.773E-04   1.861E-04   -1.101E+02   6.020E+01
     4       2.000E+04   1.540E-04   1.034E-04   -6.620E+01   1.041E+02
     5       2.500E+04   1.467E-04   9.848E-05   -7.264E+01   9.763E+01
     6       3.000E+04   1.326E-04   8.899E-05   -7.117E+01   9.911E+01
     7       3.500E+04   1.224E-04   8.219E-05   -6.222E+01   1.081E+02
     8       4.000E+04   1.342E-04   9.011E-05   -5.255E+01   1.177E+02
     9       4.500E+04   1.157E-04   7.764E-05   -4.056E+01   1.297E+02

     TOTAL HARMONIC DISTORTION =   5.262415E-01 PERCENT
```

Fig. 11.16

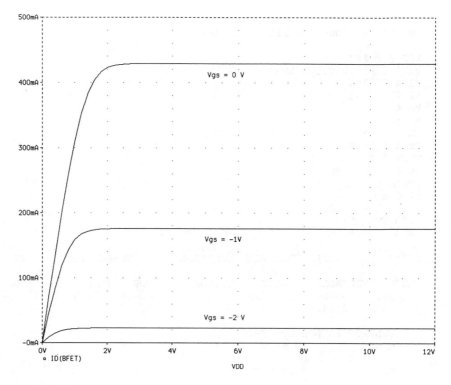

Fig. 11.17   Output Curves for GaAsFET.

in Appendix D. The default value of the pinch-off voltage V_{to} is -2.5 V. A sample input file, used to obtain the output characteristics, is shown below:

```
Output Curves for GaAsFET
VDD 2 0 12V
VGS 1 0 0V
BFET 2 1 0 B1; nodes are grain, gate, and source
.MODEL B1 GAsFET(Vto=-2.5 B=0.3 Rg=1 Rd=1 Rs=1 Vbi=0.5V)
.DC VDD 0 12V 0.2V VGS 0 -3V 1V
.END
```

Run the analysis and verify that $I_{DSS} = 429$ mA, as shown in Fig. 11.17.

PROBLEMS

Note: In SPICE, the JFET parameter *BETA* is found as

$$\beta = \frac{I_{DSS}}{V_P^2}$$

11.1 Using PSpice, determine the drain current I_D and the drain voltage V_{DS} for the JFET circuit shown in Fig. 11.18. Known values are $V_{PO} = 2$ V and $I_{DSS} = 5$ mA.

Fig. 11.18

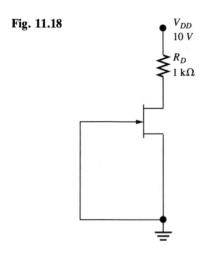

11.2 Find the Q-point values of I_D and V_{DS} for the JFET circuit shown in Fig. 11.19. The JFET has the same characteristics as in the previous problem.

Fig. 11.19

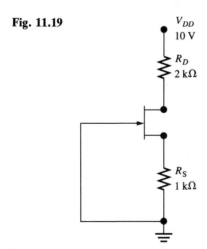

11.3 The circuit of Fig. 11.20 is for a depletion-type MOSFET with $I_{DSS} = 5$ mA and $V_{PO} = 2$ V. Find the values of I_D, V_{GS}, and V_{DS} from a PSpice analysis.

Fig. 11.20

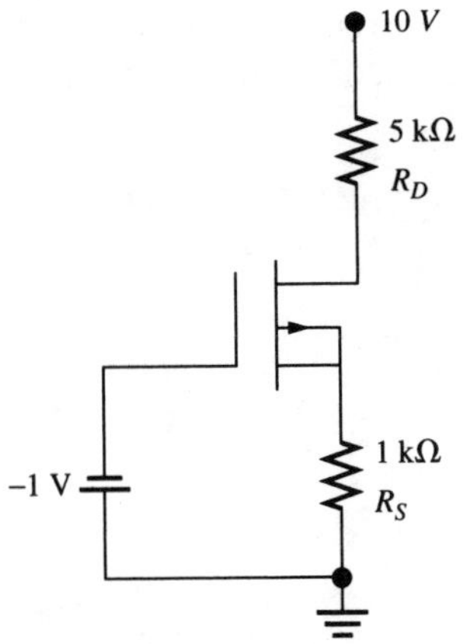

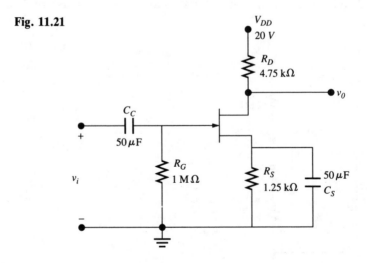

11.4 In the JFET circuit of Fig. 11.21, $I_{DSS} = 8$ mA and $V_{PO} = 5.0$ V. At the operating point $g_d = 0.3$ mS. Find the voltage gain v_o/v_i at low frequencies using a PSpice analysis.

Fig. 11.21

11.5 A JFET amplifier is shown in Fig. 11.22. Known values are $r_d = 100$ kΩ and $g_m = 2850$ μS. Use PSpice to find the voltage gain v_o/v_s.

Fig. 11.22

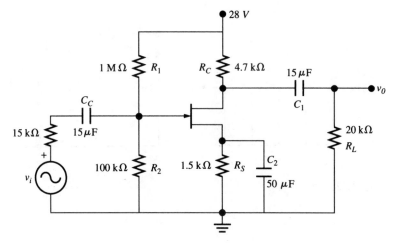

11.6 The MOSFET amplifier of Fig. 11.23 has $V_T = 2.5$ V, $\beta = 0.6$ A/V^2, and $r_d =$ 120 kΩ. Using PSpice, find the voltage gain v_o/v_s.

Fig. 11.23

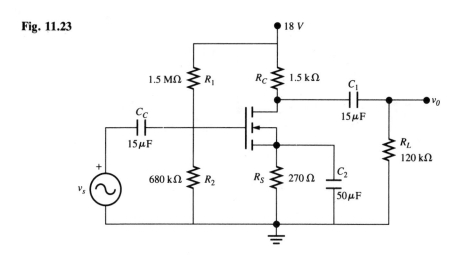

11.7 A chopper circuit is shown in Fig. 11.24. A 1-kHz sine wave is applied as v_i. Its magnitude is less than V_{PO}. The control voltage is a square wave with a frequency that is 2 kHz. Use a PSpice/Probe analysis to show the output voltage v_o.

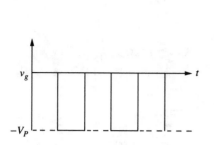

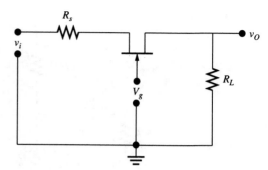

Fig. 11.24

12

Parts and User-Defined Libraries

One of the features of the Design Center of MicroSim is a program called *Parts*. It is available in the evaluation version for both DOS and Windows users.

USING PARTS TO SPECIFY NEW DEVICES

When you find it necessary or desirable to specify a new device such as a diode, BJT, JFET, power MOSFET, OpAmp, or Voltage Comparator, the task can be accomplished by using the interactive program *Parts*. The main advantage of using Parts rather than simply creating a .*MODEL* statement is that the effects of many of the parameters of the new model can be viewed graphically and modified if necessary until just the right fit is obtained.

The evaluation version of PSpice allows for modification or creation of diode models but none of the other devices. We will look at the creation of a model for a diode that is not included in the evaluation version as described in the *Analysis User's Guide*.

CREATING A NEW DIODE MODEL

While in your PSpice directory, begin by typing

```
parts
```

This initiates the program and displays a menu with the various devices listed by number. See Fig. 12.1 for the evaluation version of the menu. Assuming that you

```
0) Exit Program
1) Diode        (signal/rectifier/Zener) <- only model available in Evaluation
2) Bipolar Transistor  (gen'l purpose)
3) JFET  (small-signal, gen'l purpose)
4) Power MOSFET Transistor (all types)
5) Operational Amplifier (bipolar/FET)
6) Voltage Comparator      (bipolar OC)
7) Nonlin. Magnetic Core (ferrite/MPP)

Select: 1
```

Fig. 12.1

have the evaluation version of PSpice, the only allowed choice is 1) for Diode. Press the *Enter* key; then type the desired diode name:

1N4935

When you press *Enter*, a trace showing the voltage-current characteristic of a default diode will appear on the screen. When the session is ended, a file *D1N4935.MOD* will be created. The file will contain the parameters that you create during the session.

The menu at the bottom of the screen allows options for *Exit*, *Next_set*, *Screen_Info*, *Device_curve*, *Model_parameters*, *Trace*, *X_axis*, *Y_axis*, and *Hard_copy*.

Select *X_axis* and enter the range *0.6*, *1.6*. This will allow the selected area to conform to the desired data range as we prepare to enter voltage and current values from the manufacturer's data sheet or from the laboratory. Choose *Device_curve* so that data pairs may be entered next.

Notice that there is a box in the upper right portion of the screen with columns for *Vfwd* and *Ifwd*. These are the values that will be entered beginning with a voltage. Type *1* (for *Vfwd*) and press *Enter*, followed by *1* (for *Ifwd*), and press *Enter*.

A new menu that allows changes in data will appear, but in order to see the effect of our initial *V-I* pair, select *Exit*. The curve shown on the screen will change so that it passes through the point $V = 1$ V, $I = 1$ A. This point will be marked with a small box for easy identification. The default value of *IS* (reverse saturation current) should have changed from *10.00E−15* to *772.3E−18*.

Select *Device_curve* again from the main menu in preparation for entering a new *V-I* pair. Enter the following:

0.8 (for *Vfwd*) and press *Enter*, followed by *0.15* (for *Ifwd*), and press *Enter*.

Now select *Exit* to observe the effect of the new data. The emission coefficient *N* should have changed from its default value of *1* to *2.344*. A further change in *IS* will also occur.

Repeat the procedure for adding another data pair, using the values *1.4* for *Vfwd* and *10* for *Ifwd*.

Select *Exit* and note that *RS*, parasitic resistance, has changed from its original value of *0.1* to a new value of *19.37E−3*.

Using the Fit Command

If you now choose *Fit* from the main menu, a smoother curve fit of the data points will be obtained, and the model parameters will adjust accordingly. Look at *IS* before and after choosing *Fit*, and you should see a significant change. The final values of several parameters along with the V-I plot are shown in the printout of Fig. 12.2. Obtain a printed copy for your files.

Let us pause and think about what has been done so far. We started with a generic diode; then we shaped its *V-I* characteristic by entering three sets of data values. These values are shown in the upper right of our screen and in the printed copy. Now exit from Parts and use your text editor to load the file *D1N4935.MOD*. It should look somewhat like this:

```
*D1N4935 model created using Parts version 5.1 on mm/dd/yy at hh;mm
*
.model D1N4935 D(Is=39.14u N=3.735 Rs=19.73m Ikf=0 Xti=3 Eg=1.11 Cjo=1p
+              M=.3333 Vj=.75 Fc=.5 Isr=100p Nr=2 Bv=100 Ibv=100u Tt=5n)
```

This shows that many of the default values for the diode have been changed. These changes occurred indirectly when we gave Parts the *V-I* data values. The various parameters are now fixed at values that are in conformity with our PSpice diode model.

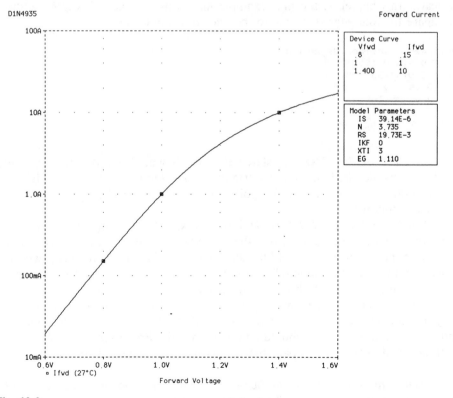

Fig. 12.2

Another file, *D1N4935.MDT*, which is a model data file, was created at the same time as the file shown above. This file is needed if you want to modify the *1N4935* diode or simply load it into Parts for further documentation.

Before we can refer to our newly created diode in a PSpice input file, it must be placed in a library. The evaluation library already contains several diodes, and we can use a text editor to paste the lines of the *D1N4935.MOD* file into this library. It would be desirable to insert the lines just after the description of the other diodes. Proceed with this if you choose to do so. It might be desirable to make a backup copy of *EVAL.LIB* on a floppy disk before you begin, just in case the file becomes damaged. When you run a PSpice analysis that refers to the diode *D1N4935*, you will see a message on the analysis screen which says

making index file for library eval.lib

This means that the index file for the library is being updated to include a reference to the new diode.

CREATING YOUR OWN LIBRARY

Another method of making use of the new diode file for the *1N4935* is instructive to learn. In Chapter 9, the first example makes use of the half-wave rectifier. Compare the following input file to the one shown in Chapter 9:

```
Half-wave Rectifier Using Our Model
v1 1 0 sin(0 12V 1000Hz)
DA 1 2 D1N4935
R 2 0 1k
.tran 0.1ms 1ms
.probe
.lib new.lib
.end
```

In place of the *.MODEL* statement, which would use the built-in diode model, we would like to use our *1N4935* diode. The input file contains a reference to the library *new.lib*, which at this point does not exist. This library can be easily created as follows:

Load the file *D1N4935.MOD* into a text editor. Without making any changes, use the *Save As* feature of the text editor to name a new file *NEW.LIB*. Then exit the editor with the old file still intact. At this point you have two files which both contain the same information about the model for the *1N4935* diode.

Now that we have a library of our own and it is referenced in the input file, we are ready to continue with the PSpice analysis. Run the analysis in the usual way. On the analysis screen you may notice that reference is made to a missing index file, but the analysis should continue until the Probe screen appears. Plot both *v(1)* and *v(2)*, and compare the results to those shown at the beginning of Chapter 9.

Exit from Probe and look at the output file which is shown in Fig. 12.3. Note the two warning statements. The first says that the index file *new.ind* was not

```
Half-wave Rectifier Using Our Model

****      CIRCUIT DESCRIPTION

v1 1 0 sin(0 12V 1000Hz)
DA 1 2 D1N4935
R 2 0 1k
.tran 0.1ms 1ms
.probe
.lib new.lib
.end

WARNING -- Unable to find index file (new.ind) for library file new.lib
WARNING -- Making new index file (new.ind) for library file new.lib
Index has 1 entries from 1 file(s).

Half-wave Rectifier Using Our Model

****      Diode MODEL PARAMETERS

            D1N4935
      IS   39.140000E-06
       N   3.735
     ISR   100.000000E-12
      BV   100
     IBV   100.000000E-06
      RS   .01973
      TT   5.000000E-09
     CJO   1.000000E-12
      VJ   .75
       M   .3333

****      INITIAL TRANSIENT SOLUTION       TEMPERATURE =   27.000 DEG C

NODE   VOLTAGE      NODE   VOLTAGE      NODE   VOLTAGE      NODE    VOLTAGE

(   1)    0.0000  (    2) 2.245E-18

    VOLTAGE SOURCE CURRENTS
    NAME          CURRENT

    v1            -2.245E-21

    TOTAL POWER DISSIPATION   0.00E+00  WATTS
```

Fig. 12.3

found; the second says that the file *new.ind* has been made and that it has one entry. Now that the index file has been created, if you run the PSpice analysis again, the message will not be shown.

If you would like to extend the library *NEW.LIB*, other model statements can be included. These can be of your own creation, or they may be copies of entries from *EVAL.LIB*.

Using the Production Version of PSpice

If you are using the production version of PSpice, there are separate libraries for diodes, BJTs, FETs, digital devices, and so forth. These libraries may be updated in the manner that has been described here. An obvious advantage of the production version is that Parts can be used for parts other than diodes.

ADDING ANOTHER DIODE TO OUR LIBRARY

We have taken data in the laboratory for the diode *1N3605*, and we would like to add this diode to our library. If you have followed the earlier discussion concerning the *1N4935* diode, you should be able to move easily through these steps.

Run the Parts program and press *Enter* to work with diodes. The program asks for

Device part number (or name): *1N3605*

Since our data values will lie within a narrow range of voltages, change the *x*-axis to *0.5 V* to *1.0 V* by selecting *X_axis, Set range*. Then type *0.5 1.0* and press *Enter*.

Next choose *Device_curve*, and note that *Vfwd* is highlighted, waiting for your entry. Type *0.7* and press *Enter*.

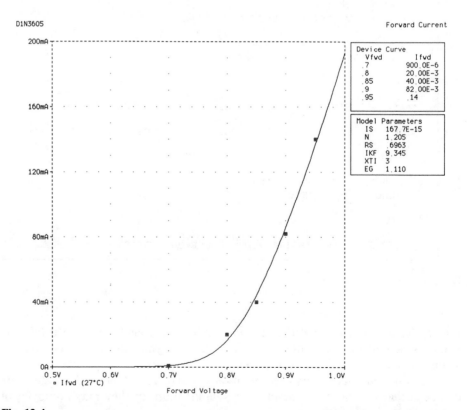

Fig. 12.4

The highlight moves to the *Ifwd*. Type *0.9m* and press *Enter*.

The other coordinate pairs may be entered in the same way. Refer to the upper right corner of Fig. 12.4 to find the other data points. After all of the data sets have been entered, *Exit* and do a curve fit by using the *Fit* command. Now change both the *x*-axis and the *y*-axis to *linear*. Obtain a *Hard_copy* of the characteristics for your files.

While you are still in Parts, select *Next_set*. The screen will be redrawn to show a plot of junction capacitance as a function of reverse voltage. Choose

```
Half-wave Rectifier Using Our Model

v1 1 0 sin(0 12V 1000Hz)
DA 1 2 D1N3605
R 2 0 1k
.tran 0.1ms 1ms
.probe
.lib new.lib
.end

WARNING -- Library file new.lib has changed since index file new.ind was
created
WARNING -- Making new index file (new.ind) for library file new.lib
Index has 2 entries from 1 file(s).

 ****        Diode MODEL PARAMETERS

               D1N3605
         IS   167.700000E-15
          N    1.205
        ISR  100.000000E-12
        IKF    9.345
         BV  100
        IBV  100.000000E-06
         RS    .6963
         TT    5.000000E-09
        CJO    1.000000E-12
         VJ    .75
          M    .3333

 ****      INITIAL TRANSIENT SOLUTION      TEMPERATURE =   27.000 DEG C

  NODE   VOLTAGE      NODE   VOLTAGE      NODE   VOLTAGE      NODE   VOLTAGE

 (   1)    0.0000  (    2)-423.3E-24

     VOLTAGE SOURCE CURRENTS
     NAME           CURRENT

     v1            4.233E-25

     TOTAL POWER DISSIPATION   0.00E+00  WATTS
```

Fig. 12.5

Next_set again and see the reverse leakage plot. Continuing to the *Next_set*, you will see reverse breakdown, and finally a reverse-recovery plot.

Now you should be ready to choose *Exit* to save the results and *Exit* to leave the program. Your two new files are *D1N3605.MOD* and *D1N3605.MDT*. Copy the *D1N3605.MOD* file into your library *NEW.LIB*. This library should now have two diode files along with anything else you might have added.

In order to test the working of the new model and the updated library, go back to your input file for the half-wave rectifier using our model, and modify it to use the device *D1N3605*. Run the PSpice analysis and plot *v(1)* and *v(2)* in Probe. Exit Probe and load the output file into a text editor. Compare the results to Fig. 12.5 on page 373. Note that the diode model parameters are for the device *D1N3605* and that they are in agreement with our model parameters as shown in *D1N3605.MOD*. The warning messages show that the library file *new.lib* has changed and that a new index file has been created.

A ZENER DIODE CASE STUDY

In this case study we will create a model file for a zener diode. We will begin with diode *1N750*, which is included with the evaluation package, by plotting its *V-I* characteristics. With the data obtained from the plot, we will use the Parts program to produce a diode that has similar *V-I* characteristics. This new diode will then become a part of our library, so that it can be used in future analyses.

Begin with the input file:

```
Characteristic Curves for 1N750
V 1 0 10V
R 1 2 5
D1 2 0 D1N750
.DC V −10V 3V 0.02V
.PROBE
.LIB EVAL.LIB
.END
```

Run the PSpice analysis, and in Probe use the cursor mode to verify (approximately) these values (or select your own):

Diode Voltage (V)	Current (mA)
−5.0	604
−4.9	−320
−4.8	−180
−4.75	−50
−4.7	−20
−4.6	−3.6
−4.5	−1.4
−3.0	−24E−3
−2.0	−1,8E−3
−1.0	−142E−6
0.1	10E−6
0.2	73E−6

Diode Voltage (V)	Current (mA)
0.3	464E−6
0.5	16E−3
0.6	96E−3
0.7	855E−3
0.8	23
0.85	82
0.9	192
0.95	334

The Probe plot of the *V-I* characteristic curve is shown in Fig. 12.6. The data from the forward and reverse characteristic curves will be used to create our own version of this diode. When you have recorded enough data pairs to give a reasonable curve fit, exit Probe and load the Parts program.

Following the procedure that has already been explained, enter the data for the forward characteristic. *Note:* In the table shown above, a value such as 96E−3 is actually 96 uA, since the column is for mA. After all points have been entered, do a curve fit and obtain hard copy for your files. This curve is shown in Fig. 12.7. Note that some of the data points are not in agreement with the actual curve.

Now we would like to give the diode its reverse-breakdown data, so that it will behave like a zener diode. While you are still in Parts, go to *Next_set* (which

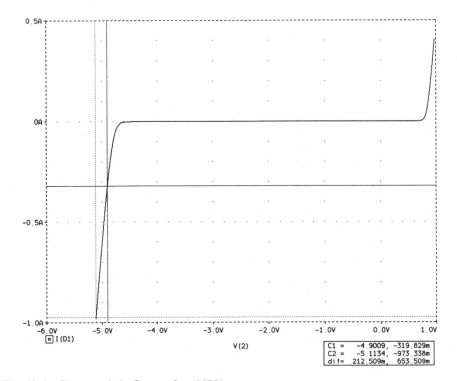

Fig. 12.6 Characteristic Curves for 1N750.

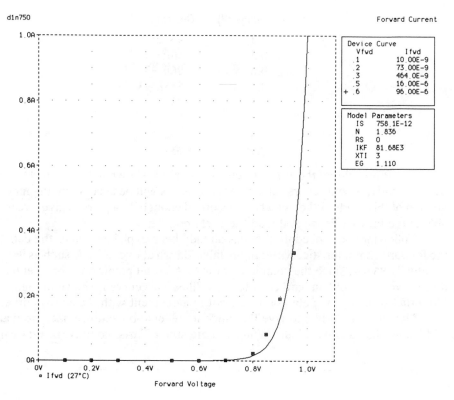

Fig. 12.7

shows junction capacitance); then *Next_set* again (which shows reverse leakage); then *Next_set* again (which shows reverse breakdown). This is where the reverse *V-I* information is to be entered. We see, however, that data pairs cannot be entered here as they were for the forward characteristics. The menu allows us to enter only three values, V_z, I_z, and Z_z. Study the reverse-characteristic curve and try to determine what pair of values to use. If you select $V_z = 4.8$ V along with its current $I_z = 108$ mA as a representative point for the zener breakdown, the new curve will bear a good resemblance to the old one.

There is another item to be entered. We must estimate Z_z. In the linear portion of the reverse-breakdown region, use the change in current from -320 mA to -20 mA and the corresponding change in voltage from -4.9 V to -4.7 V. This gives $Z_z = 0.6667$ Ω. After selecting a curve fit for these values and obtaining hard copy, we are ready to leave Parts. The reverse-breakdown curve is shown in Fig. 12.8.

Now we have produced two files, *D1N750.MOD* and *D1N750.MDT*. These files are necessary in case we need to modify the data in Parts at a later time.

Follow the procedure previously used to add the *.model D1N750* statements to the library *NEW.LIB*. Then run a sample program to see if the new model gives

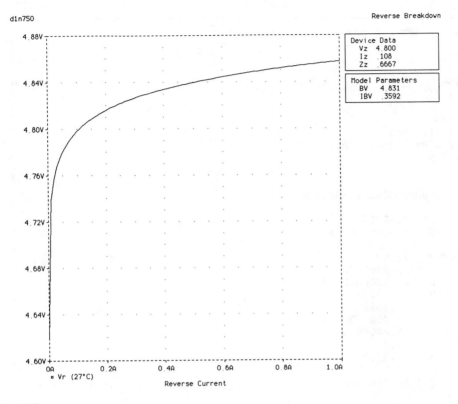

Fig. 12.8

reasonable results. Here is a zener input file you may use. See Fig. 12.9 for the circuit.

```
Testing Our Zener Diode
Vs 1 0 14V
Rs 1 2 27
RL 2 0 150
DZ 0 2 D1N750
.OP
.opt nopage
.DC Vs 14V 14V 14V
```

Fig. 12.9 Circuit for testing the zener diode.

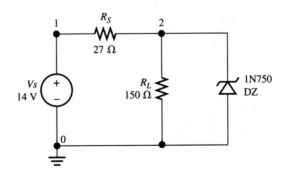

```
   Testing Our Zener Diode

Vs 1 0 14V
Rs 1 2 27
RL 2 0 150
DZ 0 2 D1N750
.OP
.opt nopage
.DC Vs 14V 14V 14V
.PRINT DC I(Rs) I(DZ) I(RL)
.LIB NEW.LIB
.END

   ****      Diode MODEL PARAMETERS


             D1N750
        IS  758.100000E-12
         N    1.836
       ISR   15.88
        NR   36.330000E+03
       IKF   81.680000E+03
        BV    4.831
       IBV    .3592
        TT    5.000000E-09
       CJO    1.000000E-12
        VJ    .75
         M    .3333


   ****     DC TRANSFER CURVES              TEMPERATURE =   27.000 DEG C

    Vs          I(Rs)        I(DZ)        I(RL)

    1.400E+01    3.404E-01   -3.084E-01    3.206E-02


   ****      SMALL SIGNAL BIAS SOLUTION      TEMPERATURE =   27.000 DEG C

  NODE   VOLTAGE      NODE   VOLTAGE     NODE   VOLTAGE      NODE    VOLTAGE

(    1)   14.0000  (    2)    4.8085


     VOLTAGE SOURCE CURRENTS
     NAME           CURRENT

     Vs           -3.404E-01

     TOTAL POWER DISSIPATION   4.77E+00  WATTS

 ****  DIODES

NAME          DZ
MODEL         D1N750
ID            -3.08E-01
VD            -4.81E+00
REQ            1.71E-01
CAP            2.93E-08
```

Fig. 12.10

```
.PRINT DC I(Rs)  I(DZ)  I(RL)
.LIB NEW.LIB
.END
```

Run the PSpice analysis and look carefully at the output file. It should give you the warning messages when the program is run for the first time. Assuming that the diode listing can be found in the library *NEW.LIB*, the diode model parameters should be like those shown in the output file of Fig. 12.10. Notice that some of these diode parameters are different from those of the *1N750* diode model found in the library *EVAL.LIB*. This is to be expected, since we changed only some of the parameters when we worked with the data points. Also, the curve fit for the forward characteristic showed that several points were off the curve.

PROBLEMS

12.1 Data for the 1N4001 diode are found from laboratory measurements as follows:

V (V)	I (mA)
0.55	1
0.6	2
0.7	10
0.75	26
0.8	90
0.81	130

Use the Parts program, selecting diode *1N4001* as a new diode, and enter these *V-I* pairs; then perform a curve fit for this diode. After you have made any necessary corrections, exit to save your new diode files for future use.

12.2 Use the file *D1N4001.MOD* from Problem 12.1 as an addition to the library *NEW.LIB*. (If you have not already created this library, do so.)

12.3 Using the *1N4001* diode from Problems 12.1 and 12.2, perform a PSpice analysis on a full-wave rectifier like the one in Chapter 9. Demonstrate that your diode was used, rather than the default diode.

12.4 For the diode whose forward characteristics are shown in Fig. 12.11, use

Fig. 12.11

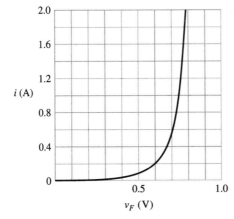

i (A)

v_F (V)

Parts to produce a model. Call this diode *1N914* and make it available in the library *NEW.LIB*.

12.5 For the Zener diode whose reverse characteristics are shown in Fig. 12.12, use Parts to produce a model. Call this diode *1N4744*. Place its model in the library *NEW.LIB*.

Fig. 12.12

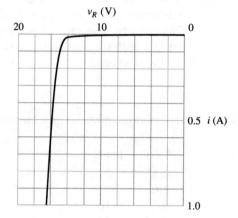

13

Two-Port Networks and Passive Filters

There are occasions when a network may be viewed as a black box, with a pair of terminals at the input and another pair of terminals at the output. The components inside the box may be either unknown or unnecessary to know in order for a circuit analysis to be performed. These networks, called *two-port networks*, might be a set of resistors, a transmission line, and filters, and they might even contain active devices.

TWO-PORT PARAMETERS

Two-port networks have two input terminals and two output terminals, with the input terminals on the source side and the output terminals on the load side of the network. These networks may be analyzed by first finding a set of parameters that define the network and then using equations that employ these parameters. This method of analysis is especially helpful when the source and load change but the network remains the same. We will look at various examples involving the y, z, h, and $ABCD$ parameters.

FINDING THE y PARAMETERS

The two-port admittance parameters are based on the equations

$$I_1 = y_{11}V_1 + y_{12}V_2$$
$$I_2 = y_{21}V_1 + y_{22}V_2$$

Fig. 13.1 Two-port network.

Figure 13.1 shows a two-port network, with the reference directions for the currents and sense directions for the voltages. By setting $V_2 = 0$, it is seen that

$$y_{11} = \left.\frac{I_1}{V_1}\right|_{V_2=0}$$

$$y_{21} = \left.\frac{I_2}{V_1}\right|_{V_2=0}$$

Thus y_{11} is found as the ratio of I_1 to V_1 with $V_2 = 0$, and y_{21} is found as the ratio of I_2 to V_1 with $V_2 = 0$. Also

$$y_{12} = \left.\frac{I_1}{V_2}\right|_{V_1=0}$$

$$y_{22} = \left.\frac{I_2}{V_2}\right|_{V_1=0}$$

PSpice can be used to find these y parameters which are called the *short-circuit admittance parameters* of the two-port network. A simple network of resistors will be used as an example.

Figure 13.2 shows a *T* network of three resistors. In order to find y_{11} and y_{21}, we will short-circuit the output terminals (on the right), making $V_2 = 0$. On the input side a 1-V dc source is used for V_1. The input file is

```
Input and Transfer Admittances
V1 1 0 1V
R1 1 2 12
R2 0 2 3
R3 2 0 6
.DC V1 1V 1V 1V
.PRINT DC I(R1) I(R2); for y11 and y21
.END
```

Fig. 13.2 *T* network.

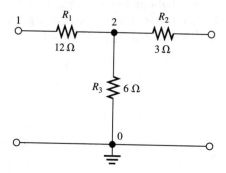

Fig. 13.3 *T* network with output shorted.

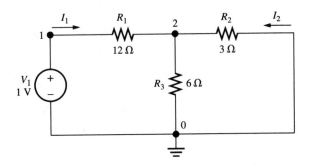

Note that the *R2* statement shows the nodes in the order *0, 2*. This will give the proper direction for the current I_2, as shown in Fig. 13.3. Run the PSpice analysis to find I_1 and I_2. The results are

$$I(R1) = 71.43 \text{ mA} \quad \text{and} \quad I(R2) = -47.62 \text{ mA}$$

Since V_1 was chosen as a 1-V source, I_1 has the same value as y_{11}, and I_2 has the same value as y_{21}. Thus

$$y_{11} = 71.43 \text{ mS} \quad \text{and} \quad y_{21} = -47.62 \text{ mS}$$

The other y parameters may be found by using a 1-V source for V_2 while making $V_1 = 0$. The latter is done by short-circuiting the input terminals. The input file is

```
Output and Transfer Admittances
V2 2 0 1V
R1 0 1 12
R2 2 1 3
R3 1 0 6
.DC V2 1V 1V 1V
.PRINT DC I(R1) I(R2); for y12 and y22
.END
```

Since V_2 was chosen as a 1-V source, I_1 has the same value as y_{12}, and I_2 has the same value as y_{22}. Run the analysis and verify that

$$y_{12} = -47.62 \text{ mS} \quad \text{and} \quad y_{22} = 142.9 \text{ mS}$$

It should be noted that $y_{12} = y_{21}$, since this is a bilateral network.

If the negative value for $y_{12} = y_{21}$ is a cause for concern, remember that the y parameters do not in themselves represent physical elements. However, it is easily shown that the π network of Fig. 13.4 is equivalent to the network contain-

Fig. 13.4 *y*-parameter equivalent circuit.

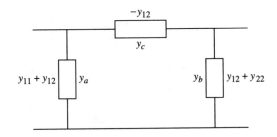

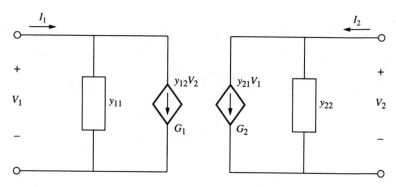

Fig. 13.5 *y*-parameter equivalent circuit with dependent sources.

ing the *y* parameters, and it is therefore equivalent to the original network, whatever it might have been. In our example

$$y_a = y_{11} + y_{12} = 23.81 \text{ mS}$$
$$y_b = y_{22} + y_{12} = 95.28 \text{ mS}$$
$$y_c = -y_{12} = 47.62 \text{ mS}$$

If these are converted to *z* values (actually *R* values in this case) by taking reciprocals, we have $z_a = 42 \ \Omega$, $z_b = 10.5 \ \Omega$, and $z_c = 21 \ \Omega$.

Another equivalent network involving the *y* parameters is shown in Fig. 13.5. This circuit employs two voltage-dependent current sources and follows directly from the original equations that define the *y* parameters. Recall that the SPICE symbol for this type of source begins with the letter *G*.

Using the y Parameters to Solve a Circuit

It is difficult to see what practical use we can find for the *y* parameters in a typical situation like that shown in Fig. 13.6. The two-port network is one for which the *y* parameters have been found, but now a voltage source and a load resistance have been added. The original equations that define the *y* parameters cannot be used directly to find the load voltage V_2 and the load current $-I_2$, since these equations would be ideally suited for the situation when V_1 and V_2 were known. Analysis shows that

$$\frac{V_2}{V_s} = \frac{-y_{21}G_s}{(y_{11} + G_s)(y_{22} + G_L) - y_{12}y_{21}}$$

Fig. 13.6 Practical circuit with source and load.

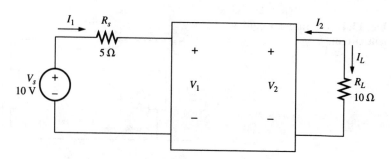

Fig. 13.7 Original T circuit with source and load.

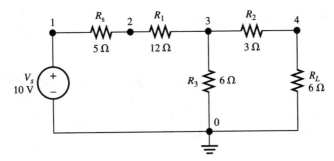

where $G_s = 1/R_s$ and $G_L = 1/R_L$. Use the y parameters from the example given above along with $V_s = 10$ V, $R_s = 5\ \Omega$, and $R_L = 10\ \Omega$ to find V_2/V_s. Verify that the result is $V_2/V_s = 0.1496$.

This can be checked by running a PSpice analysis on the original circuit, shown in Fig. 13.7. As an exercise find the ratio V(4)/Vs = 0.1496 using PSpice. Prepare an input file using the entire circuit shown in this figure.

y PARAMETERS OF NETWORK WITH DEPENDENT SOURCE

The previous example was simple enough to solve by ordinary circuit analysis, but when networks become more complicated, PSpice can be used to advantage. In the next example a dependent current source is part of the network, which is shown in Fig. 13.8. In order to find y_{11} and y_{21}, the output will be short-circuited, but a 0-V source will be included in the circuit to allow for a measurement of I_2. See Fig. 13.9 for the modified circuit. The input file is

```
Input and Transfer Admittances with Dependent Source
V1 1 0 1V
F 3 2 V1 -3
V0 0 3 0V
R1 1 2 4
R2 2 0 2
R3 2 3 2
.DC V1 1V 1V 1V
.PRINT DC I(R1) I(V0); to find I1 and I2
.END
```

Fig. 13.8 Network with dependent source.

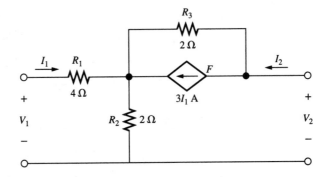

Fig. 13.9 Network with output
shorted.

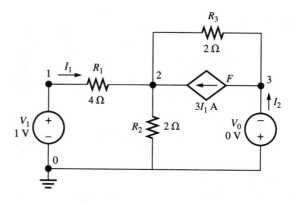

The input current I_1 will be the current through R_1, and the output current I_2 will be the current through V_0. Run the analysis and verify that

$$I(R1) = 125 \text{ mA} \quad \text{and} \quad I(V0) = 125 \text{ mA}$$

which, because of the 1-V source voltage, means that

$$y_{11} = 125 \text{ mS} \quad \text{and} \quad y_{21} = 125 \text{ mS}$$

In order to find y_{12} and y_{22}, a voltage source V_2 is connected to the output terminals, and the input terminals are shorted through a 0-V source as shown in Fig. 13.10. The input file is

```
Output and Transfer Admittances with Dependent Source
V2 2 0 1V
F 2 1 V0 -3
V0 1a 0 0V
R1 1a 1 4
R2 1 0 2
R3 1 2 2
.DC V2 1V 1V 1V
.PRINT DC I(R1) I(V2); for currents I1 and I2
.END
```

Fig. 13.10 Network with input
shorted.

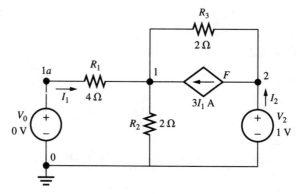

Run the analysis and verify these results:

$$I(R1) = -62.5 \text{ mA} \quad \text{and} \quad I(V2) = -187.5 \text{ mA}$$

This means that $y_{12} = -62.5$ mS and $y_{22} = 187.5$ mS.

Note that y_{22} is positive although $I(V2)$ is negative. From the circuit diagram you should be able to discover the reason for this.

THE OPEN-CIRCUIT IMPEDANCE PARAMETERS

When the currents in a two-port network are the independent variables, we may write the following equations:

$$V_1 = z_{11}I_1 + z_{12}I_2$$
$$V_2 = z_{21}I_1 + z_{22}I_2$$

from which we obtain

$$z_{11} = \left.\frac{V_1}{I_1}\right|_{I_2=0}$$

$$z_{21} = \left.\frac{V_2}{I_1}\right|_{I_2=0}$$

$$z_{12} = \left.\frac{V_1}{I_2}\right|_{I_1=0}$$

$$z_{22} = \left.\frac{V_2}{I_2}\right|_{I_1=0}$$

The simple π network of Fig. 13.11 will be used to introduce the method of solution using PSpice. The first analysis, which sets $I_2 = 0$, is used to find z_{11} and z_{21}. A source current $I_1 = 1$ A is applied to the input terminals with the output open-circuited. The input file is

```
Finding Open-Circuit Impedance Parameters z11 and z21
I1 0 1 1A
R1 1 0 42
R2 1 2 21
R3 2 0 10.5
.TF V(2) I1
.END
```

Fig. 13.11 π network.

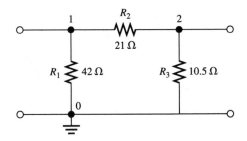

Fig. 13.12 z-parameter equivalent circuit.

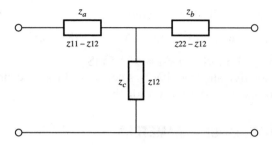

Run the analysis and obtain the following:

INPUT RESISTANCE AT I1 = 18 Ω

This gives $z_{11} = 18\ \Omega$. The output file shows V(2) = 6 V. Since the input current is 1 A, this is the same value as the transfer impedance; thus $z_{21} = 6\ \Omega$. The output file also shows

OUTPUT RESISTANCE AT V(2) = 9 Ω

Since the input source is an independent current source, PSpice considers this as an open circuit when computing the output resistance. Therefore $z_{22} = 9\ \Omega$.

The only z parameter remaining to be found is z_{12}. Since this is a bilateral network, $z_{12} = z_{21} = 6\ \Omega$.

As an exercise, verify the values of z_{12} and z_{22} by using a 1-A source as I_2 at the output terminals, while the input terminals are open-circuited.

The z parameters do not in themselves represent physical elements of an equivalent circuit. However, it is easily shown that the T network of Fig. 13.12 is equivalent to the z-parameter network and thereby equivalent to the original two-terminal network. In the present example

$$z_a = z_{11} - z_{12} = 12\ \Omega$$
$$z_b = z_{22} - z_{12} = 3\ \Omega$$
$$z_c = z_{12} = 6\ \Omega$$

Another circuit may be used to represent the equivalent of the z-parameter network. This circuit, which involves two current-dependent voltage sources, is shown in Fig. 13.13.

Fig. 13.13 z-parameter equivalent circuit with dependent sources.

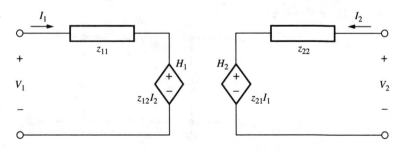

THE z PARAMETERS FOR AC CIRCUIT

The z parameters for an ac circuit, such as the one shown in Fig. 13.14, can be found using PSpice. We will find these open-circuit parameters for this circuit at a frequency $f = 500$ Hz. It is convenient to use a source current of 1 A at zero degrees to drive the circuit. The input file is

```
Find z parameters for ac circuit
I1 0 1 ac 1A
R1 1 3 20
R2 4 2 10
R3 3 0 50
L1 1 4 6.366mH
C1 3 2 12.73uF
C2 3 0 3.183uF
.ac lin 1 500Hz 500Hz
.PRINT ac v(1) vp(1) v(2) vp(2)
.END
```

Run the analysis and see that the results are

$$V(1) = 5.199E{+}01, \ VP(1) = -2.523E{+}01, \ V(2) = 5.600E{+}01,$$
$$VP(2) = -4.030E{+}01$$

from which $z_{11} = 52\underline{/-25.23°}\ \Omega$ and $z_{21} = 56\underline{/-40.3°}\ \Omega$.

Finding the other z parameters involves using a 1-A source current I_2 at the output terminals. The input file will not be shown, since it is similar to the previous one, but you should run the analysis and verify that

$$V(1) = 5.600E{+}01, \ VP(1) = -4.030E{+}01, \ V(2) = 7.325E{+}01,$$
$$VP(2) = -3.463E{+}01$$

from which $z_{12} = 56\underline{/-40.3°}\ \Omega$ and $z_{22} = 73.25\underline{/-34.63°}\ \Omega$.

Since only linear elements are used, this is a bilateral circuit, and $z_{12} = z_{21}$.

Fig. 13.14 ac network.

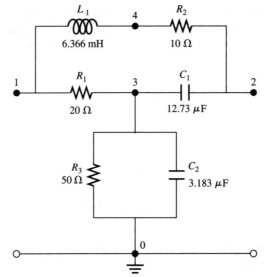

Fig. 13.15 Network with source and load.

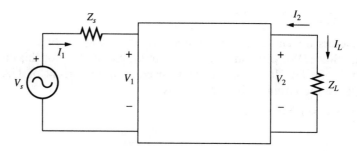

Using the z Parameters to Solve a Circuit

The typical circuit has a source with internal impedance at the sending end and a load impedance at the receiving end as shown in Fig. 13.15. It can be shown that

$$\frac{V_2}{V_s} = \frac{z_{21}Z_L}{(z_{11} + Z_s)(z_{22} + Z_L) - z_{12}z_{21}}$$

Several of the problems at the end of this chapter are related to the use of this and similar equations.

THE *ABCD* PARAMETERS

Another set of parameters, one that is widely used in power-system analysis, is the *ABCD* parameters. These are based on the equations

$$V_1 = AV_2 - BI_2$$
$$I_1 = CV_2 - DI_2$$

The minus signs are used to make these equations consistent with those of other two-port parameters, where I_2 is directed inward rather than toward the load impedance. From the defining equations it is seen that

$$A = \left.\frac{V_1}{V_2}\right|_{I_2=0}$$

$$C = \left.\frac{I_1}{V_2}\right|_{I_2=0}$$

$$B = -\left.\frac{V_1}{I_2}\right|_{V_2=0}$$

$$D = -\left.\frac{I_1}{I_2}\right|_{V_2=0}$$

Thus it is seen that *A* and *C* are open-circuit parameters while *B* and *D* are short-circuit parameters.

Fig. 13.16 *T*-section representation of transmission line.

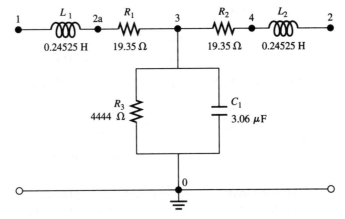

A power transmission line is often represented as a *T* section, such as that
shown in Fig. 13.16, where the series impedance is represented by resistance and
inductance and the shunt impedance is shown as resistance and capacitance. In
order to find *A* and *C*, the output will be open-circuited and the input will have a
source voltage of 1 V at 60 Hz applied. The input file is

```
V1 1 0 ac 1V
L1 1 2a 0.24525H
R1 2a 3 19.35
R3 3 0 4444
C1 3 0 3.06uF
.ac LIN 1 60Hz 60Hz
.PRINT ac v(3) vp(3) i(R1) ip(R1)
.END
```

The elements R_2 and L_2 do not appear in this file, since they are hanging
without a return path for current. Thus the voltage V_3 will be the same as the
desired voltage V_2. Run the analysis which should give

$$V(3) = 1.113E+00, \quad VP(3) = -2.750E+00,$$
$$I(R1) = 1.308E-03, \quad IP(R1) = 7.621E+01$$

Since *A* is found from the ratio V_1/V_2, use a calculator to find $A = 0.8985\underline{/2.75°}$.
The parameter *C* is found from the ratio I_1/V_2. This value is $1.175\underline{/78.95°}$ mS.

The parameters *B* and *D* are found by short-circuiting the output. Assuming
that a 1-V source is used at the input terminals, the input file is

```
Circuit to find B and D parameters
V1 1 0 ac 1V
L1 1 2a 0.24525H
R1 2a 3 19.35
R2 4 3 19.35
L2 0 4 0.24525H
R3 3 0 4444
C1 3 0 3.06uF
.ac LIN 1 60Hz 60Hz
.PRINT ac i(R2) ip(R2) i(R1) ip(R1)
.END
```

The output file gives

$$I(R2) = 5.577E{-}03, \quad IP(R2) = 1.005E{+}01,$$
$$I(R1) = 5.012E{-}03, \quad IP(R1) = -7.673E{+}01$$

Under the short-circuit conditions, B and D are found using a calculator and are

$$B = -\frac{V_1}{I_2} = 179.3\underline{/79.5^\circ}\ \Omega$$

$$D = -\frac{I_1}{I_2} = 0.8987\underline{/2.77^\circ}$$

It is seen that A and D are equal. This will always be the case when linear elements are involved and the network contains no sources.

The defining equations are used directly to find the voltage and current at the sending end of a transmission line when conditions at the load are known. Problems at the end of the chapter illustrate this method.

When conditions at the sending end of a transmission line are known, the defining equations are more useful when rearranged to solve for V_2 and I_2. It is easily shown that

$$V_2 = \frac{DV_1 - BI_1}{AD - BC}$$

$$I_2 = \frac{CV_1 - AI_1}{AD - BC}$$

It can also be shown that

$$AD - BC = 1$$

giving the final form of the equations for receiving-end voltage and current as

$$V_2 = DV_1 - BI_1$$
$$I_2 = CV_1 - AI_1$$

THE HYBRID PARAMETERS

When the input current and the output voltage are chosen as the independent variables, the two-port equations are

$$V_1 = h_{11}I_1 + h_{12}V_2$$
$$I_2 = h_{21}I_1 + h_{22}V_2$$

It is because of the mix (current and voltage) of independent variables that the hybrid parameters get their name. These are the familiar parameters that are often used to characterize bipolar-junction transistors. Although it is possible to find the h parameters for various dc networks and ac networks, they are used only to a

Fig. 13.17 *h*-parameter equivalent circuit.

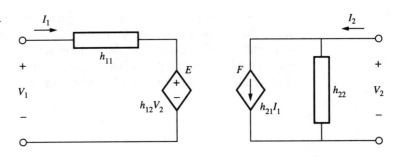

limited extent except with BJTs. From the defining equations we have

$$h_{11} = \left.\frac{V_1}{I_1}\right|_{V_2=0}$$

$$h_{21} = \left.\frac{I_2}{I_1}\right|_{V_2=0}$$

$$h_{12} = \left.\frac{V_1}{V_2}\right|_{I_1=0}$$

$$h_{22} = \left.\frac{I_2}{V_2}\right|_{I_1=0}$$

Since the *h*-parameter values for the BJT are covered in the chapter on transistor circuits, no more examples will be given here. In terms of the double-subscript notation, compare Fig. 13.17 with Fig. 3.5.

ANOTHER SET OF HYBRID PARAMETERS

Although not called hybrid parameters (to avoid confusion with the *h* parameters), another set of parameters is defined by allowing the input voltage and the output current to be the independent variables. Thus

$$I_1 = g_{11}V_1 + g_{12}I_2$$
$$V_2 = g_{21}V_1 + g_{22}I_2$$

Fig. 13.18 *g*-parameter equivalent circuit.

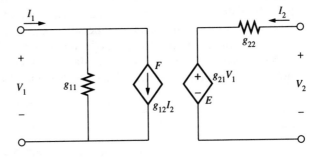

These are called the g parameters, but g does not mean that they are all conductances. In a dc network, only g_{11} represents a conductance. This is easily verified by noting the dimensions of the various terms in the defining equations. The general equivalent circuit involving the g parameters is easily obtained from these equations and is shown in Fig. 13.18 on page 393. Also it is seen that

$$g_{11} = \left.\frac{I_1}{V_1}\right|_{I_2=0}$$

$$g_{21} = \left.\frac{V_2}{V_1}\right|_{I_2=0}$$

$$g_{12} = \left.\frac{I_1}{I_2}\right|_{V_1=0}$$

$$g_{22} = \left.\frac{V_2}{I_2}\right|_{V_1=0}$$

TRANSMISSION LINES

Although there is a PSpice device called T (for transmission line), it is of limited use because it does not take line losses into account. We would prefer to use a model for the transmission line which does include losses, and involves R, L, G, and C.

A Long Telephone Line

A certain telephone line uses 104-mil-diameter copper conductors that are spaced 18 inches apart on insulators. The measured parameters for the line are given in terms of *per-loop-mile*. They are

$$R = 10.15 \ \Omega$$
$$L = 3.93 \ \text{mH}$$
$$G = 0.29 \ \mu\text{S}$$
$$C = 0.00797 \ \mu\text{F}$$

The line is 200 miles long. When operated at $\omega = 5000$ rad/s, we would like to see how the voltage and current attenuate as a function of line length when the line is terminated in its characteristic impedance Z_o.

The characteristic impedance is found as $Z_o = \sqrt{z/y}$, where $z = R + j\omega L$ and $y = G + j\omega C$. The propagation constant is found as $\gamma = \sqrt{zy} = \alpha + j\beta$. At the prescribed angular frequency these are calculated to be

$$Z_o = 445\underline{/-13.45°} \ \Omega = (724.567 - j173.285) \ \Omega$$

and

$$\gamma = 0.0297\underline{/76.13°} = 0.00712 + j0.0288$$

Fig. 13.19 Portion of a long telephone line.

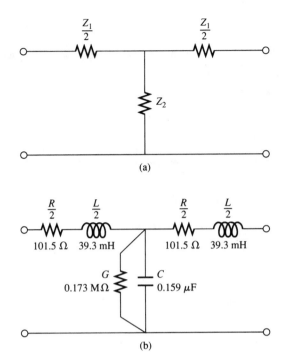

(a)

(b)

Our approach to the solution will be to decide on a reasonable segment length of line which can be represented by lumped parameters and to use the elements of this segment as a model in a subcircuit. Since the line is 200 miles long, we will choose 20 miles as the segment length, representing this segment as a lumped-parameter T section. The per-mile values for R, L, G, and C are to be multiplied by 20, and the resulting values for R and L will then be split in half for each portion of the T. The results are shown in Fig. 13.19 and are incorporated into a subcircuit. Verify the elements shown in subcircuit TLINE.

The transmission line will be fed from a 1-V source. Small sensing resistors will be included at the connections between segments of the transmission line. This will allow for voltage and current measurements to be made at these points. The line is terminated in Z_o, where the value of -173.285 Ω (capacitive reactance) is shown as 1.154 μF, based on the given frequency. Refer to Fig. 13.20 for the node designations. The input file is

```
Transmission-Line Representation
V 1 0 AC 1V
R1 1 2 0.01
R2 3 4 0.01
R3 5 6 0.01
R4 7 8 0.01
R5 9 10 0.01
R6 11 12 0.01
R7 13 14 0.01
R8 15 16 0.01
R9 17 18 0.01
```

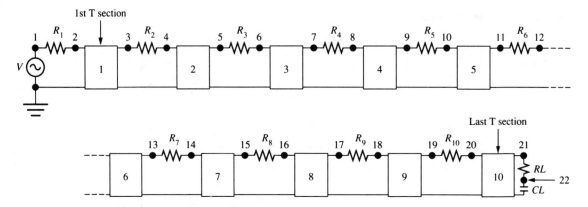

Fig. 13.20   Telephone line consisting of ten *T* sections, each representing a 20-mile length.

```
R10 19 20 0.01
RL 21 22 724.567
CL 22 0 1.154uF
X1 2 0 3 TLINE
X2 4 0 5 TLINE
X3 6 0 7 TLINE
X4 8 0 9 TLINE
X5 10 0 11 TLINE
X6 12 0 13 TLINE
X7 14 0 15 TLINE
X8 16 0 17 TLINE
X9 18 0 19 TLINE
X10 20 0 21 TLINE
.subckt TLINE 1 2 6
 R 1 3 101.5
 R1 4 5 101.5
 L 3 4 39.3mH
 L1 5 6 39.3mH
 Rs 4 2 0.172Meg
 C 4 2 0.159uF
.ends
.AC LIN 1 795.8Hz 795.8Hz
.PRINT AC I(R1) I(R2) I(R3) I(R4) I(R5) I(R6) I(R7) I(R8) I(R9) I(R10) I(RL)
.PRINT AC V(2) V(4) V(6) V(8) V(10) V(12) V(14) V(16) V(18) V(20) V(21)
.END
```

Run the analysis and from the output file verify that the sending-end current is 1.392 mA, the receiving-end current is 0.3104 mA, and the receiving-end voltage is 0.2312 V. Plot the current and voltage as functions of positions down the length of the transmission line. An exponential decay of these quantities will be evident.

You may also observe the phase shift that occurs with position along the line. Simply run the analysis and print IP(R1), IP(R2), and so forth. Or print VP(2), VP(4), and so forth. Verify that VP(4) = −33.3°. This represents a phase shift in a 20-mile segment of the telephone line and corresponds to 1.665° per mile. From $\beta = 0.0288$ rad/mi, you see that these values are in close agreement. Figure 13.21 shows the output file which includes the currents, voltages, and phases of the voltages.

```
   Transmission-Line Representation
V 1 0 AC 1V
R1 1 2 0.01
R2 3 4 0.01
R3 5 6 0.01
R4 7 8 0.01
R5 9 10 0.01
R6 11 12 0.01
R7 13 14 0.01
R8 15 16 0.01
R9 17 18 0.01
R10 19 20 0.01
RL 21 22 724.567
CL 22 0 1.154uF
X1 2 0 3 TLINE
X2 4 0 5 TLINE
X3 6 0 7 TLINE
X4 8 0 9 TLINE
X5 10 0 11 TLINE
X6 12 0 13 TLINE
X7 14 0 15 TLINE
X8 16 0 17 TLINE
X9 18 0 19 TLINE
X10 20 0 21 TLINE
.subckt TLINE 1 2 6
  R 1 3 101.5
  R1 4 5 101.5
  L 3 4 39.3mH
  L1 5 6 39.3mH
  Rs 4 2 0.172Meg
  C 4 2 0.159uF
.ends
.AC LIN 1 795.8Hz 795.8Hz
.PRINT AC I(R1) I(R2) I(R3) I(R4) I(R5) I(R6) I(R7) I(R8) I(R9) I(R10)
I(RL)
.PRINT AC V(2) V(4) V(6) V(8) V(10) V(12) V(14) V(16) V(18) V(20) V(21)
.PRINT AC VP(2) VP(4) VP(6) VP(8) VP(10) VP(12) VP(14) VP(16) VP(18 VP(20)
VP(21)
.END

      FREQ          I(R1)        I(R2)        I(R3)        I(R4)        I(R5)
      7.958E+02     1.392E-03    1.202E-03    1.038E-03    8.953E-04    7.693E-04
      FREQ          I(R6)        I(R7)        I(R8)        I(R9)        I(R10)
      7.958E+02     6.608E-04    5.709E-04    4.967E-04    4.308E-04    3.678E-04
      FREQ          I(RL)
      7.958E+02     3.104E-04
      FREQ          V(2)         V(4)         V(6)         V(8)         V(10)
      7.958E+02     1.000E+00    8.613E-01    7.412E-01    6.390E-01    5.528E-01

      FREQ          V(12)        V(14)        V(16)        V(18)        V(20)
      7.958E+02     4.784E-01    4.117E-01    3.518E-01    3.015E-01    2.626E-01
      FREQ          V(21)
      7.958E+02     2.312E-01
      FREQ          VP(2)        VP(4)        VP(6)        VP(8)        VP(10)
      7.958E+02     -1.676E-04   -3.330E+01   -6.671E+01   -1.002E+02   -1.337E+02
      FREQ          VP(12)       VP(14)       VP(16)       VP(18)       VP(20)
      7.958E+02     -1.669E+02   1.601E+02    1.268E+02    9.283E+01    5.873E+01
      FREQ          VP(21)
      7.958E+02     2.568E+01
```

Fig. 13.21

CONSTANT-*k* FILTER

The constant-*k* filter is ideally made up of pure reactances. In its simplest form, it might represent either a low-pass or a high-pass filter. A low-pass filter as a *T* section is shown in Fig. 13.22. The elements chosen for this example are $L = 0.04$ H and $C = 0.1\ \mu$F. Such a filter is usually terminated in its characteristic impedance as given by

$$Z_{oT} = \sqrt{Z_1 Z_2 + \frac{Z_1^2}{4}}$$

where $Z_1 = j\omega L$ and $Z_2 = 1/(j\omega C)$.

When a frequency of $f = 1592$ Hz is chosen, Z_{oT} is calculated to be 600 Ω as a pure resistance. In Fig. 13.22, a voltage source with an internal resistance $R = 0.01$ Ω is connected on the left, and the value of load resistance is $R_L = 600$ Ω. The PSpice analysis will find input and output currents and voltages. The input file is

```
Constant-k Filter
V 1 0 AC 1V
L 2 3 0.02H
L1 3 4 0.02H
C 3 0 0.1uF
R 1 2 0.01
RL 4 0 600
.AC LIN 1 1592Hz 1592Hz
.PRINT AC I(R) I(RL) I(C) V(2) V(3) V(4) VP(2) VP(4)
.END
```

Run the analysis and obtain a printed copy of the output file. In interpreting the results, recall that at the frequency $f = 1592$ Hz, the characteristic impedance is resistance. This frequency is in the pass band of frequencies, where there will be no attenuation ($\alpha = 0$). This simply means that the input and output currents should be the same. Verify that the input current I(R) and the output current I(RL)

Fig. 13.22 Constant-*k*, low-pass filter.

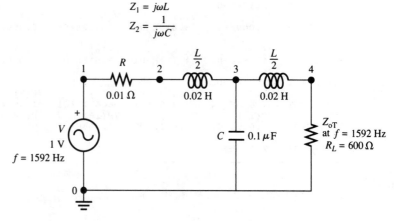

are both 1.667 mA. The phase shift for the filter is given by

$$\beta = 2 \tan^{-1}\left(\frac{\sqrt{A}}{\sqrt{1-A}}\right)$$

where

$$A = \left|\frac{Z_1}{4Z_2}\right|$$

The calculated value for the phase shift is 36.88°, which is in agreement with the PSpice results showing VP(4) = −36.88°.

The low-pass filter has a cutoff frequency given by

$$f_c = \frac{1}{\pi\sqrt{LC}}$$

which for the elements chosen gives $f_c = 5033$ Hz. The MathCAD calculations for the constant-*k* filter in the pass band are given in Fig. 13.23.

Stop-Band Behavior of the Constant-*k* Filter

We continue the example for the low-pass filter. Choosing $f = 6$ kHz will mean that this frequency should not be passed without attenuation. To keep the filter

```
MathCAD solution of Constant-k filter for f = 1592 Hz
This is a low-pass filter with a cut-off fc = 5033 Hz

L := 0.04    C := 0.1·10^{-6}    j := √-1    f := 1592

  The cut-off frequency is       fc := ────────       fc = 5.033·10^3
                                        π·√(L·C)

  ω := 2·π·f    Z1 := j·ω·L    Z1 = 400.113i

                               Z2 := ────        Z2 = -999.717i
                                     j·ω·C

  ZoT := √(Z1·Z2 + Z1²/4)    ZoT = 599.981       A := Z1/(4·Z2)    A = -0.1

  a := |A|    a = 0.1

  β := 2·atan[√a/√(1-a)]    β = 0.644    b := β·(180/π)    b = 36.881
```

Fig. 13.23

properly terminated, the value of Z_{oT} at 6 kHz must be calculated. This is found to be $Z_{oT} = j410.47\ \Omega$. This gives an inductance value for the load of $L = 10.888$ mH.

Now look at filter response for $f = 6$ kHz, which is in the stop band of frequencies. The modified input file is

```
Const.    Filter, Stop-Band
V 1 0 AC 1V
L 2 3 0.02H
L1 3 4 0.02H
C 3 0 0.1uF
R 1 2 0.01
LL 4 0 10.888mH
.AC LIN 1 6000Hz 6000Hz
.PRINT AC I(R) I(LL) I(C) V(2) V(3) V(4) VP(2) VP(4)
.END
```

Run the analysis and obtain a copy of the output file. Verify that the input current $I(R) = 2.436$ mA and that the output current $I(LL) = 0.7187$ mA. The phase shift β is found as $VP(4) = -180°$. The propagation constant is given by

$$\gamma = \ln\left(\frac{I_{in}}{I_{out}}\right) = \alpha + j\beta$$

From our results, $\gamma = \ln(3.3895\underline{/180°})$. The value of α is found as the log of the magnitude of γ. This gives $\alpha = 1.22$ nepers. The formula is

$$\alpha = 2\ln(\sqrt{B-1} + \sqrt{B})$$

where

$$B = \left|\frac{Z_1}{4Z_2}\right|$$

This gives a calculated value of $\alpha = 1.22$ nepers, in agreement with the PSpice results. The neper is a fundamental measure of attenuation and corresponds to a current ratio of 2.71728. If a conversion is appropriate, 1 neper = 8.686 dB. Figure 13.24 shows the output files for both the pass-band and the stop-band examples. The MathCAD calculations for the stop band are given in Fig. 13.25.

Lossless Transmission Line

The constant-k filter can also serve as a useful model for a lossless transmission line. Figure 13.26 shows a segment of such a line, using $L = 2$ mH and $C = 50$ nF. Assume that the model represents an actual length of line equal to 1 m. The values of L and C then become series inductance per meter and shunt capacitance per meter, respectively. The cutoff frequency for this line is readily found and is $f_c = 31.8$ kHz. The analysis will involve $f = 10$ kHz, a frequency that is well within the pass band.

In order to properly terminate the line, you must solve for Z_{oT}. Verify that $Z_{oT} = 189.874\underline{/0°}\ \Omega$. Using the method described in the previous section, you can find the phase shift β of a section of this line. Verify that $\beta = 36.62°$. Since the elements were designated as representing 1 m of line length, the phase shift represents 36.62°/m. Figure 13.27 on page 403 shows the MathCAD calculations for this constant-k filter.

```
     Constant-k Filter, Pass Band Frequency

****        CIRCUIT DESCRIPTION

V 1 0 AC 1V
L 2 3 0.02H
L1 3 4 0.02H
C 3 0 0.1uF
R 1 2 0.01
RL 4 0 600
.AC LIN 1 1592Hz 1592Hz
.PRINT AC I(R) I(RL) I(C) V(2) V(3) V(4) VP(2) VP(4)
.END

   FREQ        I(R)        I(RL)        I(C)         V(2)          V(3)

   1.592E+03   1.667E-03   1.667E-03   1.054E-03   1.000E+00   1.054E+00

   FREQ        V(4)        VP(2)        VP(4)

   1.592E+03   1.000E+00   -2.884E-08   -3.688E+01



     Constant-k Filter; Stop-band Frequency

   ****        CIRCUIT DESCRIPTION

V 1 0 AC 1V
L 2 3 0.02H
L1 3 4 0.02H
C 3 0 0.1uF
R 1 2 0.01
LL 4 0 10.8875mH
.AC LIN 1 6000Hz 6000Hz
.PRINT AC I(R) I(LL) I(C) V(2) V(3) V(4) VP(2) VP(4)
.END

   FREQ        I(R)        I(LL)        I(C)         V(2)          V(3)

   6.000E+03   2.436E-03   7.187E-04   3.155E-03   1.000E+00   8.369E-01

   FREQ        V(4)        VP(2)        VP(4)

   6.000E+03   2.950E-01   1.396E-03   -1.800E+02
```

Fig. 13.24

For a preliminary investigation, create an input file to verify some of the findings. The input file is

```
Transmission Line as Lumped Elements
v 1 0 sin(0 1 10kHz)
L 1 2 1mH
L1 2 3 1mH
C 2 0 50nF
R 3 0 189.874
.tran 1us 100us
.probe
.end
```

MathCAD solution of Constant-k filter for f = 6000 Hz
This is a low-pass filter with a cut-off fc = 5033 Hz

$L := 0.04$ $C := 0.1 \cdot 10^{-6}$ $j := \sqrt{-1}$ $f := 6000$

The cut-off frequency is $fc := \dfrac{1}{\pi \cdot \sqrt{L \cdot C}}$ $fc = 5.033 \cdot 10^{3}$

$\omega := 2 \cdot \pi \cdot f$ $Z1 := j \cdot \omega \cdot L$ $Z1 = 1.508 \cdot 10^{3} \, i$

$Z2 := \dfrac{1}{j \cdot \omega \cdot C}$ $Z2 = -265.258i$

$ZoT := \sqrt{Z1 \cdot Z2 + \dfrac{Z1^{2}}{4}}$ $ZoT = 410.474i$

$B := \dfrac{Z1}{4 \cdot Z2}$ $B = -1.421$ $b := |B|$

$b = 1.421$

$\propto := 2 \cdot \ln\left[\sqrt{b-1} + \sqrt{b}\right]$ $\propto = 1.221$

Fig. 13.25

Fig. 13.26 Section of a loss-less-transmission line.

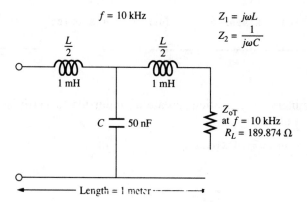

$f = 10 \text{ kHz}$

$Z_1 = j\omega L$

$Z_2 = \dfrac{1}{j\omega C}$

$\dfrac{L}{2}$ $\dfrac{L}{2}$

1 mH   1 mH

$C \rightleftharpoons 50 \text{ nF}$

Z_{oT}
at $f = 10$ kHz
$R_L = 189.874 \, \Omega$

Length = 1 meter

MathCAD solution of Lossless Transmission Line Represented as a series of T sections. This is a constant-k filter representation with a cut-off frequency of 31.83 kHz.

$L := 0.002$ $C := 50 \cdot 10^{-9}$ $j := \sqrt{-1}$ $f := 10000$

The cut-off frequency is $fc := \dfrac{1}{\pi \cdot \sqrt{L \cdot C}}$ $fc = 3.183 \cdot 10^4$

$\omega := 2 \cdot \pi \cdot f$ $Z1 := j \cdot \omega \cdot L$ $Z1 = 125.664i$ $Z2 := \dfrac{1}{j \cdot \omega \cdot C}$ $Z2 = -318.31i$

$ZoT := \sqrt{Z1 \cdot Z2 + \dfrac{Z1^2}{4}}$ $ZoT = 189.874$ $A := \dfrac{Z1}{4 \cdot Z2}$ $A = -0.099$

$a := |A|$ $a = 0.099$

$\beta := 2 \cdot atan\left[\dfrac{\sqrt{a}}{\sqrt{1 - a}}\right]$ $\beta = 0.639$ radian

$b := \beta \cdot \dfrac{180}{\pi}$ $b = 36.62$ degrees

Fig. 13.27

Run the analysis and in Probe plot v(1) and v(3). Obtain a printed copy of this plot for further study. The synthesized output-voltage wave appears to have a higher amplitude than the input-voltage wave. Move beyond the positive peaks to the negative peaks and verify that the negative peak of v(1) is at 76 μs, while the corresponding peak of v(3) is at 86 μs. Record the peak magnitude of v(3), which is 1.008 V, for future use. What interpretation can be given to the 10-μs time interval between the two waves? The wave length of a transmission line is $\lambda = 360°/\beta$, which for this example is 360/36.62 = 9.83 m. For a 10-kHz frequency, its velocity will be $v = f\lambda = 98.3$ km/s.

There is, of course, a simple relationship between time and distance for a transmission line. In our example, we conclude that a line that is 98.3 km in length is a *one-second* line. That is, it takes a wave 1 second to travel 98.3 km down the line. A 1-μs length of line would be 0.0983 m long. The 10-μs time interval between the waves v(1) and v(3) is equivalent to (10)(0.0983) = 0.983 m. Since the T section represents 1 m of line, the results are in close agreement.

Looking carefully at the plots of v(1) and v(3), the delay of v(3) in beginning its sine-wave form is explained in terms of this time delay of 10 μs. If you sketch v(3) as a true sine wave toward its beginning, you will see that it crosses the X-axis at about the 10 μs point. Probe simply uses a curve fit to produce the graph, thereby obscuring this detail. The traces of v(1) and v(3) are shown in Fig. 13.28.

Remove the plots of voltage and plot i(R) for the line segment. Verify that the peak value (use the negative peak) is 5.3 mA. The magnitude of the load impedance is $v/i = 1.008/0.0053 = 190.2\ \Omega$. (The voltage was recorded earlier.) Since these two waves are exactly in phase, the 190.2 Ω represents a pure resistance. This is in close agreement with the value of $Z_{oT} = 189.874\underline{/0°}\ \Omega$.

The graphs are plotted over a time interval of 100 μs, so that the sinusoidal forms will be more evident, but this line is only 1 m in length, which as you know corresponds to 10 μs.

An even more convincing graph is obtained by removing the current trace and plotting v(3)/i(R). Observe that this is a *flat* trace. In the cursor mode, verify that the characteristic impedance $Z_{oT} = 189.9\ \Omega$.

Now plot together i(R) and v(3)/190. What do you get, and why? See Fig. 13.29 for these traces.

Lossless Line Composed of Several Sections

We may extend the analysis of the constant-k representation of a lossless transmission line to include any number of sections. As an example, let us use five

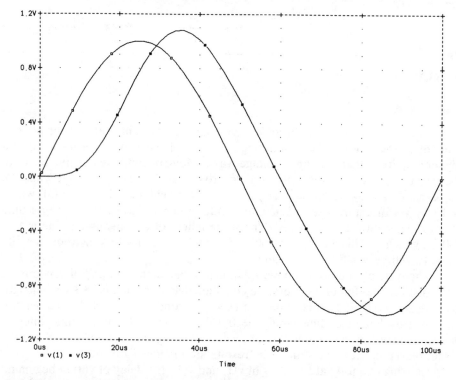

Fig. 13.28 Transmission Line as Lumped Elements.

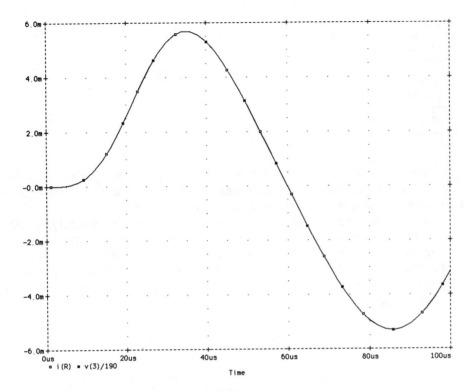

Fig. 13.29 Transmission Line as Lumped Elements.

sections, as shown in Fig. 13.30. At the sending end is a 1-V, 10-kHz source. Between the segments small resistors are inserted, so that we may find currents as well as voltages. The input file is

```
Transmission Line with 5 Sections
v 1 0 sin(0 1 10kHz)
R1 2 3 0.001
R2 4 5 0.001
R3 6 7 0.001
R4 8 9 0.001
RL 10 0 189.874
X1 1 0 2 LC
X2 3 0 4 LC
```

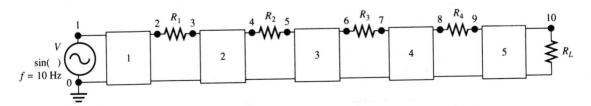

Fig. 13.30 Transmission line with 5 sections.

```
X3 5 0 6 LC
X4 7 0 8 LC
X5 9 0 10 LC
.subckt LC 1 2 3
 L 1 a 1mH
 L1 a 3 1mH
 C a 2 50nF
.ends
.tran 1us 200us
.probe
.end
```

Run the analysis, and in Probe plot v(1), v(3), v(5), v(7), v(9), and v(10). Each wave is shifted from its adjacent wave by the time required for the wave to pass through one section of line. See Fig. 13.31 for these traces.

In order to make measurements, remove these traces and plot only v(1) and v(10). Observe where each curve crosses the X-axis going negative. Verify that for v(1) this is at $t = 50\ \mu s$ and for v(10) this is at $t = 100\ \mu s$. This means that the delay of the entire line is 50 μs. Although the plots are taken for 200 μs and the line is only 50 μs in length, the sinusoidal plots give a clear picture of a traveling wave on the transmission line.

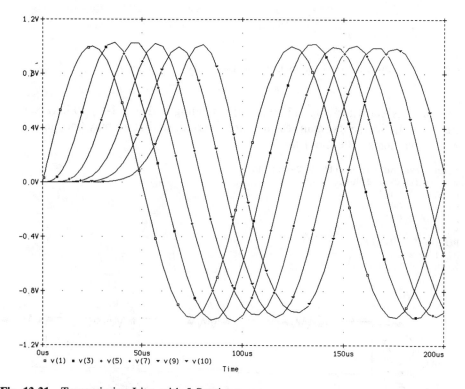

Fig. 13.31 Transmission Line with 5 Sections.

INPUT IMPEDANCE AT POINTS ALONG THE LINE

When a transmission line is properly terminated, the input impedance should be equal to its characteristic impedance regardless of whether the line is one or more sections in length. An ac analysis will readily indicate the results if we call for voltage and current magnitudes and phases. Continuing the previous example, the

```
Transmission Line with 5 Sections

****     CIRCUIT DESCRIPTION
v 1 0 ac 1
R1 2 3 0.001
R2 4 5 0.001
R3 6 7 0.001
R4 8 9 0.001
R 10 0 189.874
X1 1 0 2 LC
X2 3 0 4 LC
X3 5 0 6 LC
X4 7 0 8 LC
X5 9 0 10 LC
.subckt LC 1 2 3
  L 1 a 1mH
  L1 a 3 1mH
  C a 2 50nF
.ends
.ac lin 1 10kHz 10kHz
.print ac v(10 i(R) vp(10) ip(R)
.print ac v(9) i(R4) vp(9) ip(R4)
.print ac v(7) i(R3) vp(7) ip(R3)
.print ac v(5) i(R2) vp(5) ip(R2)
.print ac v(3) i(R1) vp(3) ip(R1)
.end

    FREQ          V(10)         I(R)          VP(10)        IP(R)

    1.000E+04     1.000E+00     5.267E-03     1.769E+02     1.769E+02

    FREQ          V(9)          I(R4)         VP(9)         IP(R4)

    1.000E+04     1.000E+00     5.267E-03     -1.465E+02    -1.465E+02

    FREQ          V(7)          I(R3)         VP(7)         IP(R3)

    1.000E+04     1.000E+00     5.267E-03     -1.099E+02    -1.099E+02

    FREQ          V(5)          I(R2)         VP(5)         IP(R2)

    1.000E+04     1.000E+00     5.267E-03     -7.324E+01    -7.324E+01

    FREQ          V(3)          I(R1)         VP(3)         IP(R1)

    1.000E+04     1.000E+00     5.267E-03     -3.662E+01    -3.662E+01
```

Fig. 13.32

input file is modified to become

```
Transmission Line with 5 Sections
v 1 0 ac 1
R1 2 3 0.001
R2 4 5 0.001
R3 6 7 0.001
R4 8 9 0.001
R 10 0 189.874
X1 1 0 2 LC
X2 3 0 4 LC
X3 5 0 6 LC
X4 7 0 8 LC
X5 9 0 10 LC
.subckt LC 1 2 3
 L 1 a 1mH
 L1 a 3 1mH
 C a 2 50nF
.ends
.ac lin 1 10kHz 10kHz
.print ac v(10) i(R) vp(10) ip(R)
.print ac v(9) i(R4) vp(9) ip(R4)
.print ac v(7) i(R3) vp(7) ip(R3)
.print ac v(5) i(R2) vp(5) ip(R2)
.print ac v(3) i(R1) vp(3) ip(R1)
.end
```

Run the analysis and look at the output file. At the receiving end, V(10) = 1 V and I(R) = 5.267 mA. The two waves are in phase with a 176.92° phase shift. This gives $Z = 189.86\underline{/0°}$ Ω, which is the characteristic impedance. The input impedance at the next section is found from V(9) = 1 V and I(R4) = 5.276 mA along with VP(9) = −146.5° and IP(R4) = −146.5°. The voltage and current magnitudes are the same as before, and since the waves are still in phase, the impedance is again equal to the characteristic impedance. Note that at each section of line the voltage values remain the same in magnitude and the current values are also the same. Also the phase shift between sections is 36.6°. Refer to Fig. 13.32 on page 407 for the output file.

The results clearly indicate a flat line, that is, a line that will have no standing waves. This is typical of transmission lines that are terminated in their characteristic impedance.

BAND-PASS FILTER USING *R* AND *C*

Passive filters containing resistors and capacitors are often used when signal loss is not a problem. The circuit of Fig. 13.33 is a band-pass filter designed for a 5-kHz bandwidth between 15 kHz and 20 kHz using

$$f_{C1} = \frac{1}{2\pi C_1 R_1}$$

$$f_{C2} = \frac{1}{2\pi C_2 R_2}$$

Fig. 13.33 Band-pass filter.

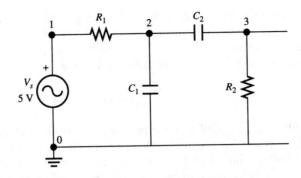

If we choose $C_1 = C_2 = 0.01 \ \mu F$, we may solve these equations for R_1 and R_2, respectively. This gives $R_1 = 1061 \ \Omega$ and $R_2 = 976 \ \Omega$. The input file is

```
Band-Pass Filter
Vs 1 0 ac 5V
R1 2 0 1061
R2 2 3 796
C1 1 2 0.01uF
C2 3 0 0.01uF
.ac DEC 50 100Hz 100kHz
.PROBE
.END
```

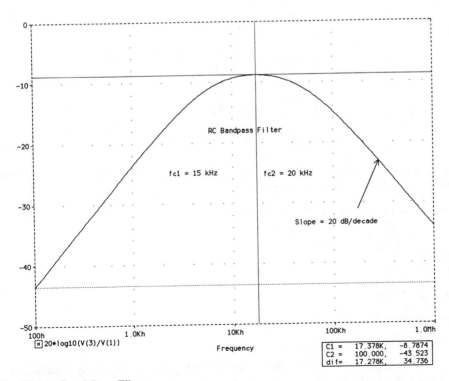

Fig. 13.34 Band-Pass Filter.

Run the Probe analysis to obtain a plot like Fig. 13.34 on page 409. Verify that in the pass band the smallest attenuation of output voltage is 8.784 dB. Note that this occurs at the geometric mean of f_{C1} and f_{C2} at $f = 17.38$ kHz. As an exercise solve for V_3 at this frequency using

$$\text{gain}_{dB} = 20 \log \frac{V_3}{V_1}$$

Verify that $V_3 = 1.81$ V in the middle of the pass band. Also find the attenuation at 100 Hz and 1 kHz to verify that the loss in the stop band is 20 dB/decade.

While you are still running the Probe analysis, remove the previous trace and plot V(3). Here you can easily check your prediction that $V(3)_{max} = 1.81$ V. As an exercise compute the voltage that corresponds to a 3 dB loss from $V(3)_{max}$; then use the cursor to find the frequencies that correspond to these voltages. Restore the original trace to verify your results.

Another Band-Pass Filter

A more elaborate passive filter is shown in Fig. 13.35. Both the series elements and the shunt elements contain capacitance and inductance. Formulas for the elements are found in Ware and Reed, *Communication Circuits*, p. 166, and are shown here for reference:

$$C_2 = \frac{1}{\pi R_o(f_o^2 - f_o^1)}$$

$$L_1 = \frac{R_o}{\pi(f_o^2 - f_o^1)}$$

$$C_1 = \frac{f_o^2 - f_o^1}{4\pi f_o^1 f_o^2 R_o}$$

$$L_2 = C_1 R_o^2 = \frac{R_o(f_o^2 - f_o^1)}{4\pi f_o^1 f_o^2}$$

The filter is matched to a 600-Ω line for a pass band between 1 kHz and 2 kHz. Our input file is designed for use with Probe:

```
Band-pass Filter Using Passive Elements
Vs 1 0 ac 1V
C1 1 2 0.1326uF
C2 4 5 0.1326uF
C3 3 0 0.53uF
L1 2 3 95.5mH
L2 3 4 95.5mH
L3 3 0 23.85mH
RL 5 0 600
.ac DEC 50 100H 10kHz
.PROBE
.END
```

In Probe, plot the attenuation as shown in Fig. 13.36. Since the attenuation is so large outside the pass band, change the Y-axis by setting the range from −50 to

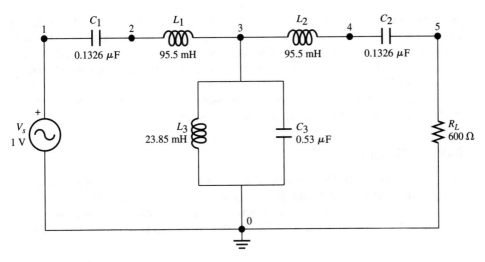

Fig. 13.35 Band-pass filter.

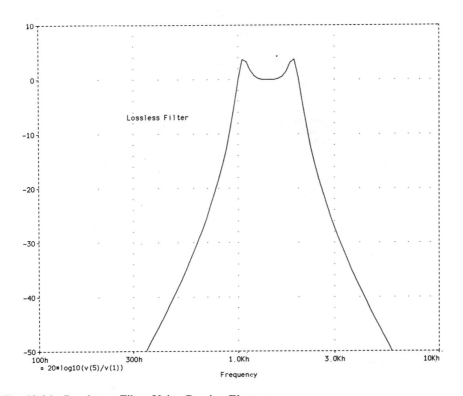

Fig. 13.36 Band-pass Filter Using Passive Elements.

10. Note that there is no attenuation near the center of the pass band and that due to the resonance properties of the circuit, the output voltage rises near the limits of the pass band. As an exercise, find the gain at each of the peaks. Verify that it is 3.89 dB at the first peak and 3.97 dB at the second peak. Also find the attenuation of $f = 2.4$ kHz.

   Practical filter elements, especially inductors, contain an amount of resistance. A problem at the end of the chapter addresses this matter.

BAND-ELIMINATION FILTER

When the T section of the previous study uses parallel inductance and capacitance in the series arms and series elements in the shunt branch, the filter may be designed for band elimination. The design is again based on formulas from Ware and Reed, with an elimination band extending from 2 kHz to 3 kHz. The equations are

$$L_1 = \frac{R_o(f_o^2 - f_o^1)}{\pi f_o^1 f_o^2}$$

$$C_1 = \frac{1}{4\pi R_o(f_o^2 - f_o^1)}$$

$$L_2 = \frac{R_o}{4\pi(f_o^2 - f_o^1)}$$

$$C_2 = \frac{f_o^2 - f_o^1}{\pi f_o^1 f_o^2 R_o}$$

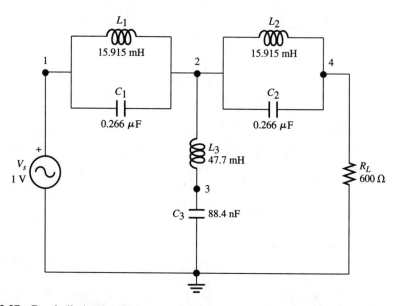

Fig. 13.37 Band-elimination filter.

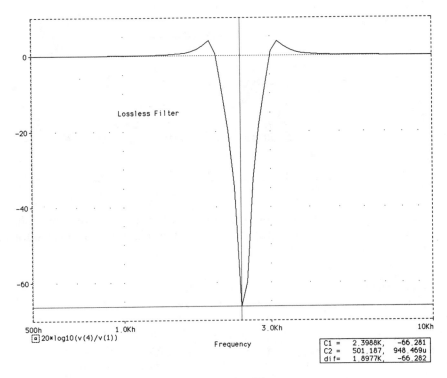

Fig. 13.38 Band-elimination Filter Using Passive Elements.

The elements are shown in Fig. 13.37 and the input file is

```
Band-elimination Filter Using Passive Elements
Vs 1 0 ac 1V
L1 1 2 15.915mH
L2 2 4 15.915mH
C1 1 2 0.266uF
C2 2 4 0.266uF
L3 2 3 47.7mH
C3 3 0 88.4nF
RL 4 0 600
.ac DEC 50 100 10kHz
.PROBE
.END
```

In Probe, obtain the dB plot of the ratio of output voltage to input voltage. Change the X range and Y range as shown in Fig. 13.38. Verify that the maximum attenuation occurs in the stop band at $f = 2.4$ kHz where the attenuation is 66.28 dB.

PROBLEMS

13.1 Use PSpice to find the y parameters of the circuit shown in Fig. 13.39. In this and other problems, plan your work so that a minimum amount of pencil-and-paper calculations will be required.

Fig. 13.39

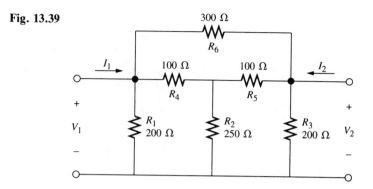

13.2   In Fig. 13.40, $R_s = 50\ \Omega$ and $R_L = 200\ \Omega$. The network in the box is that of Problem 13.1. Use the y parameters found in this problem to solve for the transfer function V_2/V_s.

Fig. 13.40

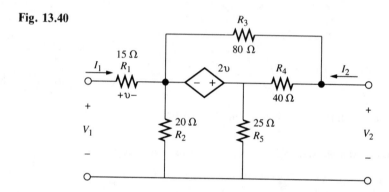

13.3   Use PSpice to find the z parameters of the circuit shown in Fig. 13.39.

13.4   In Fig. 13.6, using $R_s = 50\ \Omega$ and $R_L = 200\ \Omega$ and the z parameters of Problem 13.3, find the transfer function V_2/V_s. Compare the results to your answer to Problem 13.2.

13.5   Use PSpice to find the y parameters of the circuit shown in Fig. 13.40. Note that the circuit contains a dependent-voltage source.

13.6   The circuit in Fig. 13.41 is the π representation of a transmission line. Find its $ABCD$ parameters at $f = 60$ Hz using PSpice.

Fig. 13.41

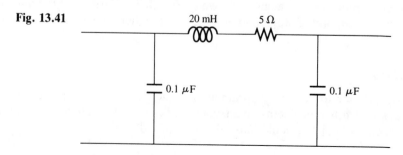

13.7 When the π circuit of Problem 13.6 is terminated with $Z_L = (20 + j20)\ \Omega$, the receiving-end current is $I_L = 3.89\underline{/-45°}$ A. Use the *ABCD* parameters to find V_1 and I_1, the voltage and current at the sending end.

13.8 In the discussion of a long telephone line, the lumped parameters were shown as a *T* section. An alternate representation is the π section as shown in Fig. 13.42. Using the same values as those given in the text example, devise a subcircuit for the 20-mile segment based on the π-section representation. Modify the input file and run the analysis. Compare the results with those obtained using the *T*-section representation.

Fig. 13.42

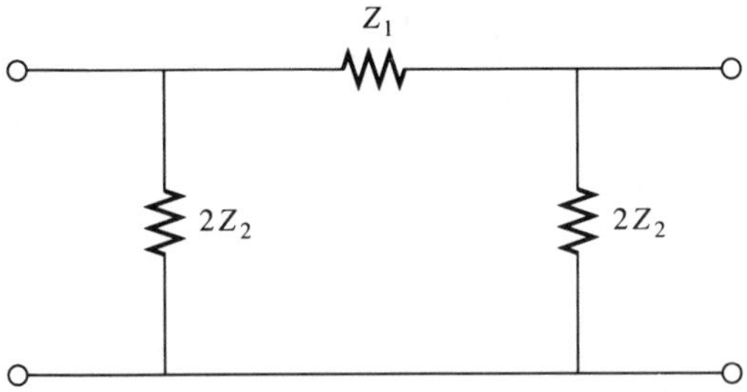

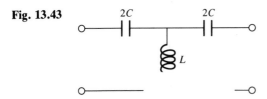

13.9 A constant-*k* high-pass filter has a cutoff frequency $f_o = 1$ kHz and $Z_o = 600$ Ω (purely resistive) at infinite frequency. The elements of the filter are shown in Fig. 13.43. The design equations give $C = 0.1326\ \mu$F and $L = 47.7$ mH. Perform an analysis like the one shown in the text at (a) $f = 2$ kHz and (b) $f = 500$ Hz.

Fig. 13.43

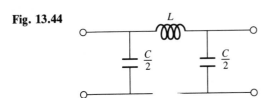

13.10 In the discussion of the lossless transmission line, a *T* section was used. An alternate form is the π section as shown in Fig. 13.44. To find the characteristic impedance of this section, you may use

$$Z_{o\pi} = \frac{Z_1 Z_2}{Z_{oT}}$$

Rework the example given in the text based on the π-section rather than the *T*-section representation of the lossless transmission line.

Fig. 13.44

Filter Synthesis and the Filter Designer

The Filter Designer is available in an evaluation version from the MicroSim Corporation. It has the complete capabilities of the production version, except that it is limited to third-order filters. The *Synthesis User's Guide* contains documentation for the design of active filters. Texts such as Van Valkenburg's *Analog Filter Design* and Daryanain's *Principles of Active Network Synthesis and Design* should be consulted for further reference.

The Filter Designer deals with classic filter approximations of the Butterworth, Bessel (or Thomson), Chebyshev, Inverse Chebyshev, and Elliptic design.

TRANSFER FUNCTIONS, POLES, AND ZEROS

The simplest filter might be a circuit such as that of Fig. 14.1 with a series element Z_1 and a shunt element Z_2. The low- and high-pass filters at the beginning of Chapter 4 are of this design. The transfer function is found by voltage division as

$$T = \frac{V_2}{V_1} = \frac{Z_2}{Z_1 + Z_2}$$

More generally, the transfer function, relating output to input voltage, is often given in terms of the complex variable s. For the low pass-filter, $Z_1 = R$ and $Z_2 = 1/sC$, and the transfer function becomes

$$T(s) = \frac{V_2}{V_1} = \frac{1/sC}{R + 1/sC}$$

The low-pass filter has a pole at $s = -1/RC$.

Fig. 14.1 A simple filter.

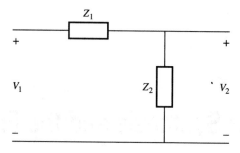

For the high-pass filter, $Z_1 = 1/sC$ and $Z_1 = R$. The transfer function is

$$T(s) = \frac{V_2}{V_1} = \frac{R}{R + 1/sC}$$

The high-pass filter has a zero at $s = 0$ and a pole at $s = -1/RC$.

A more general bilinear transfer function is shown as

$$T(s) = \frac{a_1 s + a_2}{b_1 s + b_2} = \frac{p(s)}{q(s)}$$

This may also be written as

$$T(s) = \frac{a_1}{b_1} \frac{s + a_2/a_1}{s + b_2/b_1} = K \frac{s + z_1}{s + p_1}$$

where z_1 is the zero of $T(s)$ and p_1 is the pole of $T(s)$. These may be shown in the s plane at

$$s = -z_1 \quad \text{and} \quad s = -p_1$$

For higher-order filters, the numerator and denominator of the $T(s)$ equation contain many factors, these being generally shown in terms of quadratics.

USING THE FILTER DESIGNER FOR A LOW-PASS FILTER

Using the evaluation version of the Filter Designer, we can easily implement a few designs. Let us begin with a low-pass filter. It is to have a pass band that extends from zero to 3 kHz. Ideally, the stop band would begin at the latter frequency, but practically we must allow for a transition between the pass band and the stop band. Let us specify that the stop band is to begin at 4 kHz. Ideally, there would be flat response, with zero gain throughout the pass band, but from a practical standpoint, there must be an allowed ripple in the pass band. We will specify a ripple of 2 dB in the pass band. Finally, the rate of attenuation in the stop band is needed. Let us choose a value of 6 dB per decade for this attenuation.

Now we are ready to use the Filter Designer software. In the filter directory, run the program by typing

```
filter
```

```
File  Specify  Coefficients  Circuits   Plots    Preferences
========================= Preferences =========================
   Units                   Sampled Data              S to Z Transform
   (F2)      Hz            Clock Freq.        KHz     (F3)  ► mod. bilinear
          ► KHz                                             bilinear
            MHz                                             forward Euler
            GHz                                             backward Euler
─────────────────────────────────────────────────────────────
                  Specification Input Format

   Specify all std. filters by:      Specify band pass/reject filters by:
   (F4)  ► stop band                 (F5)    center freq. & bandwidth
           order                           ► Upper/lower band limits

                     Approximation:
                     (F6)   ► normal
                             modified (for LC ladders)
─────────────────────────────────────────────────────────────
   SCF Netlist Format
   (F7)  ► PSPICE (normal)            OK - (F1)        Exit - (ESC)
           PSPICE (ideal switch)
           SWITCAP
```

Fig. 14.2   Setting preferences.

When the initial menu appears, choose *Preferences* by pressing *r*. Verify the settings shown in Fig. 14.2. Then press *F1* for *OK*. Next select *Specify* and choose *Low Pass*. The screen is now ready to accept the four values we have selected. Type each of the values (*3, 4, 2,* and *6*), following each with either *Tab* or *Enter*.

Below the entered data, a section of the screen shows the type of approximations from which you may choose:

> (F1) Butterworth
>
> (F2) Chebyshev
>
> (F3) Inv Chebyshev
>
> (F4) Elliptic
>
> (F5) Bessel

Chebyshev Approximation

Choose *Chebyshev* by pressing *F2*. In just a moment the column marked *Required Order* will display the values *3, 2, 2, 2,* and − for each of the respective filters. We see that our Chebyshev filter may be implemented with a second-order filter. Refer to Fig. 14.3 to see this screen.

At this point, press *Esc* to return on the main menu, and select *Plots*. Under this item, select *Bode* by pressing *Enter*. A plot of the filter response will appear on screen after a moment. The plot should be like the one shown in Fig. 14.4. The vertical axis shows the gain in dB, and the horizontal axis shows the logarithmic frequency range. In our example the frequencies are from 1 kHz to 10 kHz. You may wish to obtain a printed copy of this response curve by pressing *F10*.

Note that *F2* is a toggle between *Linear* and *Log* frequency. You may change to either as desired. *F1* goes through a cycle of options. There is *Horizontal*, which is currently activated, as well as *Left Vertical* and *Right Vertical*. With each

```
File  Specify  Coefficients  Circuits   Plots    Preferences
═══════════════════════════ Standard Functions ═══════════════════
    Low Pass Limits                                Pre-warp to:

Pass band cutoff       3.000_     KHz                n/a
Stop band cutoff       4.000⁻     KHz                n/a
Pass band ripple       2.000      dB
Stop band atten.       6.000      dB
──────────────────────────────────────────────────────────────────
Choose an Approximation:  Chebyshev              Required Order

        (F1)    Butterworth                          3
        (F2)    Chebyshev                            2
        (F3)    Inv Chebyshev                        2
        (F4)    Elliptic                             2
        (F5)    Bessel                               -
```

Fig. 14.3 Low-pass filter, Chebyshev approximation.

of the vertical choices, *F2* cycles through the choices, *Gain dB*, *Gain Linear*, *Phase*, *Linear Phase*, *Phase Delay*, and *Group Delay*. After choosing a particular combination, you may press the *Enter* key to see the new display. Try various combinations and observe the results.

Up to this point the filter is still a mathematical abstraction. It is possible to choose the circuit and its components by returning to the main menu. Begin by choosing *Coefficients*; then select *s Biquads*. After a brief interval, the *Filter Coefficients* will appear on the screen. For our example verify that $k_2 = 0$, $k_1 = 0$, $k_0 = 232.2919$ M, $Q = 1.1286$, and $\omega_o = 17.1009$ krad/s. Note that these coefficients are for stage number one, which is a second-order stage, and that no other stages are needed.

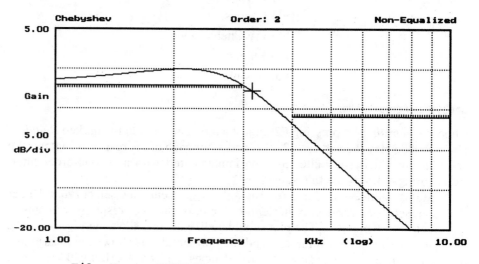

```
File Name:  FILTER
Plot Mode:  s domain
Data:       continuous
```

Fig. 14.4 Low-pass Chebyshev filter Bode plot.

These coefficients are for the second-order transfer function

$$T(s) = \frac{(k_2 s + k_1)s + k_0}{(s + \omega_o/Q)s + \omega_o^2}$$

The Tow-Thomas Low-Pass Design

The coefficients may be changed, and a certain amount of fine tuning is sometimes desirable. We will not attempt that now, however. Return to the main menu and select *Circuits*. Then choose *3 Op Amp RC Biquads*, and under this select the *Tow-Thomas* configuration. After you press *Enter*, the screen will display the components which will include resistors R_1 through R_6 and capacitors C_1 and C_2. Confirm that $R_1 = 6.60 \text{ k}\Omega$, $R_2 = 5.85 \text{ k}\Omega$, and so forth. The right arrow key may be used to allow viewing of the components that are not shown. Since the elements are not standard values, it is possible and often desirable to round them. Choose the *Round* command by pressing *F1*. A pop-up menu shows what to do next. Let us select 10% standard resistors by pressing *F2* until the 10% choice has the triangular marker beside it, and choose 5% standard capacitors by pressing *F3*. Observe the message:

> *Rounding modifies filter coefficients. Recompute coefficients before trying a different implementation.*

Press *F1* for *OK*. Note that $R_1 = 6.8 \text{ k}\Omega$, $R_2 = 5.6 \text{ k}\Omega$, and so forth, as a result of the selected rounding. Press *Esc* three times; then look at the *Coefficients for s Biquads*. Note that the values have changed, as we knew they should. Now press *Esc*.

If you now return to the Bode plot screen and look at this new implementation, you will see that the plot has been seriously affected. It is generally recommended that 1% resistors be used, but for our illustration we will continue with the 10% values. Return to the main menu and select *Circuits*; then choose *Schematic*.

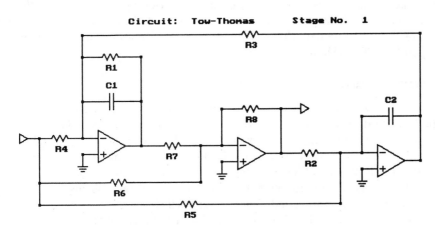

Fig. 14.5 Low-pass Chebyshev filter showing labels.

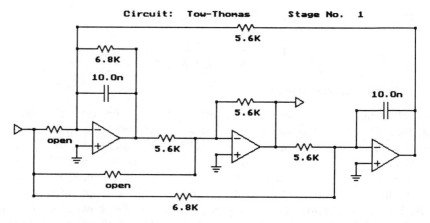

Fig. 14.6 Low-pass filter showing component values.

In keeping with our choices, the plot shows three op amps and the connections for the various resistors and capacitors that comprise the circuit. The components are labeled *R1*, *R2*, and so forth, if symbols are selected. They may be labeled with numeric values instead, depending on whether *F1* or *F2* is pressed. Symbols are shown in Fig. 14.5 on page 421, and component values are shown in Fig. 14.6.

Elliptic Approximation

There are other choices for the approximation type. When you return to *Specify, Low Pass* on the main menu, the original low-pass limits are still there. Select Elliptic (F4) for the new design, which will be second order; then return to the main menu. Choose *Circuits*, and again select *3 Op Amp RC Biquads, Tow-Thomas*. Verify that R_1 = 8.4 kΩ, R_2 = 5.42 kΩ, and so forth. Now select rounding to allow for 1% resistors and no rounding of the capacitors. Observe the slight changes in the resistor values. Look at the Bode plot, and notice that the attenuation is much greater than for the Chebyshev design. This is shown in Fig. 14.7. The circuit configuration will be the same as before, since we have not changed from Tow-Thomas.

Butterworth Approximation

Return to *Specify, Low Pass* on the main menu, and select *Butterworth*. This will result in a third-order filter design. Stay with the *3 Op Amp RC Biquads, Tow-Thomas*, with 1% resistors and no rounding of the capacitors. Look at the Bode plot and see that the response is slightly different from that of the Chebyshev. Note that the Butterworth has flat response through most of the pass band, while the Chebyshev is slightly attenuated at zero frequency, then begins to rise as the cutoff frequency is approached. The Butterworth has better attenuation in the stop band. Refer to Fig. 14.8 for the Bode plot of the Butterworth. Stages 1 and 2

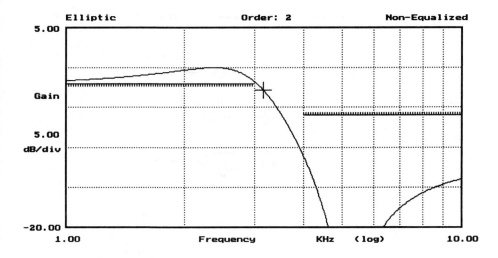

File Name: FILTER
Plot Mode: s domain
Data: continuous

Fig. 14.7 Low-pass elliptic filter Bode plot.

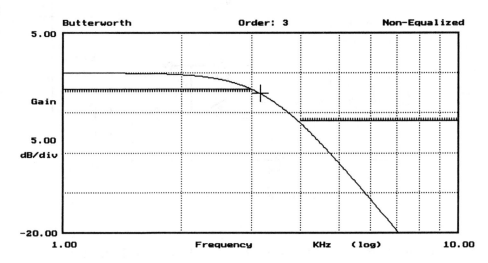

File Name: FILTER
Plot Mode: s domain
Data: continuous

Fig. 14.8 Low-pass Butterworth filter Bode plot.

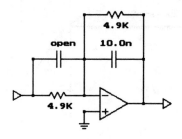

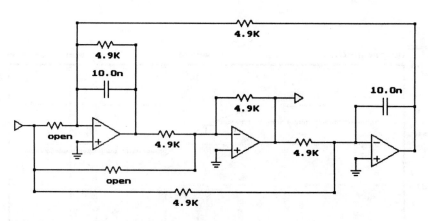

Fig. 14.9 Low-pass filter, Butterworth approximation.

of this design are shown in Fig. 14.9. The symbols of stage 2 are as shown in Fig. 14.5.

Further documentation may be obtained by returning to the main menu and selecting *File*, *Print Summary*, *Printer*. This produces the full summary of the filter design, shown in Fig. 14.10. If you wish to add notes of your own, print to a

```
                        FILTER DESIGN SUMMARY

   File Name:          FILTER
   Order:              3
   Function:           Low Pass
   Approximation:      Butterworth
   Pass band cutoff: 3.0000K
   Stop band cutoff: 4.0000K
   Pass band ripple: 2.0000
   Stop band atten.: 6.0000


   Filter Coefficients - s domain
   Stg  Stg
   No.  Ord   K2          K1          K0          Q          Wo
    1    1    0.0000e+0   0.0000e+0   2.0534e+4   0.0000e+0   2.0534e+4
    2    2    0.0000e+0   0.0000e+0   4.2164e+8   1.0000e+0   2.0534e+4

   Transfer Function Roots - s domain
               Zero1-Re       Zero1-Im        Zero2-Re        Zero2-Im
        1      infinity       infinity        infinity        infinity
        2      infinity       infinity        infinity        infinity

               Pole1-Re       Pole1-Im        Pole2-Re        Pole2-Im
        1     -2.0534e+4      0.0000e+0       infinity        infinity
        2     -1.0267e+4     -1.7783e+4      -1.0267e+4       1.7783e+4


   Circuit Implementation :    Tow - Thomas
   Stg  Sch
   No.  No.   R1          R2          R3          R4          R5
    1   10    4.87e+3     4.87e+3       open        open        open
    2   31    4.87e+3     4.87e+3     4.87e+3       open      4.87e+3

   Stg  Sch
   No.  No.   R6          R7          R8          C1          C2
    1   10      open        open        open        open     10.00e-9
    2   31      open      4.87e+3     4.87e+3    10.00e-9   10.00e-9
```

Fig. 14.10 Summary report of low-pass Butterworth filter.

file instead of to the printer. Another item of interest is the creation of a netlist file that may be used in PSpice. This is done by selecting *Netlist* from the *File* menu. The results are shown in Fig. 14.11.

Saving the File

If you wish to save this design, return to the main menu, select *File, Save as...*, and give the file a name. No extension is required.

Fig. 14.11 Netlist of low-pass Butterworth filter.

```
* *********    FILTER SUBCKTS    ********

*    FILTER DESIGNER 5.1

*    FILTER FILENAME: Filter1.cir
*    Order:          3
*    Function:       Low Pass
*    Approximation:  Butterworth
*    Pass band cutoff: 3.0000K
*    Stop band cutoff: 4.0000K
*    Pass band ripple: 2.0000
*    Stop band atten.: 6.0000


.LIB FILTER.LIB

.SUBCKT FILTER IN OUT AGND
RIN IN 1 1
 X1 1 2 AGND LINSTG
+   PARAMS:  C1VAL= 1.0e-20   C2VAL= 10.000e-9
+            R1VAL= 4.870e+3   R2VAL= 4.870e+3
 X2 2 3 AGND TTSTG
+   PARAMS:  C1VAL= 10.000e-9  C2VAL= 10.000e-9
+            R1VAL= 4.870e+3   R2VAL= 4.870e+3
+            R3VAL= 4.870e+3   R4VAL= 1.000e+20
+            R5VAL=4.870e+3   R6VAL=1.000e+20
+            R7VAL=4.870e+3   R8VAL=4.870e+3
ROUT 3 OUT 1
ROUTGND OUT AGND 1E12
.ENDS
```

A HIGH-PASS FILTER

The design of a high-pass filter is done in much the same fashion as that of the low-pass filter. As an example, for a Pass-band cutoff frequency of *18* kHz, a Stop-band cutoff frequency of *15* kHz, a Band-pass ripple of *2* dB, and a Stop-band attenuation of *6* dB, enter these data after selecting *Specify*, *High Pass*.

The Chebyshev Approximation

At this point a filter approximation is needed. Choose *Chebyshev*, and note that the required order is *3*. Since the Butterworth requires fifth order, the evaluation version of the Filter Designer cannot be used for this. Look at the *Coefficients for s biquad*. Stage one is first order with $k_1 = 1.0001$, $Q = 0$, and $\omega_o = 306.571$ krad/s. Stage two is second order with $k_2 = 1.0001$, $Q = 2.5516$, and $\omega_o = 120.1468$ krad/s. Look under *s polynomial* at the *s Domain Transfer Function Polynomials*. Verify that the numerators are zero, except for $s^3 = 1$, and the denominators are $s^0 = 4.42E15$, $s^1 = 2.89E15$, $s^2 = 3.54E5$, and $s^3 = 1$.

Look at the Bode plot, which is shown in Fig. 14.12. Note the attenuation in the stop band, the rise in response near the corner frequency, and the fall and rise in response in the pass band.

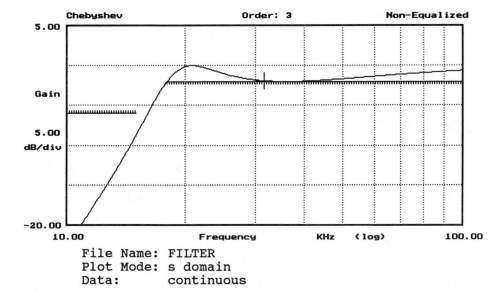

File Name: FILTER
Plot Mode: s domain
Data: continuous

Fig. 14.12 High-pass filter, Chebyshev approximation Bode plot.

Basic MLF Implementation

For implementation, choose *Circuit, 1 Op Amp RC Biquads*; then choose *Basic MLF*. The components should include for the first stage R_1 = open, $R_2 = 3.26$ kΩ, $C_1 = 1$ nF, and so forth. For the second stage, $R_1 = 1.09$ kΩ, $R_2 = 63.71$ kΩ, $C_1 =$

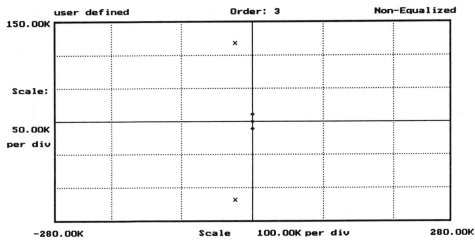

File Name: FILTER
Plot Mode: s domain
Data: continuous

Fig. 14.13 High-pass filter, Chebyshev approximation poles and zeros.

1 nF, $C_2 = 999.89$ pF, and $C_3 = 1$ nF. Round the resistor values to 1%; then return to the Bode plot and note that it is essentially unchanged.

Return to the *Plot* menu and obtain a plot of the poles and zeros which are shown in Fig. 14.13 on page 427. Finally, look at the component placement under *Schematic*, and either sketch the circuit or obtain a printed copy. Save the filter file if desired.

BAND-PASS FILTER

From the main menu use *Specify*, *Band Pass* and enter the following values: Lower stop-band cutoff *5* kHz, Lower pass-band cutoff *6.6* kHz, Upper pass-band cutoff *10* kHz, Upper stop-band cutoff *11.6* kHz, Pass-band ripple *3* dB, Stop-band attenuation *6* dB. These values were chosen to allow implementation within the limits of the evaluation version of the Filter Designer.

The Elliptic Approximation

For the approximation, choose *Elliptic* and note that the required order is 2. Obtain the Bode plot from the main menu. This should look like Fig. 14.14. Note the steep attenuation on either side of the center of the pass band. It would be desirable to specify a pass-band ripple of less than 3 dB, but this would require a fourth-order approximation.

An Akerberg-Mossberg Implementation

Now under *Circuits* choose *3 Op Amp RC Biquad*, *Akerberg-Mossberg*. For the components, round the resistors to 1%, with no rounding for the capacitors. The

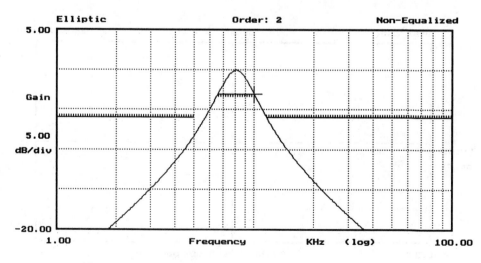

File Name: FILTER
Plot Mode: s domain
Data: continuous

Fig. 14.14 Band-pass filter, elliptic approximation Bode plot.

```
File  Specify  Coefficients  Circuits   Plots    Preferences  Scroll (L/R Arrow)
══════════════════════════════ RC Biquads ══════════════════════════════
 Stg                      Akerberg - Mossberg
 No     Sch          R1          R2          R3          R4          R5

  1     32    │     1.96K       1.96K       1.96K       4.64K        open




         (F1) Round        (F2) Resize    │        (ESC) Exit
```

Fig. 14.15 Some of the elements for the band-pass filter.

values should be $R_1 = 1.96$ kΩ, $R_2 = 1.96$ kΩ, $R_3 = 2.96$ $k\Omega$, $R_4 = 4.64$ kΩ, $R_5 =$ open, $R_6 = 4.64$ kΩ, $R_7 = 1.96$ kΩ, $R_8 =$ open, $C_1 = 10$ nF, $C_2 = 10$ nF, and $C_3 =$ open. Some of the elements are shown in Fig. 14.15. Return to the main menu and examine the coefficients which should include $k_1 = 21.5517$ k, $Q = 2.3674$, and $\omega_o = 51.0804$ krad/s. On the screen, use the right-arrow key to see the other components.

Finally, choose *Schematics* from the *Circuits* menu and look at the component placement and component values. *Warning*: The values for R_1, R_2, and R_3 may be shown incorrectly on the schematic (unless this bug has been fixed). Other resistor values are rounded to two significant digits. Therefore rely on the listing shown under *RC Biquads* rather than the values obtained from the schematic. The schematic is shown in Fig. 14.16. Save your work in a file if desired.

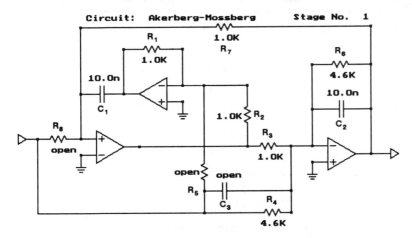

Fig. 14.16 Band-pass filter, Akerberg-Mossberg implementation.

BAND-REJECT FILTER

From the main menu, select *Specify*, *Band Reject*; then use the following values: Lower pass-band cutoff *30* kHz, Lower stop-band cutoff *31* kHz, Upper stop-band cutoff *32* kHz, Upper pass-band cutoff *33* kHz, Pass-band ripple *3* dB, and Stop-band attenuation *8* dB.

Inverse-Chebyshev Approximation

For the approximation choose *Inverse Chebyshev*, which will result in a second-order filter. Under *Circuits* choose *3 Op Amp RC Biquads*, *Tow-Thomas*. Use 1% resistors and verify the values $R_1 = 54.9$ kΩ, $R_2 = 5.11$ kΩ, $R_3 = 5.11$ kΩ, $R_4 = 54.9$ kΩ, $R_5 = 5.11$ kΩ, $R_6 = 5.11$ kΩ, $R_7 = 5.11$ kΩ, $R_8 = 5.11$ kΩ, $C_1 = 1$ nF, and $C_2 = 1$ nF.

Obtain the Bode plot shown in Fig. 14.17 by changing the horizontal axis to *Linear* (press *F2-[Rtn]*). Look at the schematic of the circuit and obtain a printed copy if it is desired.

AN *LC* LADDER FILTER

The *LC* ladder filter is an early, passive design and is included here in a low-pass implementation. To activate this style filter, go to *Preferences* on the main menu, and press *F6* to select *modified* (for *LC* ladders); then press *F1* for *OK*.

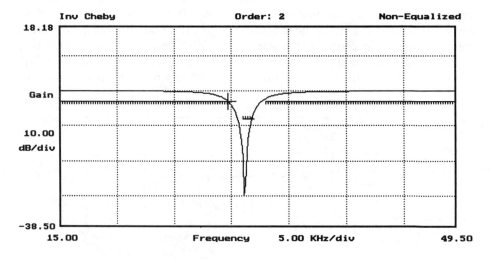

Fig. 14.17 Band-reject filter, inverse-Chebyshev implementation Bode plot.

Fig. 14.18 *LC* ladder filter.

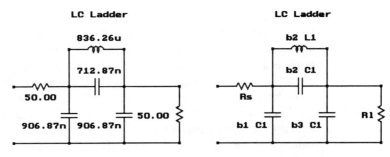

```
                    FILTER DESIGN SUMMARY

     File Name:        FILTER
     Order:            3
     Function:         Low Pass
     Approximation:    Elliptic
     Pass band cutoff: 5.0000K
     Stop band cutoff: 6.0000K
     Pass band ripple: 1.0000
     Stop band atten.: 8.0000


     Filter Coefficients - s domain
     Stg  Stg
     No.  Ord   K2          K1          K0          Q          Wo
       1   1    0.0000e+0   0.0000e+0   2.2054e+4   0.0000e+0  2.2054e+4
       2   2    6.1122e-1   0.0000e+0   1.0253e+9   3.7345e+0  3.2020e+4

     Transfer Function Roots - s domain
                 Zero1-Re     Zero1-Im      Zero2-Re     Zero2-Im
       1         infinity     infinity      infinity     infinity
       2         0.0000e+0    -4.0957e+4    0.0000e+0    4.0957e+4

                 Pole1-Re     Pole1-Im      Pole2-Re     Pole2-Im
       1         -2.2054e+4   0.0000e+0     infinity     infinity
       2         -4.2871e+3   -3.1732e+4    -4.2871e+3   3.1732e+4


     Circuit Implementation :     LC Ladder
     Branch
       No.   Type  Role     C1        C2        L1        L2
        1     1    shunt    9.07e-7   open      short     short
        2     4    series   7.13e-7   open      8.36e-4   short
        3     1    shunt    9.07e-7   open      short     short


                Filter Designer 5.1 - MicroSim Corp.
                     (c) Crescendo 1990-91
```

Fig. 14.19 Low-pass *LC* ladder filter design summary.

Under *Specify*, choose *Low Pass*; then type in the following values: Pass-band cutoff *5* kHz, Stop-band cutoff *6* kHz, Pass-band ripple *1* dB, and Stop-band attenuation *8* dB. For the approximation, choose *Elliptic*, noting that this will require a third-order filter. Then return to the main menu and under *Circuits* choose *LC Ladder*. The display shows the L and C elements required in each of three branches along with the terminating resistance value of 50 Ω. Press *Esc*; then look at the schematic under *Circuits*. Obtain a printed copy of the circuit, which is shown in Fig. 14.18 on page 431. It will be helpful to look at both the values and the symbols, where b_1 refers to branch *1*, and so forth. Also document the design with an output file like Fig. 14.19 on page 431.

PROBLEMS

Note: These problems are selected so that they may be solved using the capabilities of the evaluation version of the Filter Designer, which is limited to third-order filters.

14.1 Design a low-pass filter with a pass-band cutoff of 5 kHz, a stop-band cutoff of 6 kHz, a pass-band ripple of 2 dB, and a stop-band attenuation of 8 dB. Can the *Chebyshev* approximation be used with the evaluation package? If not, choose *Elliptic*. Look at the Bode plot. Then select the *3 Op Amp RC Biquads*, *Akerberg-Mossberg* implementation, with 10% standard resistors, and produce the final Bode plot. Show the circuit required along with the component values. Compare these elements with those shown for the Tow-Thomas implementation.

14.2 Design a high-pass filter with a pass-band cutoff frequency of 12 kHz, a stop-band cutoff frequency of 10 kHz, a band-pass ripple of 2 dB, and a stop-band attenuation of 8 dB. Use the *Chebyshev* approximation. For the implementation, select *3 Op Amp RC Biquads*, *Tow-Thomas*. Use 1% resistors, and produce the Bode plot. Find the component values required, and show the necessary circuit.

14.3 Design a band-pass filter. Attempt to use a lower stop-band cutoff of 10 kHz, a lower pass-band cutoff of 11 kHz, an upper pass-band cutoff of 15 kHz, an upper stop-band cutoff of 16 kHz, a pass-band ripple of 3 dB, and a stop-band attenuation of 8 dB. Verify that at least a fourth-order filter is required. Change the values to 10, 11.5, 15, 16.6, 3, and 6 in order to obtain a lower-order design. Select the Butterworth approximation, using *3 Op Amp RC Biquads*, *KHN* (Kerwin-Huelsman-Newcomb). With standard 10% resistors, what are the required components? Produce a Bode plot, and show the circuit with the elements labeled.

14.4 Design a band-reject filter with a lower pass-band cutoff of 12 kHz, a lower stop-band cutoff of 13.7, an upper stop-band cutoff of 16 kHz, an upper pass-band cutoff of 17.7 kHz, a pass-band ripple of 3 dB, and a stop-band attenuation of 8 dB. Select your own approximation and circuit implementation. Document your results with Bode plot and labeled circuit. Print a summary of the filter design.

14.5 Design a passive *LC* ladder filter for a band-pass application. (Don't forget to set preferences for *LC*.) Use a lower stop-band cutoff of 10 kHz, a lower pass-band cutoff of 13 kHz, an upper pass-band cutoff of 15 kHz, an upper stop-band cutoff of 17 kHz, a pass-band ripple of 4 dB, and a stop-band attenuation of 8 dB. Select the *Elliptic* approximation. Use 1% resistors; produce the Bode plot, and show the circuit with all component values.

15

Genesis and Schematics

This chapter introduces the Design Center, a CAE system. Using various components of this system, you may capture, simulate, and analyze analog and digital circuit designs. Although there are configurations of the Design Center for windows-driven, menu-driven, or direct management systems, the full implementation requires the graphical, windows-type environment.

THE DESIGN CENTER CONCEPT

MicroSim has produced a circuit design environment that may be used to define and analyze a variety of circuits. The *Design Center* is the name given to the integrated package. It consists of such items as *Genesis, Analysis,* and *Synthesis* tool kits. *Genesis* contains *Schematics*, which is a graphical circuit editor and manager. It allows the user to draw a schematic of a circuit which is suitable for creating PSpice simulations and Probe analyses. It is available for the IBM PC under Windows and also for the Sun workstations running Open Windows. Our discussion will be centered around the PC version of the packages.

GENESIS AND SCHEMATICS

If you have Microsoft Windows, version 3.0 or 3.1, and a PC compatible using the Intel 80386 or 80486 or the AMD 386 processor, you will be able to use the Genesis software. It comes in an evaluation package which should be obtained before a decision is made on the production version. At least 3 Mbytes of extended memory is essential. The production version of Genesis requires a serial port for the installation of a security plug.

Although not recommended, the Genesis evaluation version can be installed under Windows on a 286 machine running in the standard mode. But if this is done, Probe will not function properly. However, bias-point analysis and ac analysis may be performed on the 286 machine.

The installation of the software is done from within Windows from the file *setup.exe.* A directory *MSIMEVAL* is created along with its two subdirectories, *BACKUP* and *LIB.* The sample device library is like the one that comes with the DOS version of PSpice.

CREATING A CIRCUIT DRAWING

After the Design Center evaluation package has been installed, a program group containing *Schematics, PSpice, Probe, Parts,* and the *Stimulus Editor* will be available as shown in Fig. 15.1. Select *Schematics* so that the creation of a drawing may begin. The drawing may consist of symbols, attributes, connections, and text. For now, we will work with the symbols. Our goal is to create a series circuit with a dc voltage source and three resistors. Note the grid which is available for the placement of parts.

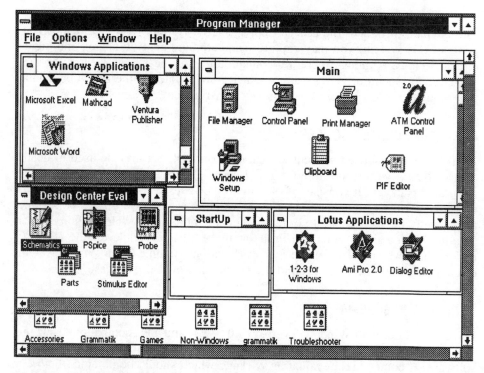

Fig. 15.1 The Design Center.

Schematics presents you with a drawing board for the placement of parts. A grid appears on the board as an aid in arranging components. This grid is shown in Fig. 15.2.

Items from the main menu are available following the usual Windows conventions. Press *Alt-d* to see the *Draw* pull-down menu. Then press *g* for *Get New Part*. Since in the beginning we do not know how the parts are named, select *Browse*. This shows a wide variety of choices depending on which symbol library is active. Using the mouse, select *source.slb* from the window on the right; then from the window on the left choose *VSRC* (for voltage source) and click on *OK*.

The cursor takes the form of the selected symbol, allowing you to move to the desired location on the grid where the source symbol will be placed. Choose a convenient spot, about an inch down from the top and an inch from the left border; then click the left mouse button. When the button is released, the cursor is free to move to another location. Since we do not need another voltage source, press *Esc* to allow the selection of another part. Note that the source is marked V_1. It is not completely defined but will be edited later.

In order to select a resistor, in the main menu choose *Draw, Get New Part, Browse*, and choose another library *analog.slb*. As you move through the available parts in this library, you will find the desired choice which is *R*. When this selection is made, the cursor turns to a resistor symbol in the horizontal position. Move the cursor several grid marks to the right and above the voltage source. Click the mouse at the desired location, and note that it is shown as R_1 with a

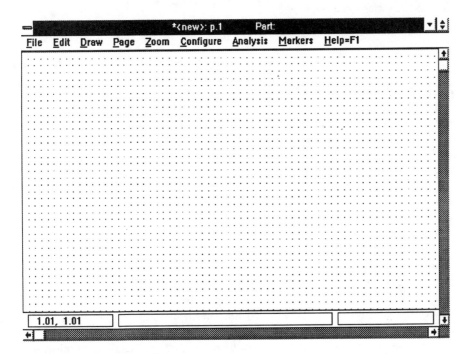

Fig. 15.2 The Schematics drawing board.

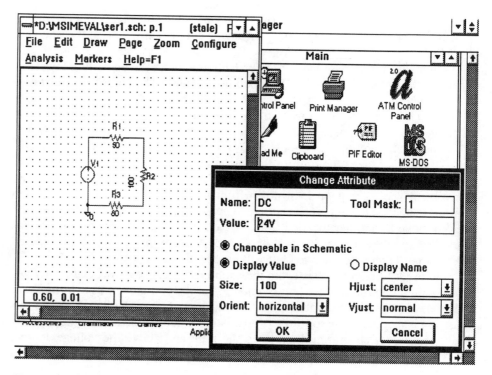

Fig. 15.3   A series circuit in Schematics.

value of 1 kΩ. Refer to Fig. 15.3 which shows the schematic in the window on the left.

Now we would like to place another resistor in the drawing. Move the cursor to a position directly to the right of the source and over several grid marks from the first resistor. Click the mouse to place the resistor; then press *Esc*, and under *Edit* choose *Rotate*. The second resistor will assume a vertical orientation and may be dragged with the mouse to the desired spot. When it is in line with the source and to the right of R_1, release the left mouse button.

Next choose *Draw, Get New Part* and see that *R* is still selected. Click on *OK* and place the third resistor directly below R_1. Click the left button; then click the right button (or press *Esc*).

The circuit will need wiring to connect the elements in series. From the *Draw* menu select *Wire (Alt-d w)* and notice that the cursor becomes a pencil. Move the tip to the top wire on the voltage source, click the left mouse button, and move up to where you will make a right turn in line with R_1. Click again; then move to the line at the left end of R_1. Click left; then click right. This same process is to be repeated until you have wired together the four elements in series. Now we must add an analog ground.

Return to *Draw, Get New Part, Browse*, and in the right window select *global.slb*. In the left window choose *GND_ANALOG*. The symbol for the ground is to be placed at the lower left of the figure at the node between V_1 and R_3. Position the ground and click left, then right.

If you find the detailed instructions on selecting parts too much bother, simply use a trial-and-error technique until you have learned the basics of creating a schematic.

With the elements in place, the next task will be to edit each element for its proper attributes. In order to specify V_1, click the left mouse button on the circle symbol of V_1. It should turn red, indicating that it has been selected.

Choose *Edit, Attributes*; then click on *DC =*; then click on *Change*. In the *Value* box type the voltage value *24V*. If you want this value displayed on the drawing, choose that option as shown in Fig. 15.3. Click on *OK* to leave the edit mode.

Resistor values are to be changed such that $R_1 = 50\ \Omega$, $R_2 = 100\ \Omega$, and $R_3 = 80\ \Omega$. Select R_1 by clicking on the zig-zag portion of the resistor. When properly selected, the resistor will turn red. If a box appears instead, try again. Next, on the main menu, choose *Edit, Attributes*; then click on *value = 1 k, Change*. In the value box type *50*; then press *Del* twice to remove the old value. Click on *OK*, then *OK* again to return to the schematic. Repeat the process and set R_2 and R_3 to their desired values. The value of R_2 (in the vertical position) may show on top of its symbol, but this can be avoided by choosing to orient the value in the vertical mode.

When you are satisfied with the schematic, you may now save the file. On the menu, select *File, Save*. Choose the name *ser1*. The file will be saved in the current directory with the extension *sch*.

Creating a Netlist

Go to the *Analysis* menu and choose *Create Netlist*. This will create a file *ser1.net*. Now choose *Examine Netlist* from the *Analysis* menu, and a notepad will appear on the screen with the file *ser1.net* loaded for viewing. In this file you will see a listing of *v_V1, R_R1, R_R2,* and *R_R3*. The underline characters are used to avoid conflict with other usual node assignments. Nodes created in Schematics are identified by *$N_0002* and other like entries. Since you had no direct control over the marking of the nodes, you may want to mark the nodes in your sketch of the circuit. Perhaps a notation such as *N2* for *$N_0002*, and so forth, will be helpful.

The netlist is shown in Fig. 15.4. You may print the netlist if desired; then close the notepad.

Running PSpice

Return to the *Analysis* menu and choose *Run PSpice*. The simulation status screen will appear in a large window. This is like the screen with which you are familiar from the DOS version of PSpice. Since this is a dc analysis, the bias-point calculations are all that will be performed. The PSpice window at the conclusion of the analysis is shown in Fig. 15.5. It has been resized to allow the circuit to be

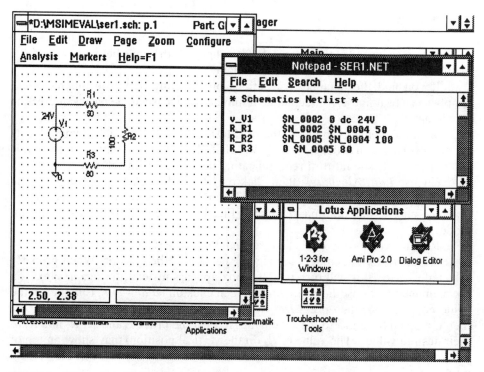

Fig. 15.4 The netlist of the series circuit.

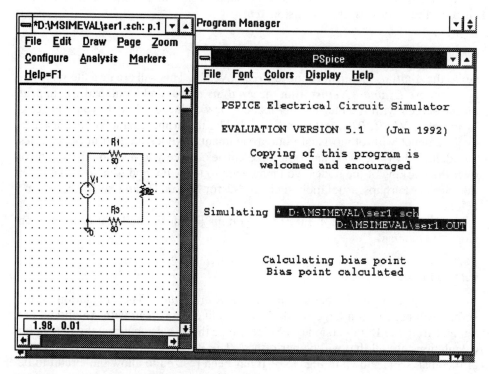

Fig. 15.5 The PSpice analysis completed.

shown on the left side of the screen. An output file *cirl.out* has been created as a result of the analysis. The contents of this file are shown in Fig. 15.6.

The Anatomy of the Output File

At the beginning of the output file there is a comment line which indicates the name of the schematic file including its path. Other comment lines give the version number of Schematics, the time, and the date. There is a *.lib* statement, although no library devices were used. Then there is an *.INC* statement which refers to the netlist. Then there is the actual netlist. This is followed by another *.INC* statement which refers to the aliases file. Aliases are a means whereby a mapping takes place between names in the schematic and names acceptable to PSpice and Probe. Each line of the *Schematics Aliases* shows a device name, following the usual first-character convention, a schematic name, and a netlist node equivalence.

After the .END statement, the results of the analysis are shown. As expected, we find the node voltages, the voltage-source currents, and the total power dissipation. Each node is identified in keeping with the netlist. For example, node $N_0004 (which we might call *N4*) is between R_1 and R_2. The voltage at this node is 18.783 V. The analysis is now complete. Had there been the need, another type of analysis could have been performed by choosing *Setup* in the analysis menu.

Altogether, five files have been created and used during the analysis of the series circuit. They are as follows:

1. *serl.sch* which contains the description of the schematic diagram. This file is coded in a form that Schematics can recognize so that the circuit can be redrawn and edited if necessary.

2. *serl.net* which is the netlist for the circuit. The contents of this file were available for viewing before the analysis was begun. It is also shown in the output listing.

3. *serl.als* which contains the aliases as previously described.

4. *serl.cir* which contains this brief set of entries:

```
* D: \MSINEVAL\ser1.sch

* Schematics Version 5.1 — January 1992
* Tue Aug 04 15:35:39 1992

* From [SCHEMATICS NETLIST] section of msim.ini:
.lib eval.lib

.INC "D:\MSIMEVAL\ser1.net"
.INC "D:\MSIMEVAL\ser1.als"

** Analysis setup **

.END
```

The circuit file contains the information required to put the analysis together, bringing in the library with the *.lib* statement and bringing in the *.net* and *.als* files with the *.INC* statements.

```
  * D:\MSIMEVAL\ser1.sch

 * Schematics Version 5.1 - January 1992
 * Tue Aug 04 15:35:39 1992

 * From [SCHEMATICS NETLIST] section of msim.ini:
 .lib eval.lib

 .INC "D:\MSIMEVAL\ser1.net"

 **** INCLUDING D:\MSIMEVAL\ser1.net ****
 * Schematics Netlist *

 v_V1      $N_0002 0 dc 24
 R_R1      $N_0002 $N_0004 50
 R_R2      $N_0005 $N_0004 100
 R_R3      0 $N_0005 80

 **** RESUMING D:\MSIMEVAL\ser1.cir ****
 .INC "D:\MSIMEVAL\ser1.als"

 **** INCLUDING D:\MSIMEVAL\ser1.als ****
 * Schematics Aliases *

 .ALIASES
 v_V1        V1(+=$N_0002 -=0 )

 R_R1        R1(1=$N_0002 2=$N_0004 )

 R_R2        R2(1=$N_0005 2=$N_0004 )

 R_R3        R3(1=0 2=$N_0005 )

 .ENDALIASES

 **** RESUMING D:\MSIMEVAL\ser1.cir ****

 ** Analysis setup **

 .END

  ****       SMALL SIGNAL BIAS SOLUTION          TEMPERATURE =   27.000 DEG C

 NODE    VOLTAGE      NODE    VOLTAGE      NODE    VOLTAGE      NODE    VOLTAGE

 ($N_0002)    24.0000                   ($N_0004)    18.7830

 ($N_0005)    8.3478

     VOLTAGE SOURCE CURRENTS
     NAME           CURRENT

     v_V1          -1.043E-01

     TOTAL POWER DISSIPATION   2.50E+00  WATTS
```

Fig. 15.6

5. *ser1.out* which is the output file that is produced when PSpice is invoked. Again refer to Fig. 15.6 for the contents of this file. Note that the reference to *D:* in the *.INC* statements means that the files are on the *D* disk drive. In many cases a single disk drive will be used, and it will be drive *C*. The node voltages are given in terms of the alias notation. The Schematics netlist shows where the elements are connected in relation to the alias nodes.

It has taken a good bit of work to produce the simple series-circuit schematic from which the PSpice analysis has been performed. Needless to say, all of this was done for the learning process rather than for its practical value.

If you want to obtain a printed copy of the schematic, this is readily done by selecting *Print* from the *File* menu as shown in Fig. 15.7. Our circuit of only four elements will print postage-stamp size with a large blank area on the rest of the page. (If you elect to use a zoom factor, the drawing will be larger, but you will obtain several extra sheets for a cut-and-paste, larger version.) Again, it will be helpful to mark the nodes N0, N2, and N4 (using the simpler reference, rather than N_0002, etc.). When the elements were placed in the schematic diagram, they were automatically numbered. The node order for each of the elements was also automatically given. In Fig. 15.8 the circuit has been shown with the node assignment added by the author, along with the reference directions for the vari-

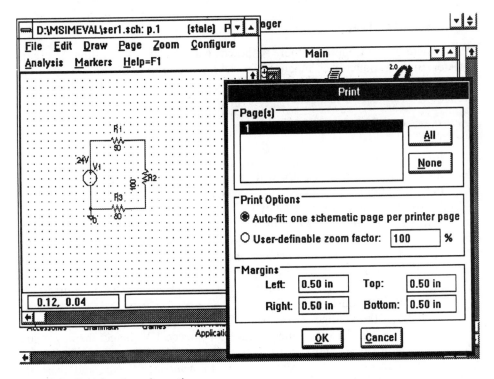

Fig. 15.7 Printing the schematic.

Fig. 15.8 The printed version of the schematic (arrows and nodes added).

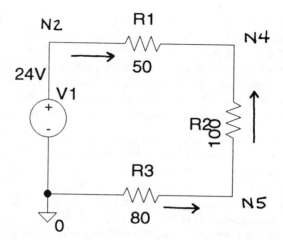

ous currents. Thus if we could ask for a current I(R1), it would be positive, but the current I(R2) or I(R3) would be negative. This may be a minor irritation, but in the drawing of the schematic, when you place an element in the horizontal position, the first node will be assigned on the left, and when the element is rotated, the first node will be at the bottom.

AN AC STEADY-STATE PROBLEM

At the beginning of Chapter 2, there is shown an ac circuit with a voltage source, a resistor, and an inductor connected in series. The circuit is to operate at 60 Hz, and we were to find the current in the loop. This required the following statements:

```
.AC LIN 1 60Hz 60Hz
.PRINT AC I(R)  IR(R)  II(R)  IP(R)
```

We would like to perform the same analysis by using Schematics to reproduce the circuit. The first attempt will be unsuccessful but instructive.

In Schematics, construct a series circuit with $V_1 = 1$ V, $R_1 = 1.5 \, \Omega$, and $L_1 = 5.3$ mH. Briefly, this is done as follows:

Select *Draw, Get New Part, Browse,* and from *source.slb* choose *VSRC* to draw the voltage source. Then from *Draw, Get New Part* choose *R* for the resistor R_1. Return to *Draw, Get New Part* and choose *L* for the inductor L_1. Rotate the inductor; then use *Draw, Wire* to complete the series circuit. Next in *Draw, Get New Part* go to the *global.slb* library, choose *GND_ANALOG,* and insert the ground beneath the source.

Now select the voltage source, then *Edit Attributes.* Select *AC =, Change.* Indicate a change on the value line to make the source 1 V. Select display value. Then in turn edit the R_1 and L_1 attributes to include their correct values.

In order to specify the frequency, go to *Analysis, Setup,* and from the submenu select *AC Sweep.* In the proper box select *Linear;* then for the *Sweep*

Parameters select *1* point starting at *60* Hz and ending at *60* Hz. This will give the desired ac sweep statement in the circuit file. At the bottom of the *AC Sweep* window, select *Enabled*, then *OK*.

There is no simple way to call for the printing of current values. We will return to this problem after we attempt to run the analysis. Create a netlist under *Analysis*, and when you are asked for a name of the file, call it *acrl*. Then examine the netlist to make sure that you have a complete list. Also make note of the labeling of the nodes. You may use a simplified notation such that $N_0002 becomes N2, and so forth.

Finally, we are ready to run the analysis. Under *Analysis* choose *Run PSpice*, and the analysis will begin. In the PSpice analysis window, the status of the run will be shown. After the ac sweep is completed, PSpice will automatically invoke *Probe*.

When Probe attempts to load the data file, the processing will be interrupted. An error message is displayed:

> *Skipping section 1. Section 1 has less than 2 rows of data.*

This problem occurs since we have gone into Probe without a true range of values for the frequency sweep. Exit Probe and look at the output file. There is a statement for the ac analysis which reads

```
.ac LIN 1 60 60
```

but there is obviously no print statement, for we did nothing to produce one. If you go back to the input file, which has been named *acrl.cir,* you may include the line

```
.print ac i(R1) ir(R1) ii(R1) ip(R1)
```

and run the analysis again. However, when you do, PSpice will give the following error message:

> *ERROR—PRINT device R is undefined*

The proper line to include (in the file *acrl.cir*) is

```
.print ac i(R_R1) ir(R_R1) ii(R_R1) ip(R_R1)
```

Put this line directly underneath the ac sweep statement. This may be done by loading the file into the Notepad, adding the desired line, saving the file, then exiting. The details of this operation are shown in Fig. 15.9.

Now run the PSpice analysis again, this time directly from the PSpice icon. Choose *File, Open* and specify the file name, including the path. When the analysis is done, the message

> *Simulation Completed. No errors occured.*

should appear. The results may be displayed on the screen by loading it into the Windows Notepad. Incidentally, it is a good idea to include the Notepad icon in the Design Center program group, so that it will be readily available for looking at netlists, circuit files, and output files. This screen display is shown in Fig. 15.10,

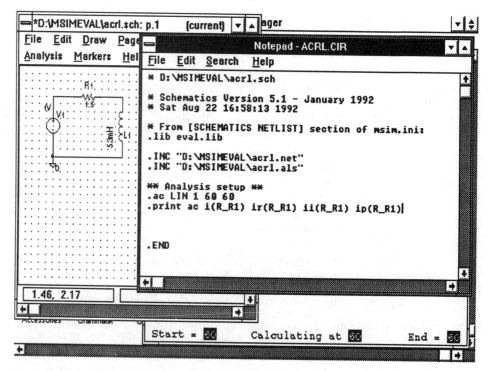

Fig. 15.9 Adding a print statement to the input file.

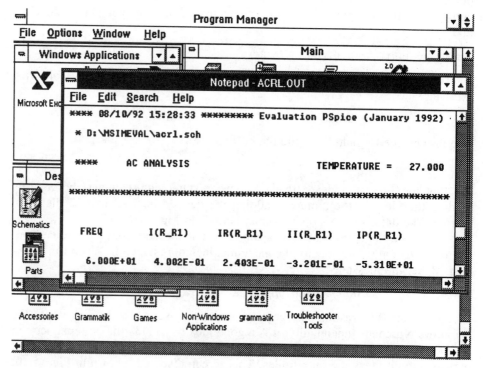

Fig. 15.10 Examining the output file.

```
 * D:\MSIMEVAL\acrl.sch

 * Schematics Version 5.1 - January 1992
 * Mon Aug 10 10:19:55 1992

 * From [SCHEMATICS NETLIST] section of msim.ini:
 .lib eval.lib

 .INC "D:\MSIMEVAL\acrl.net"

 **** INCLUDING D:\MSIMEVAL\acrl.net ****
 * Schematics Netlist *

 v_V1       $N_0002 0   ac 1V
 R_R1       $N_0002 $N_0004 1.5
 L_L1       0 $N_0004 5.3mH

 **** RESUMING acrl.cir ****
 .INC "D:\MSIMEVAL\acrl.als"

 **** INCLUDING D:\MSIMEVAL\acrl.als ****
 * Schematics Aliases *

 .ALIASES
 v_V1       V1(+=$N_0002 -=0 )
 R_R1       R1(1=$N_0002 2=$N_0004 )
 L_L1       L1(1=0 2=$N_0004 )
 .ENDALIASES

 **** RESUMING acrl.cir ****

 ** Analysis setup **
 .ac LIN 1 60 60
 .print ac i(R_R1) ir(R_R1) ii(R_R1) ip(R_R1)

 .END

  * D:\MSIMEVAL\acrl.sch

  * D:\MSIMEVAL\acrl.sch

 ****      AC ANALYSIS                  TEMPERATURE =   27.000 DEG C

   FREQ         I(R_R1)     IR(R_R1)    II(R_R1)    IP(R_R1)

   6.000E+01    4.002E-01   2.403E-01  -3.201E-01  -5.310E+01
```

Fig. 15.11

and the output file is shown in Fig. 15.11. Study this output file so that you can identify the various components of the analysis.

What do we conclude from this exercise? Simply this: if we are interested in running an ac steady-state analysis, this is not the way to do it. Try the following method instead.

RUNNING PSPICE WITHOUT SCHEMATICS ·

The Windows version of PSpice does not require that a schematic diagram be created. If you enter PSpice through Schematics, PSpice will automatically use the current files, putting together not only the information from the schematic file but also the data in the alias file, the netlist file, and the circuit file. As seen in the previous example, if you simply choose the PSpice icon from the *Design Center* program group, you may specify any circuit file of your choice. Follow the example shown here to see how this works. It deals with the same problem of the series *RL* circuit.

Use the *Windows Notepad* (which you should have placed in the *Design Center* program group) to produce the following as a new file:

```
AC Circuit with R and L in Series (Coil)
V 1 0 ac 1V
R 1 2 1.5
L 2 0 5.3mH
.ac LIN 1 60Hz 60Hz
.print ac i(R) ir(R) ii(R) ip(R)
.opt nopage
.end
```

Save the file with the name *acseries.cir*; then exit from the *Notepad*.

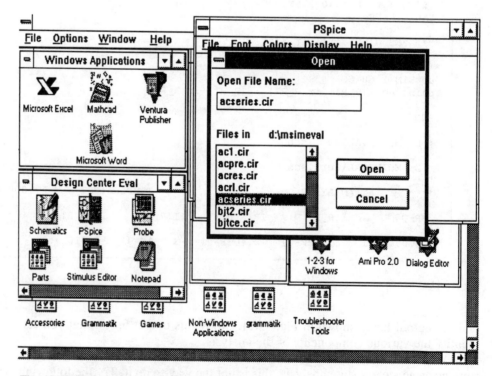

Fig. 15.12   Selecting a file to run in PSpice.

```
AC Circuit with R and L in Series (Coil)

V 1 0 ac 1V
R 1 2 1.5
L 2 0 5.3mH
.ac LIN 1 60Hz 60Hz
.print ac i(R) ir(R) ii(R) ip(R)
.opt nopage
.end

 ****     AC ANALYSIS                  TEMPERATURE =   27.000 DEG C

   FREQ        I(R)         IR(R)        II(R)        IP(R)

  6.000E+01   4.002E-01    2.403E-01   -3.201E-01   -5.310E+01
```

Fig. 15.13

From the *Design Center* program group, select the PSpice icon. In PSpice choose *File, Open,* and select the circuit file *acseries.* The analysis will begin when you select *Open* as shown in Fig. 15.12 on page 448. This time there will be no attempt to run Probe as there was when we used Schematics and specified a

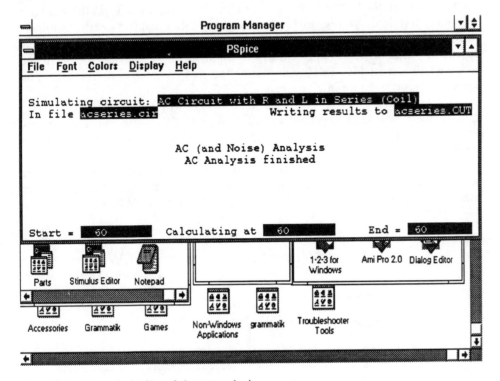

Fig. 15.14 The conclusion of the ac analysis.

sweep. The analysis will give exactly what we want, and the results can be viewed in the *Notepad*, then edited to delete the bias-point information, and then printed. See Fig. 15.13 on page 449 for the results which look like the output file using the DOS version of PSpice. Also see Fig. 15.14 on page 449 for the screen view at the conclusion of the PSpice run. For this type of problem, unless there is some compelling reason to use it, Schematics should be avoided. Also, if both the DOS and the Windows versions of PSpice are available, the simplest and fastest results will be obtained by using the DOS version.

DEPENDENT SOURCES

Beginning in Chapter 1, dependent sources are shown using a diamond symbol. This is the correct standard symbol, but it is not the one used in Schematics. The Schematics symbol is a box similar to that which is used for a two-port network.

Voltage-Dependent Voltage Source

For the first example, consider the circuit shown in Fig. 15.15. The dependent source E_1 has a value that is twice the voltage drop across R_4. In order to create the schematic diagram, begin with a new file in Schematics, and place the parts as they are shown in Fig. 15.16. Use the part *e* from the *analog.slb* library for the voltage-dependent voltage source. Note that each of the dependent sources has two terminals on the left and two on the right. The two terminals on the right are the source terminals. In Fig. 15.15, they are terminals *3* and *0*. The two terminals on the left must be connected in such a way that they show how the source is dependent on another voltage. Clearly, they are to be connected between the nodes as shown in Fig. 15.16. Give attention to the polarity markings of the sources, as you must do in all cases.

Complete the drawing by connecting the parts with wires where necessary; then use the *Edit* feature to set the attributes of each of the elements. The window on the right of Fig. 15.16 shows the attributes for the dependent source which has a gain of *two*.

Save the schematic, giving the file the name *d*. Create and examine the netlist, making note of how the nodes are assigned numbers; then run the PSpice analysis. Refer to Fig. 15.17 for the results. In this output file it is seen that the

Fig. 15.15 Circuit with dependent source.

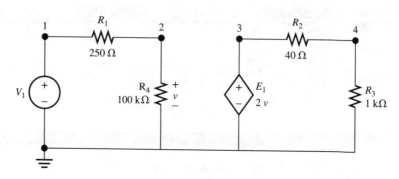

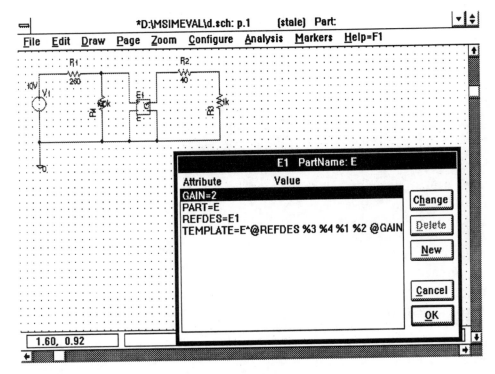

Fig. 15.16 Schematics representation of dependent source.

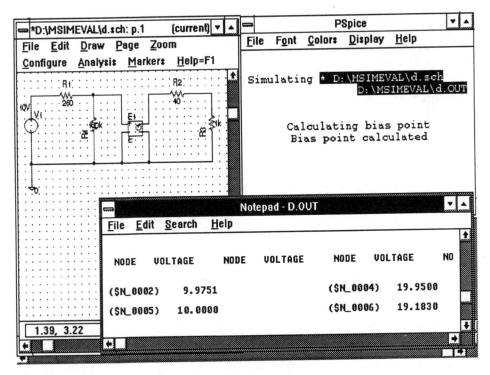

Fig. 15.17 Results of the PSpice analysis.

```
 * D:\MSIMEVAL\d.sch

* Schematics Version 5.1 - January 1992
* Sat Aug 08 19:18:02 1992

* From [SCHEMATICS NETLIST] section of msim.ini:
.lib eval.lib

.INC "D:\MSIMEVAL\d.net"

**** INCLUDING D:\MSIMEVAL\d.net ****
* Schematics Netlist *

E_E1      $N_0004 0 $N_0002 0 2
R_R1      $N_0005 $N_0002 250
R_R2      $N_0004 $N_0006 40
R_R3      0 $N_0006 1k
R_R4      0 $N_0002 100k
v_V1      $N_0005 0 dc 10V

**** RESUMING D:\MSIMEVAL\d.cir ****
.INC "D:\MSIMEVAL\d.als"

**** INCLUDING D:\MSIMEVAL\d.als ****
* Schematics Aliases *

.ALIASES
E_E1        E1(3=$N_0004 4=0 1=$N_0002 2=0 )
R_R1        R1(1=$N_0005 2=$N_0002 )
R_R2        R2(1=$N_0004 2=$N_0006 )
R_R3        R3(1=0 2=$N_0006 )
R_R4        R4(1=0 2=$N_0002 )
v_V1        V1(+=$N_0005 -=0 )
.ENDALIASES

**** RESUMING D:\MSIMEVAL\d.cir ****

** Analysis setup **

.END

 * D:\MSIMEVAL\d.sch

 ****       SMALL SIGNAL BIAS SOLUTION            TEMPERATURE =   27.000 DEG C

NODE   VOLTAGE      NODE   VOLTAGE      NODE   VOLTAGE      NODE   VOLTAGE

($N_0002)    9.9751                 ($N_0004)    19.9500
($N_0005)   10.0000                 ($N_0006)    19.1830

    VOLTAGE SOURCE CURRENTS
    NAME          CURRENT

    v_V1          -9.975E-05

    TOTAL POWER DISSIPATION   9.98E-04  WATTS
```

Fig. 15.18

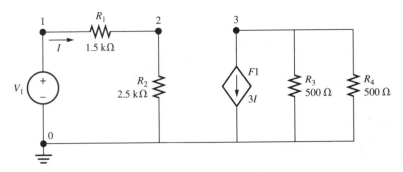

Fig. 15.19 Circuit with dependent current source.

voltage drop across R_4 (shown as N_0002) is 9.9751 V, and the voltage at node N_0004 is 19.95 V, in agreement with the gain of *two* for the dependent source. The entire output file is shown in Fig. 15.18.

Current-Dependent Current Source

A circuit with a current-dependent current source is shown in Fig. 15.19. The dependent source F_1 has a current value that is three times the current in the loop

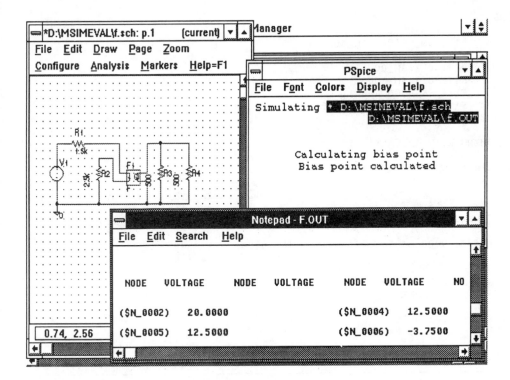

Fig. 15.20 Circuit and output from dependent current source.

```
  * D:\MSIMEVAL\f.sch

* Schematics Version 5.1 - January 1992
* Sat Aug 08 18:15:09 1992

* From [SCHEMATICS NETLIST] section of msim.ini:
.lib eval.lib

.INC "D:\MSIMEVAL\f.net"

**** INCLUDING D:\MSIMEVAL\f.net ****
* Schematics Netlist *

v_V1      $N_0002 0 dc 20V
R_R1      $N_0002 $N_0004 1.5k
R_R2      0 $N_0005 2.5k
R_R3      0 $N_0006 500
R_R4      0 $N_0006 500
F_F1      $N_0006 0 VF_F1 3
VF_F1     $N_0004 $N_0005 0V

**** RESUMING D:\MSIMEVAL\f.cir ****
.INC "D:\MSIMEVAL\f.als"

**** INCLUDING D:\MSIMEVAL\f.als ****
* Schematics Aliases *

.ALIASES
v_V1        V1(+=$N_0002 -=0 )
R_R1        R1(1=$N_0002 2=$N_0004 )
R_R2        R2(1=0 2=$N_0005 )
R_R3        R3(1=0 2=$N_0006 )
R_R4        R4(1=0 2=$N_0006 )
F_F1        F1(3=$N_0006 4=0 )
VF_F1       F1(1=$N_0004 2=$N_0005 )
.ENDALIASES

**** RESUMING D:\MSIMEVAL\f.cir ****

** Analysis setup **

.END

 NODE    VOLTAGE      NODE   VOLTAGE      NODE   VOLTAGE     NODE   VOLTAGE

($N_0002)  20.0000                    ($N_0004)   12.5000
($N_0005)  12.5000                    ($N_0006)   -3.7500

    VOLTAGE SOURCE CURRENTS
    NAME         CURRENT

    v_V1         -5.000E-03
    VF_F1         5.000E-03

    TOTAL POWER DISSIPATION   1.00E-01  WATTS
```

Fig. 15.21

on the left. In Schematics draw this circuit, using the *F* source from the library *analog.slb*. In order to show that *F* is dependent on the current in the loop on the left, break this loop and insert the leads on the left of the dependent source into this loop. The completed drawing is shown in Fig. 15.20 on page 453, along with the screen display at the conclusion of the PSpice analysis. Be sure that all of the attributes are properly set for each of the parts before the analysis is run. Name the schematic file *f*.

The output file is shown in Fig. 15.21. It is easily verified that the current in the input loop should be 5 mA (clockwise). With a gain of three for the dependent source, the current from this source becomes 15 mA (out of the terminal at the bottom). The 15 mA divides equally between the two resistors R_3 and R_4, giving 7.5 mA as each current. This produces a voltage drop of 3.75 V, giving the voltage at node $N_0006 a value of -3.75 V.

Current directions as shown in PSpice can be confusing. The convention is worth looking at in this example. Note that the current for the voltage source v_V1 is shown as -5 mA. This means that inside this source the current from the positive terminal to the negative terminal is -5 mA. Externally this produces a positive current from the positive terminal, that is, a clockwise current in the loop on the left. In the output file, the current in VF_F1 is given as 5 mA (positive). This is a positive value associated with a dummy source in the loop on the left. Internally the current is from the positive to the negative terminal. On the right

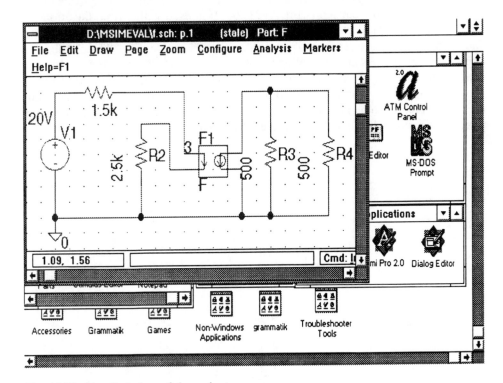

Fig. 15.22 Detailed view of dependent source.

side of the drawing the dependent source has a positive current in the direction of its current arrow (which is downward). See Fig. 15.22 on page 455 for the detail of the dependent source.

A BJT BIASING SCHEMATIC

A BJT circuit is shown in Fig. 15.23. Construct this circuit in Schematics. Obviously, if you place the resistors in a different order, they will have different numbers in your drawing. This is of little consequence, and the choice should be based on convenience. It will be easier to follow the discussion, however, if the resistors are inserted in the order shown in the figure. Note that there are two voltage sources. The ac source and the capacitor will be used later in an ac analysis. They are included now since it is usually easier to put all components in the circuit at the same time. The ac source comes from the library *source.slb*. The part is *VSIN*. Its attributes are

$$AC = 10 \text{ mV}$$
$$voff = 0$$
$$vamp = 10 \text{ mV}$$
$$freq = 5 \text{ kHz}$$

This ac voltage is specified as a sine wave to allow for a transient analysis later. The library transistor chosen is the *2N2222A*. It is used without modification.

After all of the parts are inserted and their attributes have been set, create and examine the netlist, saving the circuit file with the name *bjtce*. The circuit as it appears in Schematics and the netlist are shown in Fig. 15.24. Next run the PSpice analysis for the bias-point solution. An abbreviated version of the output file is shown in Fig. 15.25. On your hard copy of the circuit, based on the information in the output file, it is now possible to identify the nodes in the circuit. Compare your node designation with that shown in Fig. 15.26. The various node voltages should

Fig. 15.23 BJT circuit for Schematics.

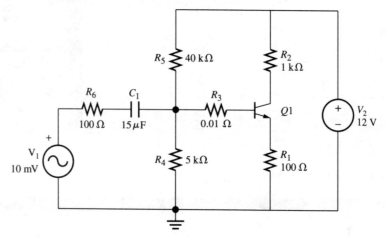

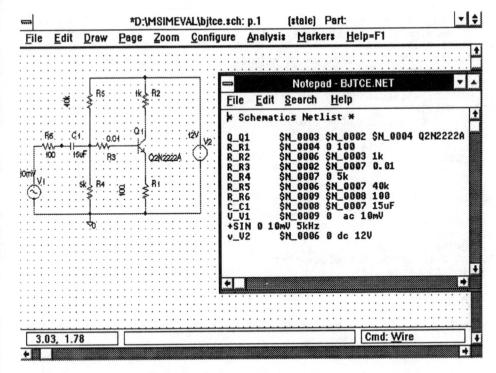

Fig. 15.24 Schematic of BJT circuit.

be examined to verify that they are reasonable values. For example, verify that the collector voltage is 6.8373 V.

A Transient Analysis

The BJT circuit we have just analyzed has a sinusoidal-voltage source $V_1 = 10$ mV at $f = 5$ kHz. If these values have been set in Schematics, you may continue with the transient analysis by choosing *Transient* analysis from *Analysis Setup*. Since the period is $T = 200$ μs, select a *Print step* of 2 μs and a *Final time* of 200 μs. This will allow a full cycle of the sine wave to be displayed in Probe. After these values have been selected and the sweep has been enabled, run PSpice from *Analysis* on the main menu. The bias-point calculations will be completed; then the transient solution will begin. At the conclusion of the analysis, the screen will show the familiar Probe oscilloscope grid. Now you may select *Add_trace* and select from the items listed on the screen.

Select $V(Q1{:}c)$, which represents the voltage at the collector of Q_1, and plot this in Probe. The display shows a cycle of the 5-kHz sine wave as it appears at the collector. Note that the first half cycle is down-going, as it should be. The Schematics window, the PSpice window, and the Probe window are shown in Fig. 15.27. The voltage has a peak of 6.9289 V and a valley of 6.7456 V. The ac axis is at 6.8373 V.

```
   * D:\MSIMEVAL\bjtce.sch
 * Schematics Version 5.1 - January 1992
 * Thu Aug 13 12:09:14 1992
 * From [SCHEMATICS NETLIST] section of msim.ini:
 .lib eval.lib
 .INC "D:\MSIMEVAL\bjtce.net"
 **** INCLUDING D:\MSIMEVAL\bjtce.net ****
 * Schematics Netlist *
 Q_Q1        $N_0003 $N_0002 $N_0004 Q2N2222A
 R_R1        $N_0004 0 100
 R_R2        $N_0006 $N_0003 1k
 R_R3        $N_0002 $N_0007 0.01
 R_R4        $N_0007 0 5k
 R_R5        $N_0006 $N_0007 40k
 R_R6        $N_0009 $N_0008 100
 C_C1        $N_0008 $N_0007 15uF
 V_V1        $N_0009 0   ac 10mV
 +SIN 0 10mV 5kHz
 v_V2        $N_0006 0 dc 12V
 **** RESUMING D:\MSIMEVAL\bjtce.cir ****
 .INC "D:\MSIMEVAL\bjtce.als"
 **** INCLUDING D:\MSIMEVAL\bjtce.als ****
 * Schematics Aliases *
 .ALIASES
 Q_Q1        Q1(c=$N_0003 b=$N_0002 e=$N_0004 )
 R_R1        R1(1=$N_0004 2=0 )
 R_R2        R2(1=$N_0006 2=$N_0003 )
 R_R3        R3(1=$N_0002 2=$N_0007 )
 R_R4        R4(1=$N_0007 2=0 )
 R_R5        R5(1=$N_0006 2=$N_0007 )
 R_R6        R6(1=$N_0009 2=$N_0008 )
 C_C1        C1(1=$N_0008 2=$N_0007 )
 V_V1        V1(+=$N_0009 -=0 )
 v_V2        V2(+=$N_0006 -=0 )
 .ENDALIASES
 **** RESUMING D:\MSIMEVAL\bjtce.cir ****
 ** Analysis setup **
 .END
  ****      BJT MODEL PARAMETERS
              Q2N2222A
              NPN
         IS   14.340000E-15
         BF  255.9
 NODE    VOLTAGE       NODE    VOLTAGE       NODE    VOLTAGE       NODE     VOLTAGE
 ($N_0002)    1.2062                     ($N_0003)     6.8373
 ($N_0004)     .5191                     ($N_0006)    12.0000
 ($N_0007)    1.2062                     ($N_0008)     0.0000
 ($N_0009)    0.0000
    VOLTAGE SOURCE CURRENTS
    NAME       CURRENT
    V_V1        0.000E+00
    v_V2       -5.433E-03
    TOTAL POWER DISSIPATION   6.52E-02  WATTS
```

Fig. 15.25

Fig. 15.26 BJT schematic as printed.

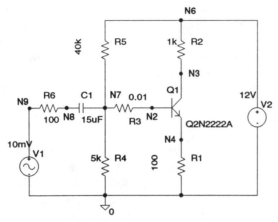

You may also want to observe the base and emitter voltages. In order to remove a trace, select its descriptive label with the mouse; then choose *Trace Delete*. The available traces include such descriptions as I(R6), and so forth. This signifies the current through R_6. The reference direction for this current can be found in the netlist entry

R_R6 $N_0009 $N_0008 100

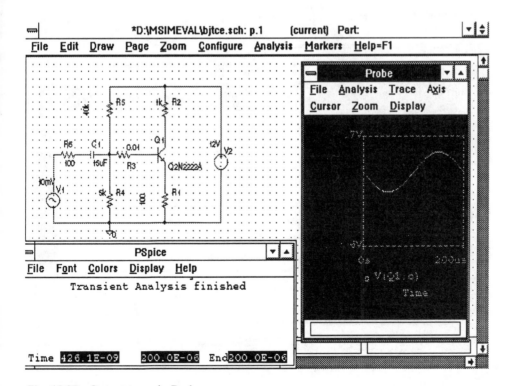

Fig. 15.27 Output trace in Probe.

This shows that the current is from *N9* (the first node) to *N8* (the second node). The plot of this current is up-going for the first half cycle. This emphasizes once more the importance of identifying the nodes (as assigned by Schematics) in your circuit drawing. Also include reference directions for currents, especially if currents are to be plotted in Probe.

Other items on the *Add Traces* list include such descriptions as

V(R6:1)

This notation refers to the voltage at one node of R_6 with respect to ground. The question is whether it refers to the node on the left or right of R_6. The colon followed by *1* means that the reference is to the first node in the netlist entry for *R_R6*. In this example, this is node *N9*. Since this is the node between V_1 and R_6, this is the voltage that will be displayed on this trace. Plot this voltage and you will see that it is the same as the input sine wave.

If we plot V(C1:2), this will give the voltage from *N7* to ground, since the second node of C_1 is shown in the netlist as *N7*. Plot this voltage and observe that it is the same as V(Q1:b). The suggested traces are shown in Fig. 15.28.

The Windows version of Probe allows for adjustments to be made in the size

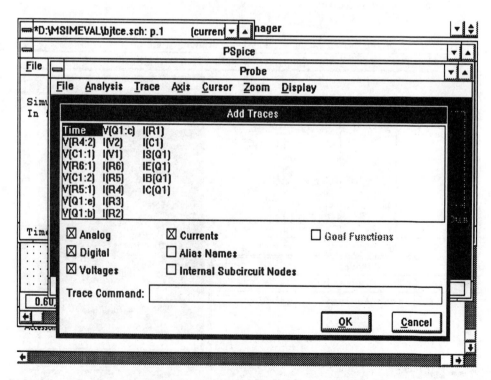

Fig. 15.28 Suggested traces for Probe analysis.

of the Probe window. It also is convenient to keep the Probe window open while some of the setup features are altered. As an example, in order to show two full cycles of the 5-kHz sine wave, shrink the Probe window and the PSpice window; then in the Schematic window make the necessary change under *Analysis Setup Transient*. Change the final time to 400 μs. Then run PSpice again. You will be asked if you want to abandon the current Probe file, and the answer is yes. After the PSpice analysis is done, display the voltage V(Q1:c) and see two cycles of the sine wave as shown in Fig. 15.29.

When hard copy is chosen from Probe, the choice can be made between portrait and landscape, as would be expected in other Windows applications. Several features are available in the DOS version of Probe, however, that are not found in the current Windows version of Probe. For example, using the DOS version, text can be written, arrows drawn, and boxes inserted; the cursor moves in response to the right- and left-arrow keys; and a cursor version of hard copy is available. These features are missing in the current Windows version of Probe.

Return to the *Analysis* option on the main menu, and under *Setup* note that there are several choices such as *AC Sweep, DC sweep,* and *Transient*. See Fig. 15.30 for the full list of options. The menus for some of the options are shown in Figs. 15.31 through 15.33.

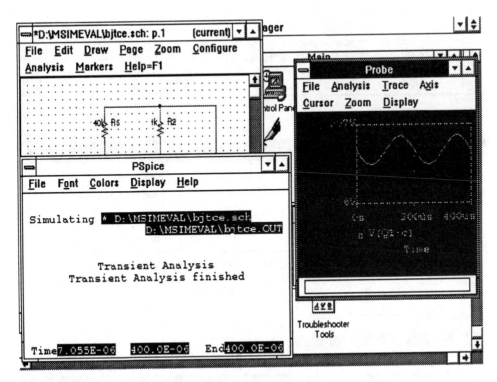

Fig. 15.29 Revised analysis in Probe.

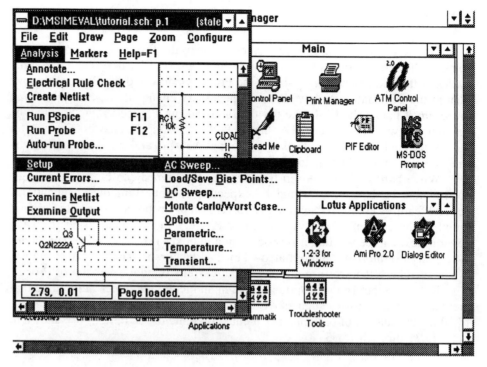

Fig. 15.30 Setup options.

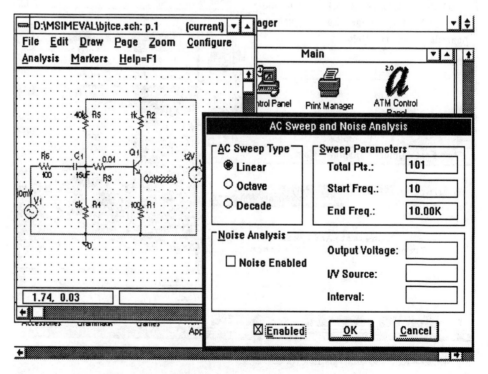

Fig. 15.31 ac sweep and noise analysis.

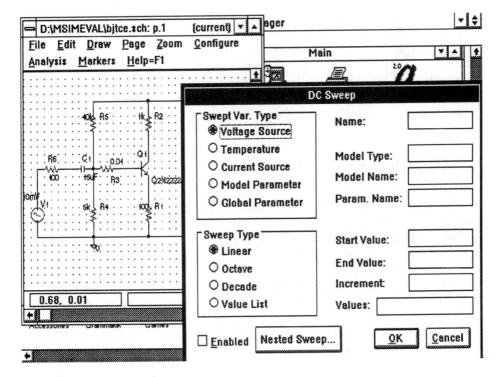

Fig. 15.32 dc sweep options.

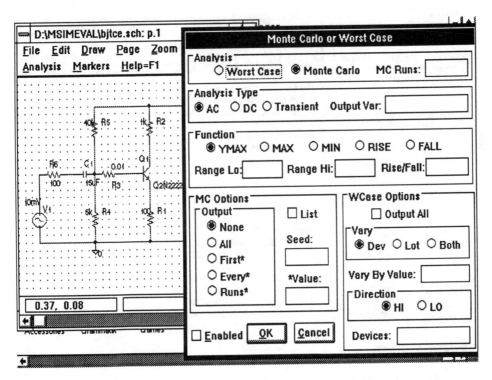

Fig. 15.33 Options for Monte Carlo or worst case.

GETTING HELP

As is typical for Windows applications, on-screen help is available. In the menu, select *Help*, or press the *F10* key. Some of the items available in *Help* are shown in Fig. 15.34.

THE SCHEMATICS SYMBOL LIBRARIES

In the evaluation version of Schematics, the libraries available under *Draw, Get New Part* include *abm.slb* (analog-behavioral modeling blocks symbol library), *analog.slb* (analog-symbol library), *breakout.slb* (breakout symbol library), *eval.slb* (evaluation-devices symbol library), *global.slb* (global symbol library), *marker.slb* (Probe-marker symbol library), *source.slb* (voltage- and current-source symbol library), and *special.slb* (special-device symbol library).

The analog symbol library is shown in Fig. 15.35, with the devices shown in the order in which they appear on the pull-down menu. If you use the descriptions such as C, E, F, and so forth (not C_1, E_1, F_1), under *Draw, Get New Part*, you will not have to use the more lengthy process of browsing. The voltage- and current-source library is shown in Fig. 15.36. Note that this library includes stimulus

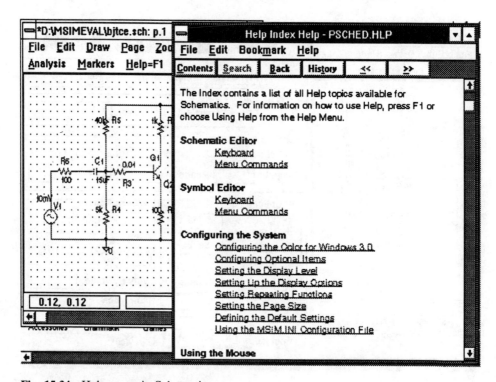

Fig. 15.34 Help menu in Schematics.

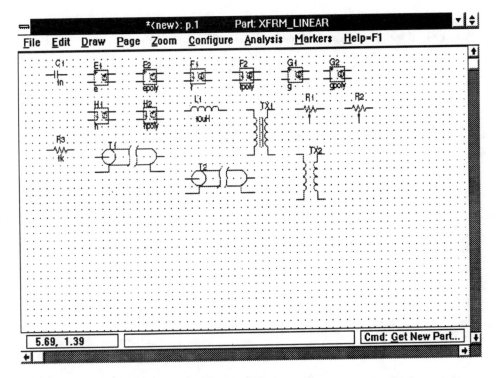

Fig. 15.35 Analog components in Schematics.

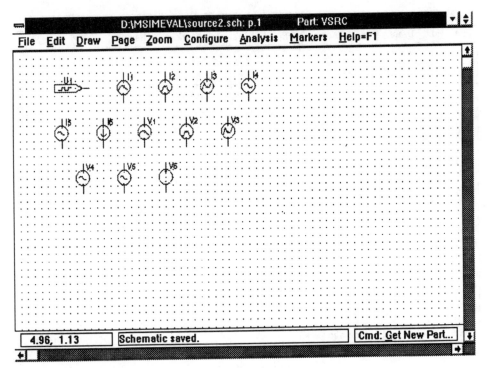

Fig. 15.36 Independent sources in Schematics.

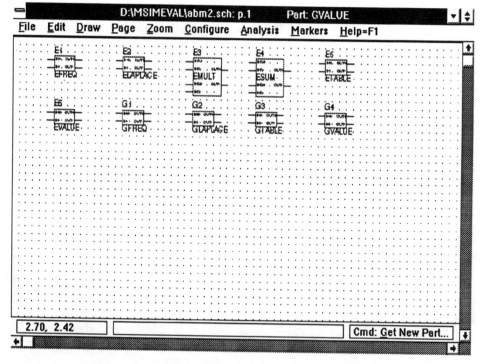

Fig. 15.37 Analog-behavioral modeling blocks.

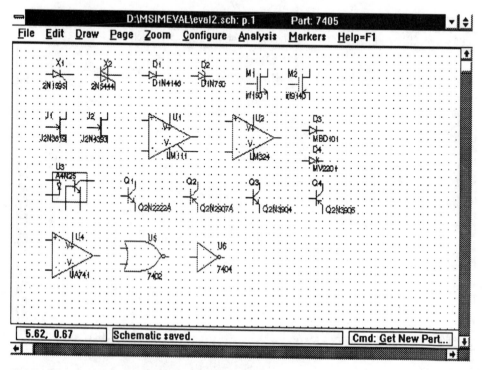

Fig. 15.38 Some devices in the evaluation library.

devices. The analog-behavioral modeling blocks library is shown in Fig. 15.37, and some of the devices in the evaluation library are shown in Fig. 15.38.

PARTS AND STIMULUS EDITOR

In the Windows version of the *Design Center*, icons are available for *Parts* and the *Stimulus Editor*. These are not Windows applications; they are the DOS programs that are simply available in the *Design Center* program group.

16

Nonlinear Devices

In many applications, the elements of an electrical circuit may behave in a non-linear fashion. The obvious examples include diodes and transistors, which have already been used in numerous examples. There are nonlinear resistors, such as the filament of a lamp, and there are nonlinear magnetic circuits, such as iron-core transformers, and so forth. Can PSpice handle these? The answer is a qualified *yes*.

THE NONLINEAR RESISTOR

A nonlinear resistor or other passive element can be simulated by using a dependent source. In Fig. 16.1 the basic circuit consists of a voltage source and two resistors R_i and R_{L1}. The current through R_{L1} is to be a function of the voltage across it, so that the basic form of Ohm's law, $v_1 = R_{L1}i$, does not apply unless you assume that R_{L1} is not constant. By placing a copy of R_{L1} in a loop with a dependent source, the volt-ampere response of this resistor can be of various shapes. If you use the polynomial form for the dependent source, you can let the response curve be almost anything that you can predict in terms of a curve fit. Recall that the polynomial is of the form

$$k_0 + k_1 x + k_2 x^2 + k_3 x^3 + \cdots$$

If you can assign values to the coefficients, you can predict the response. This is not always easy to do, but for some cases the relationship might be simple to state. Figure 16.1 contains two types of dependent sources. One is E, and you can let its value be determined by the voltage v_2 across R_{L1}. The other is F, and you can let

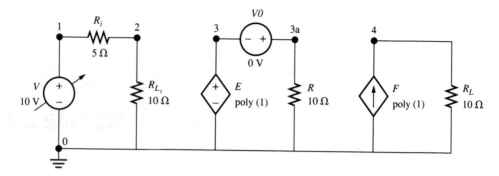

Fig. 16.1 Variable-resistor simulator.

its value be a function of the current through some branch of the circuit. Choose to let F be a function of the current through R. This example introduces no new concepts, and its input file is

```
Variable Resistor Simulator
V 1 0 10V
Ri 1 2 5
```

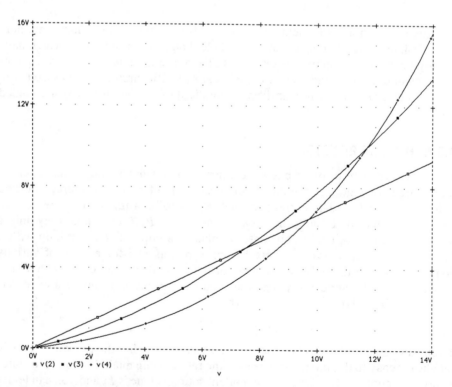

Fig. 16.2 Variable Resistor Simulator.

```
RL1  2  0  10
E  3  0  POLY(1)  2,0  0  0.5  0.1
VO  3A  3  0V
R  3A  0  10
F  0  4  POLY(1)  VO  0  -0.5  0.5
RL  4  0  10
.DC  V  0  14V  1V
.PROBE
.END
```

Run the analysis and in Probe plot v(2), v(3), and v(4). Because the resistors are actually linear, the current plots will obviously have the same shape as the voltage plots. Note that only the plot of v(2) is linear, while the other two have shapes that are determined by their respective polynomials. See Fig. 16.2 for these traces.

The same polynomial technique will apply to other elements such as capacitors and inductors as well.

THE IRON-CORE INDUCTOR

In any electric circuit where there is a current, there is also a magnetic field. The property of the magnetic field that is directly proportional to the current I is the magnetic field intensity H. The two quantities are related by a constant that is a function of the circuit configuration. An example is a coil of wire of a certain size and shape. The simple equation $H = kI$ applies, although it is often difficult to find an exact value for k.

The magnetic flux density is related to the field intensity by the equation $B = \mu H$. In free space the permeability is given the designation μ_o and is equal to $4\pi \times 10^{-7}$ newtons/ampere^2.

When the magnetic field is in a medium other than free space, a relative permeability u_r is introduced such that $\mu = \mu_o \mu_r$. The relative permeability is often not constant but is a function of the current. As the magnetic material begins to reach saturation, further increases in H produce little increase in B. When the current begins to decrease, the retentivity of the magnetic material prevents the B and H relationship from following the same pattern as when the current was increasing. This produces the familiar BH characteristic curve with its hysteresis cycle.

PSpice uses a model for the iron-core inductor based on the Jiles-Atherton treatment for the magnetic domains. It is beyond the scope of this text to describe this model completely; however, we can investigate the BH curve for various conditions and look at what happens to voltage and current waves in transformers when saturation occurs.

The BH Curve

The circuit of Fig. 16.3 contains a ferromagnetic coil with $R_L = 10 \ \Omega$ containing 20 turns. This is based on the sample program in MAGNETIC.LIB from the PSpice diskettes. Note that the statement for the inductor is

```
L1  1  0  20
```

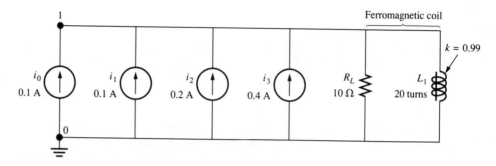

Fig. 16.3 Circuit to drive an iron-core inductor.

where the *20* represents *turns* and not 20 *henries*. This is in connection with the model statement containing the key word *core*. If the model were not used, the *20* would represent 20 henries. The four current generators are used to produce low-frequency (1-Hz) sine waves. The first current has a peak value of 0.1 A beginning at $t = 0$. Then at $t = 1$ s the next current is introduced. It also has a peak value of 0.1 A. Then a third and a fourth sine wave are added during the second and third seconds, respectively. The sine waves are of increasing amplitude in order to demonstrate the effects of saturation. The .MODEL statement uses the key word *core* and allows for the nonlinear magnetic parameters of the core model to take effect. The entire input file is

```
This is the sample magnetic core problem
i0 0 1 sin(0 0.1A 1Hz 0) ; no time delay for i0
i1 0 1 sin(0 0.1A 1Hz 1) ; another current added at 1 s
i2 0 1 sin(0 0.2A 1Hz 2) ; larger current added at 2 s
i3 0 1 sin(0 0.4A 1Hz 3) ; larger current added at 3 s
RL 1 0 10
L1 1 0 20 ; turns=20, not 20 henries since a model is used
K1 L1 0.99 KT ; coefficient of coupling
.model KT Core(MS=420E3 ALPHA=2E-5 A=26 K=18
+ C=1.05 AREA=1.17 PATH=8.49) ; the plus signs are for
+ continuation of statement
.options itl5=0
* Iteration parameter ITL5=0 gives ITL5=infinity
.tran 0.1 4
.probe
.end
```

Run the Probe analysis; then plot B(K1) as a function of time. This shows the nonlinear magnetic flux density in the core over the period from 0 to 4 s. Note that the first time periods show little nonlinearity compared to the later times. Verify that the first peak of B is at 995 oersteds, the second is for $B = 2000$ oersteds, the third is for $B = 3374$ oersteds, and the final peak is at $B = 4444$ oersteds. Refer to Fig. 16.4 for this trace.

In order to produce the standard *BH* curve, change the *X*-axis to represent H(K1). This is the magnetic field intensity in the core, which is directly proportional to the current. The *Y*-axis still represents B(K1). See Fig. 16.5 for these

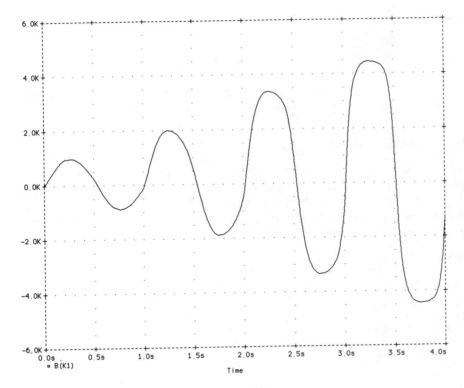

Fig. 16.4 This is the sample magnetic core problem.

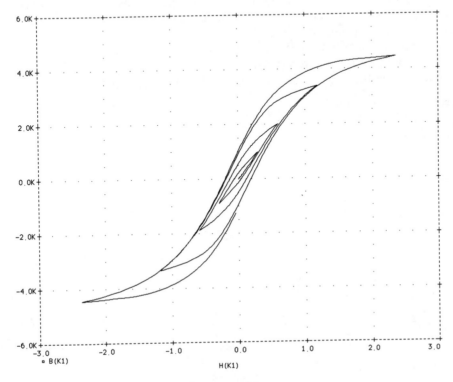

Fig. 16.5 This is the sample magnetic core problem.

traces. The four hysteresis loops are for the four levels of current. Verify where the peak values shown on the previous trace appear on this plot.

   You might want to try changing the number of turns on the coil and running the analysis again for comparison with the previous results.

THE IRON-CORE TRANSFORMER

When using a transformer with a magnetic core, the presence of the iron core will affect the shape of the current wave in the secondary. To illustrate this, consider the circuit of Fig. 16.6, which shows a current-driven transformer. The resistor R_1 is a resistor in parallel with the current source as required. The number of turns on the primary and secondary windings is to be 150 each. The input file is

```
Iron-Core Transformer
i 0 1 sin(0 1A 60Hz)
R1 1 0 1k
L1 1 0 150 ; turns = 150 on primary
L2 2 0 150 ; turns = 150 on secondary
R2 2 0 1
K1 L1 L2 0.9999 KT
.model KT core ; use default values for the core model
.options ITL5=0
.tran 1ms 16.67ms
.probe
.end
```

   Run the analysis and in Probe plot i(R2) and i(L1). Observe that although the primary current is a true sine wave, the secondary current is badly distorted. These traces are shown in Fig. 16.7.

   You may want to change the value of R_2 and/or the number of turns on each winding and compare the results with those obtained here. To demonstrate what happens when there is less saturation, use the following input file:

```
Iron-Core Transformer
i 0 1 sin(0 1A 60Hz)
R1 1 0 1k
L1 1 0 10
L2 2 0 150
R2 2 0 1
K1 L1 L2 0.9999 KT
```

Fig. 16.6 An iron-core transformer.

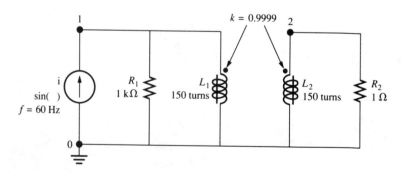

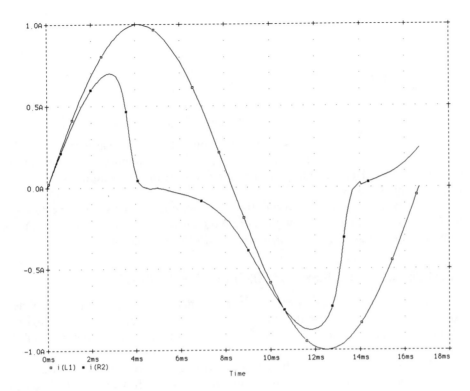

Fig. 16.7 Iron-Core Transformer.

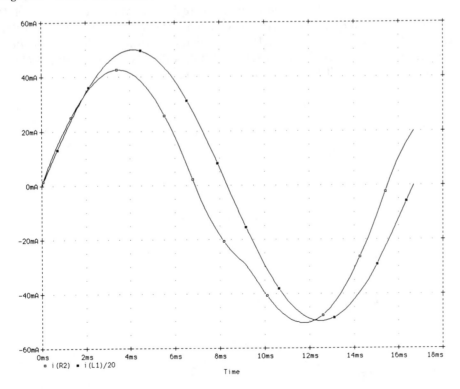

Fig. 16.8 Iron-Core Transformer.

```
.model KT core
.options ITL5=0
.tran 1ms 16.67ms
.probe
.end
```

Run the analysis and in Probe plot i(R2) and i(L1)/20. The latter plot is scaled to be comparable to the former plot. Here you will see that both curves are almost true sine waves. The secondary current still shows some distortion, however. See Fig. 16.8 for these curves.

You may want to try several other combinations of resistance and turns to become more familiar with how the core model works. You will discover that for some combinations, PSpice is unable to complete the analysis.

VOLTAGE-CONTROLLED SWITCH FOR VARIABLE RESISTOR

Another way to achieve a variation in resistance is by using either a voltage-controlled switch or a current-controlled switch. The switch can be made to open or close, depending on the value of voltage or current in another part of the circuit.

Consider Fig. 16.9, which shows the voltage-controlled switch in a series circuit with $V = 10$ V, $R_i = 50$ Ω, and $R_L = 50$ Ω. If you choose V as the controlling voltage, you can allow the switch to close when the voltage reaches a certain value. The switch will have nominal resistance values of $RON = 1$ Ω (switch closed) and $ROFF = 1$ MΩ (switch open). The latter value is necessary to prevent a floating node at the switch. The model requires the model type *vswitch*. In order to see the effects of the switch in both the off and on states, select $VON = 3$ V and use the default value $VOFF = 0$. The required input file is

```
Voltage-Controlled Switch
v 1 0 10V
Ri 1 2 50
```

Fig. 16.9 Voltage-controlled switch.

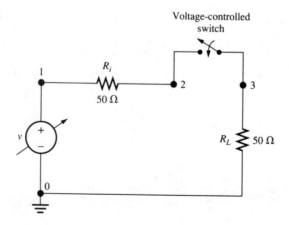

```
RL  3  0  50
S  2  3  1  0  S1 ;  vcs must begin with S
.model S1 vswitch(RON=1 ROFF=1E6 VON=3V VOFF=0)
.dc v 0 10V 0.05V
.probe
.end
```

Run the analysis and plot i(Ri). Notice that the curve shows almost no current until the input voltage nears 2 V. Then by the time the input voltage has reached $VON = 3V$, the slope of the curve is correct for the closed-switch condition. The total loop resistance with the switch closed is 101 Ω, which includes the resistance of the switch. Refer to Fig. 16.10 for this plot.

Change the value of VON to 8 V and run the analysis again. Observe the variations of the vi curve in the lower regions. The current begins to rise in the vicinity of 4 V. Remember that the slope of the curve is inversely proportional to the loop resistance. Note that the slope of the curve gradually changes; there is no abrupt change at the value of VON. This must be taken into account if you are to use this type of switch in your circuits. It is advisable to obtain a vi curve such as the one given here before employing the voltage-controlled switch in an elaborate circuit. Refer to Fig. 16.11 for this trace.

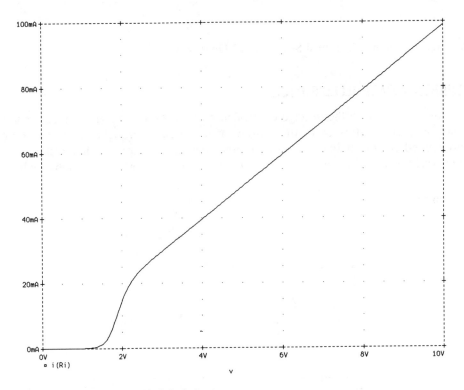

Fig. 16.10 Voltage-Controlled Switch.

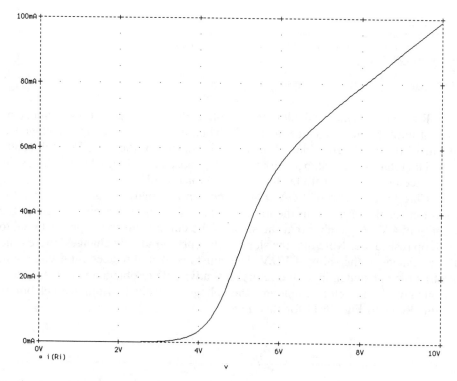

Fig. 16.11   Voltage-Controlled Switch with VON = 8 V.

CURRENT-CONTROLLED SWITCH

As an alternative to the voltage-controlled switch, the current-controlled switch may be used. In this case, the closing of a switch is brought about by some predefined current being present in another part of the circuit. This is shown in Fig. 16.12, where there is a current source supplying two branches, each of which

Fig. 16.12   Current-controlled switch.

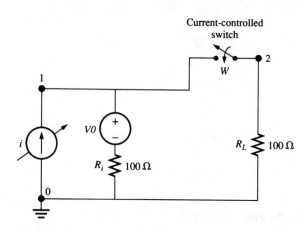

contains a 100-Ω resistor. The branch on the right contains the switch *W*. This branch has high resistance until the switch closes, with *ROFF* = 1 MΩ. When the predefined current *ION* = 10 mA is present in the branch on the left, the switch closes. *ION* uses the default value of zero. Beyond the 10-mA point, the resistance of the branch on the right is 101 Ω, since *RON* = 1 Ω. Current-controlled switches must have names beginning with *W*. The .MODEL statement must use the model type *ISWITCH*. The input file is

```
Current-Controlled Switch
i 0 1 40mA
VO 1 1A 0V
Ri 1A 0 100
RL 2 0 100
W 1 2 VO W1
.MODEL W1 ISWITCH(ION=10mA RON=1 ROFF=1E6)
.DC i 0 40mA 1mA
.PROBE
.END
```

Run the analysis and plot i(RL) as a function of *i*. Since a smooth change of resistance takes place in the circuit, the current through R_L does not have a linear slope until it is slightly greater than 15 mA. Thus for currents in this branch greater than 15 mA, the resistance of the branch is 101 Ω. See Fig. 16.13 for this trace.

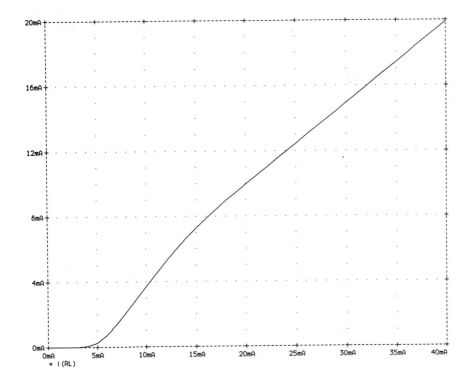

Fig. 16.13 Current-Controlled Switch.

NEW PSPICE STATEMENTS USED IN THIS CHAPTER

S[*name*] **<+ *switch node*>** **<−*switch node*>** **<+*controlling node*>**
                    <−*controlling node*> **<*model name*>**

For example,

```
S 2 3 1 0 S1
```

means that there is a voltage-controlled switch between nodes *2* and *3*. The switch is normally open (by default), but when the controlling voltage (between nodes *1* and *0*) reaches a certain value, the switch will close. The *S* statement requires a .MODEL statement to account for *on* resistance, *off* resistance, and controlling-voltage values. In this case, the model is identified as *S1*. This identification must begin with the letter *S*. Refer to Appendix D for a full description.

W[*name*] **<+*node*>** **<−*node*>** **<*vname*>** **<*model*>**

For example,

```
W 1 2 V0 W1
```

means that there is a current-controlled switch between nodes *1* and *2*. The switch is normally open (by default), but when the controlling current passing through the source voltage V0 reaches a certain value, the switch will close. The *W* statement requires a .MODEL statement similar to that required for the voltage-controlled switch described above. In the model statement the model must begin with the letter *W*. Refer to Appendix D for a full description.

NEW DOT COMMAND

.MODEL **<*name*>** **<*type*>** **([<*parameter name*>=<*value*>]*)**

For example,

```
.MODEL KT core
```

When a model is for inductor coupling, the name begins with *K*. If the model statement also contains the word core, a nonlinear model is used. The *BH* curve for this nonlinear device will be similar to that shown in Fig. 16.5.

PROBLEMS

16.1 In the discussion of the nonlinear resistor, we pointed out that the resistors are not actually nonlinear, but that the dependent sources were the nonlinear elements. Modify the circuit shown in Fig. 16.1 to produce a voltage $V(3)$ that is approximately equal to the voltage V raised to the 1.5 power.

16.2 For the circuit shown in Fig. 16.14, $L_1 = L_2 = 25$ mH, $C_1 = C_2 = 50$ nF, $R_s = 1\,\Omega$, $R_L = 1$ kΩ, and $M = 1$ mH. The statement for the transformer has the form

```
K1 L1 L2 value
```

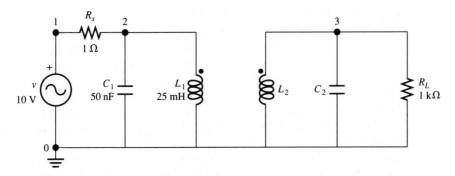

Fig. 16.14

where *value* is the coefficient of coupling. Create an input file to produce a plot of voltage across R_L in the vicinity of the resonant frequency. From the plot determine whether the coupling is less than, equal to, or greater than critical. Justify your answer.

16.3 What is the value of critical coupling in Problem 16.2? Run an analysis to demonstrate that critical coupling gives maximum power transfer to R_L.

16.4 The circuit shown in Fig. 16.15 uses an iron-core transformer (default model). The inductor L_1 contains 150 turns, while L_2 has 300 turns. Perform an analysis at $f = 4.5$ kHz to determine the load-resistor voltage and current.

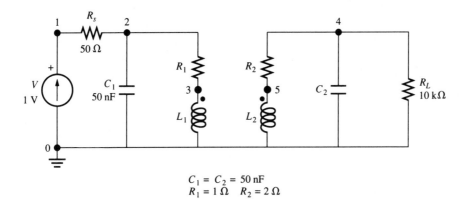

$$C_1 = C_2 = 50 \text{ nF}$$
$$R_1 = 1\,\Omega \quad R_2 = 2\,\Omega$$

Fig. 16.15

16.5 Using the same figure as in Problem 16.4, remove the .model statement and use inductive values for L_1 and L_2 instead. In order to produce approximately the same results as previously obtained, begin with $L_1 = 5$ mH and $L_2 = 10$ mH. Run an analysis and print the ac voltages V(2) and V(4). The results should be V(2) = 0.978 V and V(4) = 1.367 V. Select other values for L_1 and L_2 to obtain results that are in closer agreement with those obtained in Problem 16.4. Run the analysis several times using selected values.

16.6   A voltage-controlled switch is shown in Fig. 16.9. A design is required such that the switch will appear to have resistance of 1 kΩ when open and 1 Ω when closed. Let R_i = 50 Ω and R_L = 100 Ω. In the beginning the switch should be closed. The switch is to open when the input voltage v reaches 5 V. Prepare the input file, run the analysis, and verify your results with a Probe plot.

16.7   Again refer to Fig. 16.9. Use the resistance values given in the figure. The switch should have a resistance of 1 kΩ when open and 1 Ω when closed. In the beginning the switch should be closed. The switch is to open when the voltage across R_i reaches 0.25 V. Prepare the input file, run the analysis, and verify your results with a Probe plot.

Note that the switch statement must give the dependent nodes in the proper order for the analysis to be correct. Run the analysis to determine the proper order of these nodes. You may be surprised! In Probe verify that the current through R_i becomes significant when the source voltage v exceeds 0.55 V.

Brief Summary of PSpice Statements

The statements shown here are given in quick-reference form. This will prove helpful if you need to look up a statement that you have seen or used previously. More detailed information is given in Appendix B and Appendix D and at the end of each chapter.

General notes:

Upper-case and lower-case alphabetic characters may be used interchangeably.

< > means required information.

[ ] means optional information.

* indicates a comment line in the PSpice input file.

; shows a comment, generally following a statement on the same line.

.AC[LIN][OCT][*dec*] <*points*> <*f start*> <*f end*> ;for ac sweep

B[*name*] <*drain*> <*gate*> <*source*> <*model name*> [*area value*] ;for GaAsFET

C[*name*] <*+node*> <*−node*> [*model name*] <*value*> [IC = initial value] ;for capacitor

D[*name*] <*+node*> <*−node*> <*model name*> [*area value*] ;for diode

.DC [LIN][OCT][DEC] <*sweep variable*> <*start*> <*end*> <*increment*> [LIST] ;for dc sweep

E[*name*] <+*node*> <−*node*> <+*controlling node*>
  <−*controlling node*> <*gain*> ;VCVS

E[*name*] <+*node*> <−*node*> POLY(*value*) <+*controlling node*>
  <−*controlling node*> <*polynomial coefficient values*> ;VCVS

.END ;indicates the end of the circuit input file

.ENDS <*subcircuit name*> ;indicates the end of a subcircuit

F[*name*] <+*node*> <−*node*> <*controlling V device name*>
  <*gain*> ;CCCS

F[*name*] <+*node*> <−*node*> POLY(*value*)
  <*controlling V device name*> <*polynomial coefficient values*>
  ;CCCS

.FOUR <*frequencies*> <*output variable*> ;for Fourier analysis

G[*name*] <+*node*> <−*node*> <+*controlling node*>
  <−*controlling node*> <*transconductance*> ;VCCS

G[*name*] <+*node*> <−*node*> POLY(value) <+*controlling node*>
  <−*controlling node*> <*polynomial coefficient values*> ;VCCS

H[*name*] <+*node*> <−*node*> <*controlling V device name*>
  <*transresistance*> ;CCVS

H[*name*] <+*node*> <−*node*> POLY(*value*)
  <*controlling V device name*> <*polynomial coefficient values*>
  ;CCVS

I[*name*] <+*node*> <−*node*> [[DC] <*value*>] [AC <*magnitude*>
  [*phase*]] [*transient specification*] ; for independent current source

.IC <*Vnode* = *value*> ;shows an initial node voltage for transient
  analysis

.INC <*file name*> ;inserts another file

J[*name*] <*drain node*> <*gate node*> <*source node*>
  <*model name*> [*area*] ;for JFET

K[*name*] L[*name*] [L[*name*]] <*coupling value*> ;for inductor
  coupling

K[*name*] L[*name*] [L[*name*]] <*coupling value*> <*model name*>
  [*size value*] ;for inductor coupling model

L[*name*] <+*node*> <−*node*> [*model name*] <*value*> [IC = *value*];
  for inductor

.LIB <*file name*> ;references a model or subcircuit library in
  another file. The default file is NOM.LIB.

.MC [#runs] [DC] [AC] [TRAN] [*output variable*] YMAX ;for
  Monte Carlo analysis

.MODEL [*name*] [*type*] ;describes a built-in model

M[*name*] <*drain node*> <*gate node*> <*source node*> <*bulk/substrate node*> <*model name*> [*parameter value*] ;for MOSFET

.NODESET <*Vnode=value*> ;gives initial guess for node voltage

.NOISE <*Vnode*> [*,node*] <*name*> [*internal value*] ;gives noise analysis along with ac analysis

.OP ;gives detailed bias-point information

.OPTIONS <*option name*> ;sets options for analysis

.PLOT [DC] [AC] [NOISE] [TRAN] [*output variable*]; gives printer-type plot

.PRINT [DC] [AC] [NOISE] [TRAN] [*output variable*] ;produces a listing in the output file

.PROBE [*output variable*] ;creates PROBE.DAT file for graphics analysis

Q[*name*] <*collector node*> <*base node*> <*emitter node*> [*substrate node*] <*model name*> [*area value*] ;for BJT

R[*name*] <*+node*> <*−node*> [*model name*] <*value*> ;for resistor

S[*name*] <*+switch node*> <*−switch node*> <*+controlling node*> <*−controlling node*> <*model name*> ;for voltage-controlled switch

.SENS <*output variable*> ;used for the sensitivity analysis

.SUBCKT <*name*> [*node(s)*] ;marks the beginning of a subcircuit

T[*name*] <*+A port node*> <*−A port node*> <*+B port node*> <*−B port node*> <*ZO=value*> [TD=*value*] [F=*value*] [NL=*value*] ;for ideal transmission line

.TEMP <*value*> ;sets the temperature for the analysis in Celsius

.TF <*output variable*> <*input source*> ;for transfer function

.TRAN <*step value*> <*final value*> [*step ceiling value*] [UIC] ;for transient analysis

V[*name*] <*+node*> <*−node*> [[DC <*value*>]] [AC <*magnitude*> [*phase*]] [*transient specification*] ; for independent voltage

W[*name*] <*+switch node*> <*−switch node*> <*controlling V device name*> <*model name*> ;for current-controlled switch

.WIDTH=<*value*> ;sets the number of characters per line of output

X[*name*] <*[nodes]*> <*subcircuit name*> ;to specify a subcircuit

PSpice Devices and Statements as Listed in PSPICE.HLP

PSPICE DEVICES

B device—GaAsFET

General form:

B*<name><d><g><s><model>*[*<area>*]

Examples:

```
BIN 100  1  0 GFAST
B13   22 14 23 GNOM 2.0
```

C device—Capacitor

General form:

C*<name><+node><−node>*[*<model>*]*<value>*[IC=*<initial>*]

Examples:

```
CLOAD 15   0 20pF
CFDBK  3 33 CMOD 10pF IC=1.5v
```

D device—Diode

General form:

D*<name><+node><−node><model>*[*area*]

Examples:

```
DCLAMP 14   0 DMOD
D13      15 17 SWITCH 1.5
```

E device—(Voltage-) Controlled Voltage Source

General forms:

E*<name><+node><−node><+control><−control><gain>*

E*<name><+node><−node>*POLY(*<value>*)<<+*control><−control>>**
<<*coeff>>**

E*<name><+node><−node>*VALUE={*<exp>*}

E*<name><+node><−node>*TABLE{*<exp>*}<(*inval*),(*outval*)>*

E*<name><+node><−node>*LAPLACE{*<exp>*}{*<sexp>*}

E*<name><+node><−node>*FREQ{*<exp>*}<(*freq,magdb,phasedeg*)>*

Examples:

```
EBUFF    1    2 10 11 1.0
EAMP     13   0 POLY(1)  26 0 500
ENLIN  100 101 POLY(2)  3 0 4 0 0.0 13.6 0.2 0.005
ESQRT   10   0 VALUE = {SQRT(V(5))}
ETAB    20   5 TABLE {V(2)} (-5v,5v)  (0v,0v)  (5v,-5v)
E1POLE  10   0 LAPLACE {V(1)} {1 / (1 + s)}
EATTEN  20   0 FREQ {V(100)} (0,0,0 10,-2,-5 20,-6,-10)
```

F device—Current-Controlled Current Source

General forms:

F*<name><+node><−node><vname><gain>*

F*<name><+node><−node>*POLY(*<value>*)*<vname>>**<*coeff>>**

Examples:

```
FSENSE   1    2 VSENSE 10.0
FAMP     13   0 POLY(1) VIN 500
FNLIN  100 101 POLY(2) V1 v2 0.0 0.9 0.2 0.005
```

G device—(Voltage-) Controlled Current Source

General forms:

G*<name><+node><−node><+control><−control><gain>*

G*<name><+node><−node>*POLY(*<value>*)<<+*control><−control>>**
   <<*coeff>>**

G*<name><+node><−node>*VALUE=[*<exp>*]

G*<name><+node><−node>*TABLE[*<exp>*]=<(*inval*),(*outval*)>*

G*<name><+node><−node>*LAPLACE[*<exp>*][*<sexp>*]

G*<name><+node><−node>*FREQ[*<exp>*]<(*freq,magdb,phasedeg*)>*

Examples:

```
GBUFF     1    2 10 11 1.0
GAMP      13   0 POLY(1) 26 0 500
GNLIN    100 101 POLY(2) 3 0 4 0 0.0 13.6 0.2 0.005
GSQRT     10   0 VALUE = {SQRT(V(5))}
GTAB      20   5 TABLE{V(2)} = (−5v,5v)  (0v,0v)  (5v,−5v)
G1POLE    10   0 LAPLACE [V(1)] [1 / (1 + s)]
GATTEN    20   0 FREQ [V(100)] (0,0,0 10,−2,−5 20,−6,−10)
```

H device—Current-Controlled Voltage Source

General forms:

H<*name*><*+node*><*−node*><*vname*><*gain*>

H<*name*><*+node*><*−node*>POLY(<*value*>)<<*vname*>>*<<*coeff*>>*

Examples:

```
HSENSE    1    2 VSENSE 10.0
HAMP      13   0 POLY(1) VIN 500
HNLIN    100 101 POLY(2) V1 v2 0.0 0.9 0.2 0.005
```

I device—Current Source

General form:

I<*name*><*+node*><*−node*>[[DC]<*value*>][AC<*mag*>[<*phase*>]][<*transient*>]

Transient specifications:

     EXP(i1 ipk rdelay rtc fdelay ftc)

     PULSE(i1 i2 td trise tfall pw per)

     PWL(t1 i1 t2 i2 . . . tn fn)

     SFFM(ioff iampl fc mod fm)

     SIN(ioff iampl freq td df phase)

Examples:

```
IBIAS   13   0 2.3mA
IAC      2   3 AC .001
IACPHS   2   3 AC .001 90
IPULSE   1   0 PULSE(−1mA 1mA 2ns 2ns 2ns 50ns 100ns)
I3 26   77     DC .002 AC 1 SIN(.002 .002 1.5MEG)
```

J device—Junction FET

General form:

J<*name*><*d*><*g*><*s*><*model*>[<*area*>]

Examples:

```
JIN 100  1  0 JFAST
J13   22 14 23 JNOM 2.0
```

K device—Inductor Coupling

General forms:

K<*name*>L<*name*><L<*name*>>*<*coupling*>

K<*name*><L<*name*>>*<*coupling*><*model*>[<*size*>]

Examples:

```
KTUNED  L3OUT L4IN  .8
KXFR1   LPRIM LSEC  .99
KXFR2   L1 L2 L3 L4 .98 KPOT_3C8
```

L device—Inductor

General form:

L<*name*><+*node*><−*node*>[*model*]<*value*>[IC=<*initial*>]

Examples:

```
LLOAD   15  0 20mH
L2       1  2 .2e−6
LCHOKE   3 42 LMOD .03
LSENSE   5 12 2uH IC=2mA
```

M device—MOSFET

General form:

M<*name*><*d*><*g*><*s*><*sub*><*mdl*>[L=<*value*>][W=<*value*>]
[AD=<*value*>][AS=<*value*>][PD=<*value*>][PS=<*value*>]
[NRD=<*value*>][NRS=<*value*>]
[NRG=<*value*>][NRB=<*value*>]

Examples:

```
M1   14 2  13    0 PNOM L=25u W=12u
M13  15 3   0    0 PSTRONG
M2A   0 2 100  100 PWEAK L=33u w=12u
+ AD=288p AS=288p PD=60u PS=60u NRD=14 NRS=24 NRG=10
```

N device—Digital Input

General form:

N<*name*><*inode*><*lonode*><*hinode*><*model*>[SIGNAME=<*name*>][IS=<*init.*

Examples:

```
NRESET  7 15  16 FROM_TTL
N12       18  0 100 FROM_CMOS SIGNAME=VCO_GATE IS=0
```

O device—Digital Output

General form:

O<*name*><*iface*><*ref*><*model*>[SIGNAME=<*name*>]

Examples:

```
OVCO 17   0 TO_TTL
O5    22 100 TO_CMOS SIGNAME=VCO_OUT
```

Q device—Bipolar Transistor

General form:

Q<*name*><*c*><*b*><*e*>[<*subs*>]<*model*>[<*area*>]

Examples:

```
Q1   14 2 13 PNPNOM
Q13 15 3  0 1 NPNSTRONG 1.5
Q7  VC 5 12 [SUB] LATPNP
```

R device—Resistor

General form:

R<*name*><+*node*><-*node*>[<*model*>]<*value*>

Examples:

```
RLOAD 15 0 2k
R2     1 2 2.4e4
```

S device—Voltage-Controlled Switch

General form:

S<*name*><+*node*><-*node*><+*control*><-*control*><*model*>

Examples:

```
S12 13 17 2 0 SMOD
SRESET 5 0 15 3 RELAY
```

T device—Transmission Line

General form:

T<*name*><A+><A-><B+><B-><Z0=*value*>[TD=<*val*>]
  [F=<*val*>[NL=<*val*>]]

Examples:

```
T1 1 2 3 4 Z0=220 TD=115ns
T2 1 2 3 4 Z0=50 F=5MEG NL=0.5
```

U device—Digital

General form:

U<*name*><*type*><*parms*><*node*>*[<*parm*>=<*val*>]*

Types: BUF, INV, AND, NAND, OR, NOR, XOR, NXOR, BUF3, INV3, AND3, OR3, NOR3, XOR3, NXOR3, JKFF, DFF, PULLUP, PULLDN, STIM.

STIM Syntax:

U*<name>*STIM(*<width>*,*<radices>*)*<node>***<iomodel>*
  [TIMESTEP=*stepsize*]<<*time>*,*<value>*>|
  <<*time>*GOTO*<label>*<*n*>TIMES>|
  <<*time>*GOTO*<label>*<*rv>*<*val>*>|
  <<*time>*INCR BY*<val>*>>
  <<*time>*DECR BY*<val>*>>**<rv>*=UNTIL GT|GE|LT|LE

Examples:

```
U7 XOR() INA INB OUTXOR DEFGATE DEFIO
U101 STIM( 1, 1 ) IN1 STMIO TIMESTEP=10ns
+ (LABEL=STARTLOOP) (+10c, 0)(+5ns, 1)
+ (+40c GOTO STARTLOOP 1 TIMES)
```

V device—Voltage Source

General form:

V*<name><+node><−node>*[[DC]*<value>*][AC*<mag>*[*<phase>*]]][*<transient>*]

Transient specifications:

    EXP(iv vpk rdelay rtc fdelay ftc)
    PULSE(v1 v2 td trise tfall pw per)
    PWL(t1 v1 t2 v2 . . . tn vn)
    SFFM(voff vampl fc mod fm)
    SIN(voff vampl freq td df phase)

Examples:

```
VBIAS   13   0 2.3mV
VAC      2   3 AC .001
VACPHS   2   3 AC .001 90
VPULSE   1   0 PULSE(−1mV 1mV 2ns 2ns 2ns 50ns 100ns)
V3      26  77 DC .002 AC 1 SIN(.002 .002 1.5MEG)
```

W device—Current-Controlled Switch

General form:

W*<name><+node><−node><vname><model>*

Examples:

```
W12 13 17 VC WMOD
WRESET 5 0 VRESET RELAY
```

X device—Subcircuit Call

General form:

X*<name>*[*<node>*]**<sname>*[PARAMS:<*<par>*=*<val>**>]

Examples:

```
X12 100 101 200 201 DIFFAMP
XBUFF 13 15 UNITAMP
```

PSPICE STATEMENTS AS LISTED IN PSPICE.HLP

.AC—AC Analysis

General form:

.AC[LIN][OCT][DEC]<*points*><*start*><*end*>

Examples:

```
.AC LIN 101 10Hz 200Hz
.AC OCT  10 1KHz 16KHz
.AC DEC  20 1MEG 100MEG
```

.DC—DC Analysis

General forms:

.DC[LIN]<*varname*><*start*><*end*><*incr*>[<*nest*>]

.DC[OCT][DEC]<*varname*><*start*><*end*><*points*>[<*nest*>]

.DC<*varname*>LIST<*value*>*[<*nest*>]

Examples:

```
.DC VIN −.25 .25 .05
.DC LIN I2 5mA −2mA 0.1mA
.DC VCE 0v 10v .5v IB 0mA 1mA 50uA
.DC RES RMOD(R) 0.9 1.1 .001
.DC DEC NPN QFAST(IS) 1e−18 1e−14 5
.DC TEMP LIST 0 20 27 50 80
```

.DISTRIBUTION—User-Defined Dist

General form:

.DISTRIBUTION<*name*><<*dev*><*prob*>>*

Example:

```
.DISTRIBUTION bimodal (−1,1)(−.5,1)(−.5,0)(.5,0)(.5,1)(1,1)
```

.END—End Circuit

.ENDS—End Subcircuit

General forms:

.END

.ENDS[<*name*>]

Examples:

```
. END
. ENDS
. ENDS 741
```

.FOUR—Fourier Analysis

General form:

.FOUR*<freq><output var>**

Examples:

```
. FOUR 10KHz v(5) v(6,7)
```

.FUNC—Define Function

General form:

.FUNC*<name>*([*arg**])*<body>*

Examples:

```
. FUNC DR(D) D/57.296
. FUNC E(X) EXP(X)
. FUNC APBX(A,B,X) A+B*X
```

.IC—Initial Transient Conditions

General form:

.IC*<<vnode>=<value>>**

Examples:

```
. IC V(2)=3.4 V(102)=0
```

.INC—Include File

General form:

.INC*<name>*

Examples:

```
. INC SETUP.CIR
. INC C:\PSLIB\VCO.CIR
```

.LIB—Library File

General form:

.LIB[*<name>*]

Examples:

```
.LIB
.LIB OPNOM.LIB
.LIB C:\\PSLIB\\QNOM.LIB
```

.MC—Monte Carlo Analysis

General form:

.MC<*#runs*>[DC][AC][TRAN]<*opvar*><*func*><*option*>*

Examples:

```
.MC 10 TRAN V(5) YMAX
.MC 50   DC IC(Q7) MIN LIST
.MC 20   AC VP(13,5)RISE_EDGE(1.0) LIST OUTPUT ALL
```

.WCASE—Worst-Case Analysis

General form:

.WCASE<*analysis*><*opvar*><*func*><*option*>*

Examples:

```
.WCASE DC V(4,5) YMAX
.WCASE TRAN V(1) FALL_EDGE(3.5v) VARY BOTH BY RELTOL DEVICES RL
```

.MODEL—Model

General form:

.MODEL<*name*><*type*>[<*param*>=<*value*>[<*tol*>]]*

Typename	Devname	Devtype
CAP	Cxxx	capacitor
IND	Lxxx	inductor
RES	Rxxx	resistor
D	Dxxx	diode
NPN	Qxxx	*npn* bipolar
PNP	Qxxx	*pnp* bipolar
LPNP	Qxxx	lateral PNP
NJF	Jxxx	*n*-channel JFET
PJF	Jxxx	*p*-channel JFET
NMOS	Mxxx	*n*-channel MOSFET
PMOS	Mxxx	*p*-channel MOSFET
GASFET	Bxxx	*n*-channel GaAsFET
CORE	Kxxx	nonlinear core
VSWITCH	Sxxx	v/c switch
ISWITCH	Wxxx	c/c switch
DINPUT	Nxxx	digital i/p
DOUTPUT	Oxxx	digital o/p

Examples:

```
.MODEL RMAX RES (R=1.5 TC=.02 TC2=.005)
.MODEL QDRIV NPN (IS=1e-7 BF=30)
.MODEL DLOAD D (IS=1e-9 DEV 5% LOT 10%)
```

.NODESET—Nodeset

General form:

.NODESET<<*node*>=<*value*>>*

Examples:

```
.NODESET V(2)=3.4 V(3)=-1V
```

.NOISE—Noise Analysis

General form:

.NOISE<*opvar*><*name*>[<*ival*>]

Examples:

```
.NOISE V(5) VIN
.NOISE V(4,5) ISRC 20
```

.OP—Bias Point

General form:

.OP

Examples:

```
.OP
```

.OPTIONS—Options

General form:

```
.OPTIONS[<fopt>*][<vopt>=<value>*]
```

Flag Options:

ACCT summary & accounting
EXPAND show subcircuit expansion
LIBRARY list lines from library files
LIST output summary
NODE output netlist
NOECHO suppress listing
NOMOD suppress model param listing
NOPAGE suppress banners
OPTS output option values

Value Options:

ABSTOL	best accuracy of currents
CHGTOL	best accuracy of charges
CPTIME	CPU time allowed
DEFAD	MOSFET default AD
DEFAS	MOSFET default AS
DEFL	MOSFET default L
DEFW	MOSFET default W
GMIN	min conductance, any branch
ITL1	DC & bias pt blind limit
ITL2	DC & bias pt guess limit
ITL4	transient per-point limit
ITL5	transient total, all points
LIMPTS	max for print/plot
NUMDGT	#digits output
PIVREL	rel mag for matrix pivot
PIVTOL	abs mag for matrix pivot
RELTOL	rel accuracy of V's and I's
TNOM	default temp
TRTOL	transient accuracy adjustment
VNTOL	best accuracy of voltages
WIDTH	output width

Examples:

```
.OPTIONS NOECHO NOMOD RELTOL=.01
.OPTIONS ACCT DEFL=12u DEFW=8u
```

.PARAM—Global Parameter

General form:

.PARAM<<*name*>=<*value*>>*

Examples:

```
.PARAM pi=3.14159265
.PARAM RSHEET=120, VCC=5V
```

.PLOT—Plot

General form:

.PLOT[DC][AC][NOISE][TRAN][[<*opvar*>*][(<*lo*>,<*hi*>)]]*

Examples:

```
.PLOT DC V(3) V(2,3) V(R1) I(VIN)
.PLOT AC VM(2) VP(2) VG(2)
.PLOT TRAN V(3) V(2,3) (0,5V) ID(M2) I(VCC) (-50mA,50mA)
```

.PRINT—Print

General form:

.PRINT[DC][AC][NOISE][TRAN][<opvar>*]

Examples:

```
.PRINT DC V(3) V(2,3) V(R1) IB(Q13)
.PRINT AC VM(2) VP(2) VG(5) II(7)
.PRINT NOISE INOISE ONOISE DB(INOISE)
```

.PROBE—Probe

General forms:

.PROBE[/CSDF]

.PROBE[/CSDF][<opvar>*]

Examples:

```
.PROBE
.PROBE v(2) I(R2) VBE(Q13) VDB(5)
```

.PROBE/CSDF

.SENS—Sensitivity Analysis

General form:

.SENS<opvar>*

Examples:

```
.SENS V(9) V(4,3) I(VCC)
```

.STEP—Stepped Analysis

General forms:

.STEP[LIN]<varname><start><end><incr>

.STEP[OCT][DEC]<varname><start><end><points>

.STEP<varname>LIST<value>*

.STEP PARAM X 1 5 0.1

Examples:

```
.STEP VIN -.25 .25 .05
.STEP LIN I2 5mA -2mA 0.1mA
```

```
.STEP RES RMOD(R) 0.9 1.1 .001
.STEP TEMP LIST 0 20 27 50 80
```

.SUBCKT—Subcircuit Definition

General form:

.SUBCKT<*name*>[<*node*>*][PARAMS:<*par*>[=<*val*>]*]

Examples:

```
.SUBCKT OPAMP 1 2 101 102
.SUBCKT FILTER IN OUT PARAMS: CENTER, WIDTH=10 KHz
```

.TEMP—Temperature

General form:

.TEMP<*value*>*

Examples:

```
.TEMP 125
.TEMP 0 27 125
```

.TF—Transfer Function

General form:

.TF<*opvar*><*ipsrc*>

Examples:

```
.TF V(5) VIN
.TF I(VDRIV) ICNTRL
```

.TRAN—Transient Analysis

General form:

.TRAN[/OP]<*pstep*><*ftime*>[<*noprint*>[<*ceiling*>]]][UIC]

Examples:

```
.TRAN 1ns 100nS
.TRAN/OP 1nS 100nS 20ns UIC
.TRAN 1nS 100nS 0nS .1nS
```

.WIDTH—Width

General form:

.WIDTH OUT=<*val*>

Example:

```
.WIDTH OUT=80
```

OUTPUT VARIABLES

This section describes the types of output variables that can be used in both the .PRINT and .PLOT statements. Each such statement may have up to eight output variables.

DC Sweep and Transient Analysis

V(<*node*>)
V(<+*node*>,<−*node*>)
V(<*name*>)
Vx(<*name*>)
Vxy(<*name*>)
Vz(<*name*>)
I(<*name*>)
Ix(<*name*>)
Iz(<*name*>)

The following is an abbreviated list of the two-terminal device types for which the dc sweep and transient analysis applies.

Devtypes: C/D/E/F/G/H/I/L/R/V

For the Vx, Vxy, Ix forms, <*name*> must be a three- or four-terminal device and *x* and *y* must each be a terminal abbreviation. In abbreviated form these are

xy: D/G/S (B)
xy: D/G/S (J)
xy: D/G/S/B (M)
xy: C/B/E/S (Q)
z: A/B

AC Analysis

Suffixes:

M magnitude
DB magnitude
P phase
G group delay
R real
I imaginary

The following is an abbreviated list of devices through which currents are available:

Devtypes: C/I/L/R/T/V

For other devices, you must put a zero-valued voltage source in series with the device (or terminal) of interest.

Noise Analysis

INOISE

ONOISE

DB(INOISE)

DB(ONOISE)

COMMON SOURCES OF ERROR IN PSPICE INPUT FILES

Floating nodes have no dc path to ground. There are three frequent causes of trouble:

1. The two ends of a transmission line do not have a dc connection between them.
2. Voltage-controlled sources do not have a dc connection between their controlling nodes.
3. There is an error in the circuit description.

Assuming that this circuit is correct, the solution is to connect the floating node to ground via a large-value resistor.

PSpice checks for zero-resistance loops. These may be caused by independent voltage sources (V), controlled voltage sources (E and H), and inductors (L); or there may be an error in the circuit description.

Assuming the circuit to be correct, the solution is to add series resistance into the loop.

Convergence problems may occur in the dc sweep, in bias-point calculations, and in transient analysis:

DC Sweep—The most frequent problem is attempting to analyze circuits with regenerative feedback (e.g., Schmitt trigger). Try doing a Transient Analysis instead of the dc Sweep. Use a piecewise-linear voltage source to generate a slow ramp. You can sweep up and down again in the same analysis.

Bias Point—Use the .NODESET statement to help PSpice find a solution. Nodes such as the outputs of op amps are good candidates for .NODESET.

Transient Analysis–Unrealistic modeling of circuits with switches but no parasitic capacitance can cause problems, for example, circuits containing diodes and inductors but no parasitic resistance or capacitance.

It may be necessary to relax RELTOL from .001 to .01.

Using the "uic" modifier causes the Transient Bias-Point calculation to be skipped, causing Transient Analysis convergence problems. Use .IC or .NODE-SET instead.

With high voltages and currents, it may be necessary to increase VNTOL and ABSTOL. For voltages in the kV range, raise VNTOL to 1 mV. For currents in the amps range, raise ABSTOL to 1 nA. For currents in the kA range, raise ABSTOL to 1 uA.

PSpice's accuracy is controlled by the RELTOL, VNTOL, ABSTOL, and CHGTOL parameters of the .OPTIONS statement. The most important is RELTOL, which controls the relative accuracy of all voltages and currents that are calculated. RELTOL defaults to 0.1%. VNTOL sets best accuracy for voltages. ABSTOL sets best accuracy for currents. CHGTOL sets best accuracy for charge/flux.

Global nodes begin with the prefix "$G_". Examples are $G_VCC $G_COMMON.

Predefined digital nodes are $D_HI,$D_LO,$D_NC,$D_X.

PSpice will accept expressions in most places where a numeric value is required. This includes component values, model parameter values, subcircuit parameters, initial conditions, and so forth. An expression is contained within { } and must fit on one line.

Components of an expression include numbers, operators $+-*/$, parameter names, and functions (sin, cos, exp, etc.) For example, a resistor value could be defined in terms of a global parameter RSHEET:

```
re1 20 21 {rsheet*1.10}
```

Expressions may be used for global parameter values, but these expressions may not contain parameter names.

Expressions used in the extended controlled sources may additionally refer to node voltages, currents, and the swept variable "time."

Global parameters are defined with the .PARAM statement. They can then be used in expressions for device values, and so forth. For example,

```
.param pi=3.14159265
c1 2 0 {1/(2*pi*10khz*10k)}
```

Subcircuit parameters supply default values for subcircuits. The defaults can be overridden when the subcircuit is called. The values given to subcircuit parameters can be expressions. In addition to the normal components of an expression, subcircuit parameter expressions may refer to the names of a subcircuit's own parameters (if any).

If a global parameter and a subcircuit parameter have the same name, the subcircuit parameter definition is used. For example, here is the definition of a parasitic node:

```
. subckt para 1 params:r=1meg,c=1pf
r1 10{r}
c1 10{c}
. ends
```

and here is the subcircuit being used:

```
xparal 27 para params: c=5pf
```

Making PSpice Work

NOTES ON INSTALLING PSPICE ON THE COMPUTER

The MicroSim Corporation offers several versions of PSpice for a variety of machines. It would be impractical to include even a brief description of all of the combinations available in their Design Center systems. The information given here assumes that you will be using the evaluation version of PSpice for the IBM PC and its clones.

In preparation for installing PSpice on your computer, you should read the file README.DOC. This contains valuable information on the installation of PSpice and extended information of some of its features. Newer versions of PSpice come on packed disks and use the program INSTALL to unpack the various files in the PSpice package.

Some of the files are as follows:

EVAL.DOC which contains the models of various devices such as diodes, transistors, operational amplifiers, and logic devices.

PARTS.HLP which contains information used in the parts program PARTS.EXE to provide *Screen_info*.

PSPICE.HLP which contains the information on devices and model parameters (shown in Appendix D) along with other information used in the shell program PS.EXE.

STMED.HLP which contains information used in the STMED.EXE (stimulus editor) program to provide *Display_help* information.

Using the Probe Post-Processor

Before using Probe for the first time, you must use the setup program SETUP-DEV.EXE to create a PSPICE.DEV file. The latter file will contain a pair of statements on your display and hard-copy devices. Since SETUPDEV is menu driven, you will be able to use it without further help. Typically, the PSPICE-.DEV file contains something like this:

```
Display=IBMEGA
Hard-copy=LPT1,HPLJ100
```

for the 640×350 color adaptor and the Hewlett Packard LaserJet II printer, giving 100 dots per inch.

```
Hard-copy=LPT1:, Epson
```

is used for an Epson FX-86 or similar printer.

The author used an HP LaserJet II printer for the Probe plots which appear in this book. Although HPLJ300 would give 300 dots per inch resolution, the LaserJet II would require more than the standard 0.5 Mbyte of memory to give a full page in this much detail and the printed portion of the graphs would be smaller.

If you have the January, 1989 (or earlier), release, you will find a PARTS-.DEV file which can be modified to accommodate the display and printer in much the same manner. The Parts option is activated by typing

```
parts
```

This program is menu driven, allowing the user to customize diodes, transistors, and so forth. The evaluation version of Parts is very limited, however, working only with diodes. It is only useful to give the user a feel for what might be done with the production version of the program.

The earlier versions of PSpice contained separate libraries for devices; for example, BIPOLAR.LIB, DIODE.LIB, and so forth. The master library, NOM.LIB, called the other libraries as necessary. The latest release does not contain the separate libraries; the various devices are all placed in the master library instead.

USING EDLIN TO CREATE A PSPICE INPUT FILE

First, be sure that you either have copied EDLIN.COM (which is a part of DOS) into the SPICE directory or have set up your DOS path statement to reach EDLIN. If you are in the SPICE directory and are using the pg statement in AUTOEXEC.BAT, you will see the prompt

C:\SPICE>_

and you may create your file. Let us assume that you choose the name DC1.CIR for your file name. Type

```
edlin dc1.cir
```

This creates a working buffer for your input. The EDLIN prompt is

New file

*

Type

i

(to go into the input mode), and EDLIN responds with

1:*_

This means that you can now type the first line of your input file. End the line by pressing the Enter key [*Rtn*]. You will see a similar prompt for the second line, and so forth.

Continue typing until you have entered all of the statements, including the .END statement, closing each line with [*Rtn*]. Now press [*Ctrl*]*c*, and you will be in the edit mode.

If you want to correct any line, list it first; then correct it. For example, if line 6 has an error, type

6[Rtn]

and line 6 is listed. Now you can use the right arrow key (or the *F1* key) to display each character on the line until you find the spot where the correction is to be made. Use [*Ins*] or [*Del*] to add or remove characters, or simply type over characters that need to be changed. When you press the *F3* key, it will copy any remaining characters on the current line. Press [Rtn] to complete the line entry.

Editing commands (press [*Rtn*] after each command):

1l is used to list lines 1 through 23 of the file. (Or simply type l to do the same thing.)

12l will list 23 lines beginning with line 12.

12 lists line 12, placing the pointer at the beginning of the line to allow for editing.

5d is used to delete line 5.

5,10d is used to delete lines 5 through 10.

5i is used to insert a line ahead of line 5 and continue to insert more lines until you press F6[*Rtn*].

After all corrections are made, if you are in the edit mode, type *e* to exit and save the file.

Then run the PSpice analysis by typing

pspice dc1

This will automatically create an output file DC1.OUT. You will always be in the edit mode when you first go into EDLIN. To view the file, type

edlin dc1.out
1,201

and you will see the first 20 lines. (Do not attempt to see 23 lines at once, since any lines that exceed 69 characters will wrap-around on the screen, taking up more than one line.)

You will want to check the output file for errors if the analysis was aborted. When you see what needs to be changed, exit EDLIN by typing *e*. If there are no errors and you would like a printed copy of the output file, you should remove any unnecessary lines from the file. These will generally be lines indicating the temperature, extra blank lines, and lines that contain a form-feed. Be sure to catch these so that you will not waste paper. The form-feeds are marked with ^L at the beginning of the line.

After all extra lines have been removed and you are satisfied with the on-screen listing of the file, exit EDLIN in the usual way (by typing *e*); then obtain your printed copy, if desired, by using the DOS command

```
print fname
```

where *fname* is the name of your output file (in this example dc1.out).

USING INTERACTIVE PSPICE

The evaluation version of PSpice includes an interactive control shell. This program allows the user to carry out the various tasks of PSpice without leaving the interactive environment. Using this approach, you can create and edit an input file, decide which type of analysis to run, modify parameters for another run, and so forth.

You may enter this shell by typing

```
ps[Rtn]
```

after which the main menu appears on the screen. From this point on, options are chosen from pull-down menus. The user is never required to return to the DOS environment until the session is ended.

Before using this method for operating PSpice, you may need some experience in the more conventional method where an input file is created and saved using a full-screen editor, and then the SPICE analysis is run with the *PSpice fname* command. You will find that the latter method is faster and simpler, requiring only a minimal knowledge of DOS.

The PSpice shell program, although menu driven, requires a certain amount of additional effort to learn. The overhead involved in using this shell represents a trade-off between convenience and speed. After some experimentation, you will be able to decide which features of the shell program, if any, you want to use. Generally, short input files can be handled more easily outside the shell. Incidentally, the shell environment was not used in preparation of the files in this book although it was available.

One of the most useful features of the shell program is the on-line help, which is available at a keystroke. The function key *F1* shows the main help features; *F3* takes the user into an on-line manual. When using the Editor, you need to press *F3*, then [*Rtn*][*Rtn*] to go to the manual listing for the device or command dealing with your current input line. For example, if you were describing a voltage, the entry would begin with a *V*. This is what the program looks at to determine that you desire help in describing a voltage.

Creating an Input File

Select *Files* (press [*Rtn*]); then select *Current*. For the circuit file name, choose a new name, for example, ps1a (the *.cir* extension is not needed). Select *Edit* (by pressing *e*). Then in response to "Create?", type *y*.

The top of the screen (the second line) should now show *Circuit File Editor*. The blinking prompt indicates that you may create an input file. Remember to use a title for the first statement in the file. Continue until the entire file has been entered, finishing with the usual *.END* statement.

The editor allows the use of [*Ctrl*] (shown as ^) and [*Alt*] to aid in the editing of text as follows:

^*a* moves the cursor left one word

^*f* moves the cursor right one word

^*z* scrolls up

^*w* scrolls down

^*y* deletes a line

[*Esc*] is used to exit the editor

[*Alt*]*m* begins marking a block

[*Alt*]*c* copies marked text to the scrap

[*Alt*]*x* cuts marked text to the scrap

[*Alt*]*p* pastes from the scrapboard at the current cursor position

[*Alt*]*s* searches for a designated pattern

[*Ins*] toggles between the insert and overstrike mode

[*Del*] and [*Bk Sp*] are used in the conventional way

Arrow keys are used to move up, down, right, and left

Use the mouse to

Click left or right to begin marking

Click both left and right for *Esc*

(The mouse may also be used to select menu items.)

After checking your input file for errors, press [*Esc*] to exit the Editor.

In response to "Save or Discard edits?" press [*Rtn*] to save the input file. It will be saved to disk in the current directory, after which you will see the main menu on the screen. Available menu items are shown in bold at the top of the screen. Press [*Esc*] if necessary to leave the Files menu; then select <u>A</u>nalysis to start the analysis and create an output file.

Note that if <u>A</u>nalysis is not an available option, select <u>C</u>ircuit and check for errors.

When the analysis is done, you may select Browse to see the output file. However, if the input file contains the .PROBE statement, you will automatically be taken to the Probe post-processor.

Working with Existing Files

To use the interactive shell with files that have already been created and stored on disk, after entering interactive PSpice, select Files, Current; then press the function key *F4*. This will give a display of all the files in the current directory with the *.CIR* extension. Use the arrow keys to select the desired file, then Edit to see the contents of the file. To run the analysis, use [*Esc*] to exit the Editor.

A CONTROL SHELL EXAMPLE

The *Control Shell* is a DOS program that provides the PSpice user with access to the various features that are also available as separate programs working directly in DOS. This shell program allows access to *PSpice, Probe,* the *Stimulus Editor*, a *Text Editor*, and so forth. Enter the *Control Shell* from the PSpice directory by typing

ps

The main menu, which is shown in Fig. C.1, will be displayed. Press [*Rtn*] to select the *Files* menu. The pull-down menu is shown in Fig. C.2. In the beginning, only two of the options are available, *Current File. . .* and *Display/Prn Setup. . .* choose *Current File....*

```
 ┌──────────── PSpice Control Shell - ver 5.1 ─────────────┐
 │ Files      Circuit     StmEd     Analysis    Display      Probe      Quit │
 ├──────────────────────────────────────────────────────────┤
 │                                                                          │
 │                                                                          │
 │                                                                          │
 │                                                                          │
 │                                                                          │
 │                                                                          │
 │                                                                          │
 │                                                                          │
 │                                                                          │
 │                                                                          │
 │                                                                          │
 │              Current File:                      (New)                    │
 │   F1=Help  F2=Move  F3=Manual  F4=Choices  F5=Calc  F6=Errors  ESC=Cancel│
 └──────────────────────────────────────────────────────────┘
```

Fig. C.1 Control Shell main menu.

```
┌──────────────── PSpice Control Shell - ver 5.1 ────────────────┐
│ Files      Circuit     StmEd     Analysis    Display    Probe      Quit  │
├──────────────────────────┐
│ Edit                     │
│ Browse Output            │
├──────────────────────────┤
│ Current File...          │
│ Save File                │
├──────────────────────────┤
│ X - External Editor      │
│ R - External Browser     │
├──────────────────────────┤
│ Display/Prn Setup...     │
└──────────────────────────┘




                    Current File:                        (New)
       F1=Help  F2=Move  F3=Manual   F4=Choices   F5=Calc  F6=Errors   ESC=Cancel
```

Fig. C.2   The pull-down Files menu.

A box appears on the screen, as seen in Fig. C.3, with a space for entering the circuit file name. In this example, we want to create a new file, so type the file name

hparam

The extension *.cir* is not needed and will be assigned by the shell program. The *Current File* shown at the bottom of the screen is now *hparam.cir*. From the main menu, select *File Edit*. The Circuit Editor screen will appear, and on the top line a blinking cursor will show that you may begin typing on line 1, column 1.

```
┌──────────────── PSpice Control Shell - ver 5.1 ────────────────┐
│ Files      Circuit     StmEd     Analysis    Display    Probe      Quit  │
├──────────────────────────┐
│ Edit                     │
│ Browse Output            │
├──────────────────────────┤
│ Current File...          │
│ S ┌───────────Define Input File───────────┐
│ X │                                        │
│ R │      Circuit File Name:                │
│   │                                        │
│ D │                                        │
│   └────────────────────────────────────────┘




                    Current File:                        (New)
       F1=Help  F2=Move  F3=Manual   F4=Choices   F5=Calc  F6=Errors   ESC=Cancel
```

Fig. C.3   Box for defining input file.

Enter the listing as shown in Fig. C.4 (begin with *Small-Signal Analysis. . .*). Do not press the [*Rtn*] key after the *.END* statement, but rather press [*Esc*]. You will see the message:

Save changes (to temporary working file) or Discard (S/D)?

Press [*Rtn*] to save the circuit file.

If there are errors, such as illegal device entries, the message *Errors* appears at the bottom right on the screen, and a box appears on the upper right. The error message shows the line number and a brief description of the error. Press [*Esc*], then [*Rtn*], to edit the lines containing errors. After correcting the errors, you must save the file again in order to update it.

Figure C.4 also shows a box with the *Help* screen, which is available when you press the *F1* key. Some of the *Editor* commands are shown in this box. For example, ^*A* (for move left one word) indicates that by pressing [*Ctrl*]*A* in the *Editor*, the cursor will move back one word.

When the file has been updated and there are no more errors, the *Error* message disappears, and the word *Loaded* is shown at the lower right of the screen. You may *Esc* back to the main menu, and under *Circuit* choose *Devices, Models,* or *Parameters.*

Choose *Devices* and the screen will look like Fig. C.5. The message at the top is

Select Device to Change

which gives you an opportunity to edit each device line. This is done by selecting the desired line and then entering a new value. Assuming that no changes are to be made, *Esc* back to the main menu and prepare to run the analysis.

```
┌──────────────────── PSpice Control Shell - ver 5.1 ─────────────────┐
│ ═════════ Circuit Editor    line:   1 col:  1      [Insert] ═════════│
│Small-Signal Analysis of Transistor Circuit Using h Parameters       │
│VS 1 0 1mV                                                           │
│VO 3 3A 0                                                            │
│E 3A 0 4 0 2.5E-4                                                    │
│F 4 0 VO 50                                                          │
│RS 1 2 1k                                                            │
│RI 2 3 1.1k                                                          │
│RO 4 0 40k                                                           │
│RL 4 0 10k              ┌══════════════HELP SCREEN══════════════┐    │
│.OP                     │ ----- Editor commands -----           │    │
│.TF V(4) VS             │                                       │    │
│.END                    │ ^A - move left 1 word;                │    │
│                        │ ^F - move right 1 word;               │    │
│                        │ ^Z - scroll up;  ^W - scroll down;    │    │
│                        │ ^Y - delete a line.                   │    │
│                        │                                       │    │
│                        │ ESC - exit the editor.                │    │
│                        │ Arrow keys - move horiz. and vert.    │    │
│                        │                           more...     │    │
│                        └───────────────────────────────────────┘    │
│            Current File: hparam.cir              Loaded             │
│   F1=Help   F2=Move  F3=Manual  F4=Choices  F5=Calc  F6=Errors  ESC=Cancel │
└─────────────────────────────────────────────────────────────────────┘
```

Fig. C.4 Sample input file for PSpice; edit help screen.

```
┌──────────────── PSpice Control Shell - ver 5.1 ────────────────┐
│                    ─Select Device to Change─                   │
│ Device    Type                                                 │
│                                                                │
│ E         linear  3A     0        250.000u                     │
│ F         linear  4      0        50                           │
│ RL                4      0        10.000k                       │
│ RO                4      0        40.000k                       │
│ RI                2      3        1.100k                        │
│ RS                1      2        1.000k                        │
│ VO                3      3A       0         0        0          │
│ VS                1      0        1.000m    0        0          │
│                                                                │
│                                                                │
│                                                                │
│                                                                │
│                                                                │
│                                                                │
├────────────────────────────────────────────────────────────────┤
│              Current File: hparam.cir           Loaded         │
│     F1=Help  F2=Move  F3=Manual  F4=Choices  F5=Calc  F6=Errors   ESC=Cancel │
└────────────────────────────────────────────────────────────────┘
```

Fig. C.5 Display of circuit devices.

Running PSpice

The various options under *Analysis* are shown in Fig. C.6. Select *Run PSpice* from the pull-down menu. The PSpice analysis will be performed, and you will be returned to the Control Shell.

Select *Browse Output* from the *Files* menu; then page down to see the Output file.

```
┌──────────────── PSpice Control Shell - ver 5.1 ────────────────┐
│ Files    Circuit    StmEd    Analysis    Display    Probe    Quit │
│                            ┌─────────────────────┐              │
│                            │ Run PSpice          │              │
│                            ├─────────────────────┤              │
│                            │ AC & Noise...       │              │
│                            │ DC Sweep...         │              │
│                            │ Transient...        │              │
│                            │ Parametric...       │              │
│                            ├─────────────────────┤              │
│                            │ Specify Temperature...│            │
│                            ├─────────────────────┤              │
│                            │ Monte Carlo...      │              │
│                            │ Change Options...   │              │
│                            └─────────────────────┘              │
│                                                                │
│              Current File: hparam.cir           Loaded         │
│     F1=Help  F2=Move  F3=Manual  F4=Choices  F5=Calc  F6=Errors   ESC=Cancel │
└────────────────────────────────────────────────────────────────┘
```

Fig. C.6 Analysis options.

Using an External Editor

One of the commands available under the menu item *Files* is *X-External Editor*. If you select this feature, the default editor *EDLIN* (the primitive DOS editor) will be invoked. The default file is the current file (which in the example is *hparam-.cir*).

If you are using DOS 5.0, you will want to use the editor *EDIT*, rather than *EDLIN*. The External Editor may be changed to invoke *EDIT* as follows:

Choose the *Quit* option if you are in the *Control Shell*, and return to the DOS prompt. Give this command:

```
set psedit = edit %f
```

The *f* is a special code for the entire file name. This command sets the environment variable *psedit* to allow for the running of the external editor.

Using an External Browser

Another option under the menu item *Files* is *R - External Browser*. Recall that browsing the output is one of the *Files* options also. If you prefer to use another editor for external browsing, an environment variable is needed. Set this with this DOS command:

```
set psbrowse = edit %f
```

There are several other codes which may be used in setting either of the environment variables. They are as follows:

Code	Description	Example
%c	current working directory	spice
%e	name without extension	spice\hparam
%f	entire name	spice\hparam.cir
%n	name without directory	hparam
%p	path to PSpice	spice

On the main menu you will see that other features of Pspice are directly available. For example, the *Stimulus Editor* and *Probe* are shown as menu items. Thus the *Control Shell* offers an alternative to using DOS commands to perform various tasks. It should be mentioned, however, that the quicker method is to work directly under DOS rather than the *Control Shell*. It is suggested that the PSpice user be familiar with doing things in DOS rather than relying on the *Control Shell* as a crutch.

The On-Line Manual

When you press the *F3* key in the *Control Shell*, you will see a pop-up menu listing several choices: *Devices, Commands, Hints,* and *Exprs/Params*. As an example, when you select *Devices*, all of the PSpice devices are listed, as shown in Fig. C.7. You may then select one of the devices to obtain a *Help Screen*. Choosing the device *Q* gives the *Help Screen* shown in Fig. C.8.

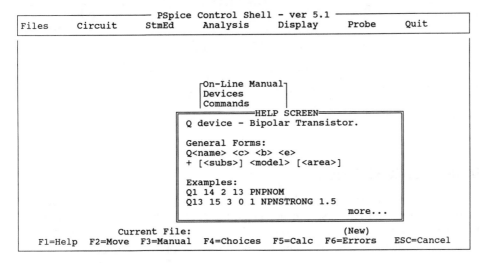

Fig. C.7 The on-line manual for devices.

Alternately, you might select *Commands* in the *On-Line Manual* to obtain a list of all the *dot* commands. These are shown in Fig. C.9. If you desire to see examples of a certain command, simply select it to obtain a *Help Screen*. When you select *AC*, the screen of Fig. C.10 is shown.

After you have worked in the *Control Shell* awhile, you will find that you have several files with the extensions *.cbk* (circuit back-up) and *.cfg* (configuration). You may want to delete these in order to save disk space after an analysis is completed.

Fig. C.8 Help screen for the BJT.

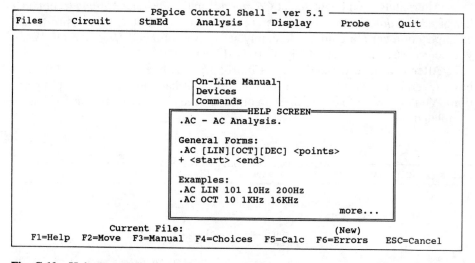

Fig. C.9 Help screen listing of PSpice commands.

Fig. C.10 Help screen for ac analysis.

USING THE DOS 5.0 EDITOR

If you are using DOS 5.0 or a later version, instead of using the primitive line editor *EDLIN*, you may use the newer editor *EDIT*. This editor has pull-down menus that are activated by using the [*Alt*] key. To create a new file *ex1.cir* for use with PSpice, at the DOS command (in the directory of your choice) type

```
edit ex1.cir
```

The text of the input file, such as that shown in Fig. C.11, is then entered. Remember to not press [*Rtn*] after the *.END* statement. Save the input file by pressing [*Alt*]*fx* followed by [*Rtn*]. If you prefer to save the input file without leaving the editor, press [*Alt*]*fs* as illustrated in Fig. C.12.

If you are in the PSpice directory, after leaving the editor, you may now run the PSpice analysis by using the command

```
pspice ex1
```

Other features of the DOS editor *EDIT* may be found by using the *Help* option on the main menu. To access *Help*, press [*Alt*]*h*. This will give the pull-down menu as shown in Fig. C.13.

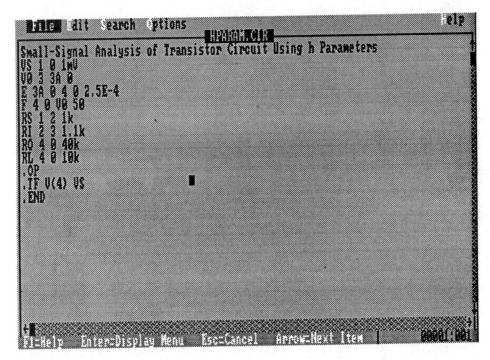

Fig. C.11 Using the DOS editor EDIT.

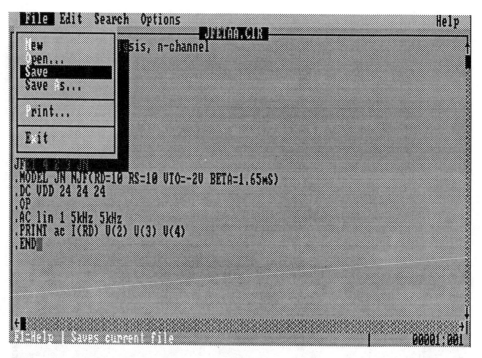

Fig. C.12 Saving a file in the DOS editor EDIT.

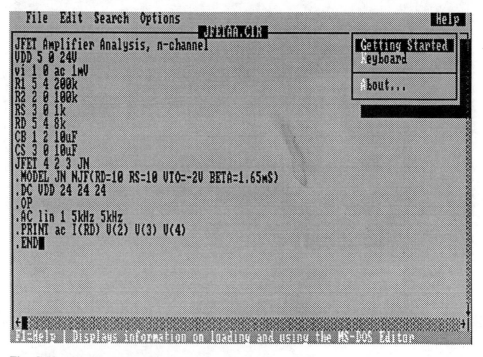

Fig. C.13 Getting help in the DOS editor EDIT.

PSpice Devices and Model Parameters

$<$ $>$* indicates that an item may be repeated.

B GaAsFET

B*<name><drain node><gate node><source node><model name>*[*area*]

Model Parameters		Default Value	Units
LEVEL	model type(1=Curtice, 2=Raytheon)	1	
VTO	threshold voltage	−2.5	volts
ALPHA	tanh constant	2	volts^{-1}
B	doping tail extending parameter	0.3	
BETA	transconductance coef.	0.1	A/V^2
LAMBDA	channel-length modulation	0	volt^{-1}
RG	gate ohmic resistance	0	ohm
RD	drain ohmic resistance	0	ohm
RS	source ohmic resistance	0	ohm
IS	gate *pn* saturation current	1E−14	ampere
M	gate *pn* grading coefficient	0.5	
N	gate *pn* emission coefficient	1	
VBI	gate *pn* potential	1	volt
CGD	gate-drain zero-bias *pn* cap.	0	farad
CGS	gate-source zero-bias *pn* cap.	0	farad
CDS	drain-source capacitance	0	farad
TAU	transit time	0	sec
FC	forward-bias dep. cap. coef.	0.5	
VTOTC	VTO temperature coef.	0	volt/°C
BETATCE	BETA exponential temp. coef.	0	%/°C
KF	flicker noise coef.	0	
AF	flicker noise exponent	1	

[*area*] is the relative device area and defaults to 1.

Fig. D.1   GaAsFET model.

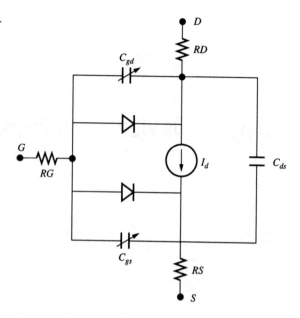

The GaAsFET, as shown in Fig. D.1, is modeled as an intrinsic FET with ohmic resistance RD in series with the drain. Another ohmic resistance RS is in series with the source, and another ohmic resistance RG is in series with the gate.

C Capacitor

C<*name*><+*node*><−*node*>[*model name*]<*value*>[IC=<*initial value*>]

Model Parameters		Default Value	Units
C	capacitance multiplier	1	
VC1	linear voltage coefficient	0	volts^{-1}
VC2	quadratic voltage coefficient	0	volt^{-2}
TC1	linear temperature coefficient	0	°C^{-1}
TC2	quadratic temperature coefficient	0	°C^{-2}

If [*model name*] is left out, then <*value*> is the capacitance in farads. If [*model name*] is given, then the capacitance is

$$<value> \cdot C(1+VC1 \cdot V+VC2 \cdot V^2)(1+TC1(T\text{-}T_{nom})+TC2(T\text{-}T_{nom})^2)$$

T_{nom} is the nominal temperature which is set with the TNOM option.

D Diode

D<*name*><+*node*><−*node*><*model name*>[*area*]

Model Parameters		Default Value	Units
IS	saturation current	1E−14	ampere
N	emission coefficient	1	
RS	parasitic resistance	0	ohm
CJO	zero-bias *pn* capacitance	0	farad
VJ	*pn* potential	1	volt
M	*pn* grading coefficient	0.5	
FC	forward-bias depletion cap. coef.	0.5	
TT	transit time	0	s
BV	reverse breakdown voltage	infinite	volts
IBV	reverse breakdown current	1E−10	ampere
EG	bandgap voltage (barrier height)	1.11	eV
XTI	IS temperature exponent	3	
KF	flicker noise coefficient	0	
AF	flicker noise exponent	1	

Fig. D.2 Diode model.

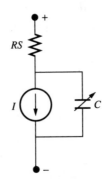

The diode, which is shown in Fig. D.2, is modeled as an ohmic resistance RS in series with an intrinsic diode.

E Voltage-Controlled Voltage Source

E*<name><+node><−node><+controlling node><−con. node><gain>*

E*<name><+node><−node>*POLY(*<value>*)
  <<+controlling node><−con. node>>*<<polynomial coefficient value>>*

F Current-Controlled Current Source

F*<name><+node><−node><controlling V device name><gain>*

F*<name><+node><−node>*POLY(*<value>*)*<controlling V device name>*<<polynomial coefficient value>>**

G Voltage-Controlled Current Source

G*<name><+node><−node><+controlling node>*
 <−controlling node><transconductance>
G*<name><+node><−node>*POLY(*<value>*)
 <<+controlling node><−controlling node>**
 <<polynomial coefficient value>>**

H Current-Controlled Voltage Source

H*<name><+node><−node><controlling V device name>*
 <transresistance>
H*<name><+node><−node>*POLY(*<value>*)
 <controlling V device name>**<<polynomial coefficient value>>***

I Independent Current Source

I*<name><+node><−node>*[[DC]*<value>*]
 [AC*<magnitude value>*[*phase value*]][*transient specification*]

If present, [*transient specification*] must be one of these:

EXP<>,PULSE<>,PWL<>,SFFM<>,or SIN<>

J Junction FET

J*<name><drain node><gate node><source node><model name>*[*area*]

Model Parameters		Default Value	Units
VTO	threshold voltage	−2.0	volts
BETA	transconductance coefficient	1E−4	A/V^2
LAMBDA	channel-length modulation	0	volt^{-1}
RD	drain ohmic resistance	0	ohm
RS	source ohmic resistance	0	ohm
IS	gate *pn* saturation current	1E−14	ampere
PB	gate *pn* potential	1	volt
CGD	gate-drain zero-bias *pn* capac.	0	farad
CGS	gate-source zero-bias *pn* capac.	0	farad
FC	foward-bias depletion cap. coef.	0.5	
VTOTC	VTO temperature coefficient	0	V/°C
BETATCE	BETA exponential temp. coef.	0	%/°C
KF	flicker noise coefficient	0	
AF	flicker noise exponent	1	

The JFET, as shown in Fig. D.3, is modeled as an intrinsic FET with an ohmic resistance RD in series with the drain. Another ohmic resistance RS is in series with the source.

Fig. D.3 JFET model.

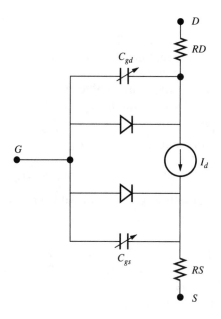

K Inductor Coupling (Transformer Core)

K<*name*>L<*inductor name*><L<*inductor name*>>*<*coupling value*>
K<*name*><L<*inductor name*>>*<*coupling value*><*model name*>[*size*]

Model Parameters (for Nonlinear Only)		Default Value	Units
AREA	mean magnetic cross section	0.1	cm²
PATH	mean magnetic path length	1	cm
GAP	effective air-gap length	0	cm
PACK	pack (stacking) factor	1	
MS	magnetization saturation	1E+6	A/m
ALPHA	mean field parameter	0.001	
A	shape parameter	1000	A/m
C	domain wall flexing constant	0.2	
K	domain wall pinning constant	500	

K<*name*> couples two or more inductors. Use the dot convention to give a dot
on the first(positive) node of each inductor.

If <*model name*> is given, then (a) the inductor is a nonlinear device with a
magnetic core; (b) the *BH* characteristics are based on the Jiles-Atherton model;
(c) the *L* values indicate windings, with the value indicating the number of turns;
and (d) a model statement is needed to specify model parameters.

L Inductor

L<*name*><+*node*><−*node*>[*model name*]<*value*>[IC=<*initial value*>]

Model Parameters		Default Value	Units
L	inductance multiplier	1	
IL1	linear current coefficient	0	ampere^{-1}
IL2	quadratic current coefficient	0	ampere^{-2}
TC1	linear temperature coefficient	0	°C^{-1}
TC2	quadratic temperature coefficient	0	°C^{-2}

If [*model name*] is omitted, then <*value*> is the inductance in henries. If [*model name*] is present, then the inductance is

$$<value> \cdot L(1+IL1 \cdot I+IL2 \cdot I^2)(1+TC1(T-T_{nom})+TC2(T-T_{nom})^2)$$

T_{nom} is the nominal temperature which is set with the TNOM option.

M MOSFET

M<*name*><*drain node*><*gate node*><*source node*><*bulk/substrate node*>
  <*model name*>[L=<*value*>][W=<*value*>][AD=<*value*>][AS=<*value*>]
  [PD=<*value*>][PS=<*value*>][NRD=<*value*>][NRS=<*value*>]
  [NRG=<*value*>][NRB=<*value*>]

Model Parameters		Default Value	Units
LEVEL	model type (1, 2, or 3)	1	
L	channel length	DEFL	meter
W	channel width	DEFW	meter
LD	lateral diffusion (length)	0	meter
WD	lateral diffusion (width)	0	meter
VTO	zero-bias threshold voltage	0	volt
KP	transconductance	2E$-$5	A/V^2
GAMMA	bulk threshold parameter	0	volt$^{1/2}$
PHI	surface potential	0.6	volt
LAMBDA	channel-len. mod.(LEVEL 1 or 2)	0	volt^{-1}
RD	drain ohmic resistance	0	ohm
RS	source ohmic resistance	0	ohm
RG	gate ohmic resistance	0	ohm
RB	bulk ohmic resistance	0	ohm
RDS	drain-source shunt resistance	infinite	ohms
RSH	drain-source diff. sheet res.	0	ohm/sq.
IS	bulk *pn* saturation current	1E$-$14	A
JS	bulk *pn* sat. current/area	0	A/m^2
PB	bulk *pn* potential	0.8	volt
CBD	bulk-drain zero-bias *pn* cap.	0	farad
CBS	bulk-source zero-bias *pn* cap.	0	farad
CJ	bulk *pn* zero-bias bot. cap./area	0	F/m^2
CJSW	bulk *pn* zero-bias perimeter cap./length	0	F/m
MJ	bulk *pn* bottom grading coefficient	0.5	
MJSW	bulk *pn* sidewall grading coefficient	0.33	
FC	bulk *pn* forward bias capacitance coefficient	0.5	
CGSO	gate-source overlap capacitance/channel width	0	F/m
CGDO	gate-drain overlap capacitance/channel width	0	F/m

Model Parameters		Default Value	Units
CGBO	gate-bulk overlap capacitance/channel length	0	F/m
NSUB	substrate doping density	0	cm^{-3}
NSS	surface state density	0	cm^{-2}
NFS	fast surface state density	0	cm^{-2}
TOX	oxide thickness	infinite	meter
TPG	gate material type	+1	
	+1 = opposite of substrate		
	−1 = same as substrate		
	0 = aluminum		
XJ	metallurgical junction depth	0	meter
UO	surface mobility	600	cm^2/Vs
UCRIT	mobility degradation critical field (LEVEL=2)	1E4	V/cm
UEXP	mobility degradation exponent (LEVEL=2)	0	
UTRA	(not used) mobility degradation transverse field coef.		
VMAX	maximum drift velocity	0	m/s
NEFF	channel charge coefficient (LEVEL=2)	1	
XQC	fraction of channel charge attributed to drain	1	
DELTA	width effect on threshold	0	
THETA	mobility modulation (LEVEL=3)	0	$volt^{-1}$
ETA	static feedback (LEVEL=3)	0	
KAPPA	saturation field factor (LEVEL=3)	0.2	
KF	flicker noise coefficient	0	
AF	flicker noise exponent	1	

The MOSFET, which is shown in Fig. D.4, is modeled as an intrinsic MOSFET with ohmic resistance RD in series with the drain, ohmic resistance RS in series

Fig. D.4 MOSFET model.

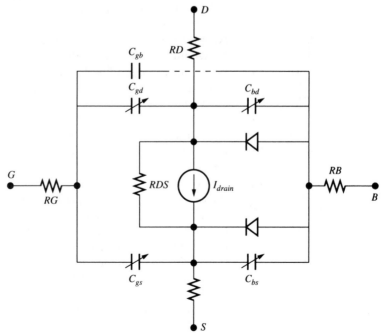

with the source, ohmic resistance RG in series with the gate, and ohmic resistance RB in series with the (bulk) substrate. A shunt resistance RDS is in parallel with the (drain-source) channel.

Q Bipolar Transistor

Q*<name><collector node><base node>*
<emitter node><[substrate node]<model name>[area value]

Model Parameters		Default Values	Units
IS	*pn* saturation current	1E−16	A
BF	ideal maximum forward beta	100	
NF	forward current emission coefficient	1	
VAF (VA)	forward Early voltage	infinite	V
IKF (IK)	corner for fwd beta high-cur roll off	infinite	A
ISE (C2)	base-emitter leakage saturation current	0	A
NE	base-emitter leakage emission coefficient	1.5	
BR	ideal maximum reverse beta	1	
NR	reverse current emission coefficient	1	
VAR (VB)	reverse Early voltage	infinite	V
IKR	corner for rev beta hi-cur roll off	infinite	A
ISC (C4)	base-collector leakage saturation current	0	A
NC	base-collector leakage emission coefficient	2.0	
RB	zero-bias (maximum) base resistance	0	ohm
RBM	minimum base resistance	RB	ohm
RE	emitter ohmic resistance	0	ohm
RC	collector ohmic resistance	0	ohm
CJE	base-emitter zero-bias *pn* capacitance	0	F
VJE (PE)	base-emitter built-in potential	0.75	V
MJE (ME)	base-emitter *pn* grading factor	0.33	
CJC	base-collector zero-bias pn capacitance	0	F
VJC (PC)	base-collector built-in potential	0.75	V
MJC (MC)	base-collector *pn* grading factor	0.33	
XCJC	fraction of Cbc connected int to Rb	1	
CJS (CCS)	collector-substrate zero-bias pn capacitance	0	F
VJS (PS)	collector-substrate built-in potential	0.75	
MJS (MS)	collector-substrate *pn* grading factor	0	
FC	forward-bias depletion capacitor coefficient	0.5	
TF	ideal forward transit time	0	s
XTF	transit time bias dependence coefficient	0	
VTF	transit time dependency on Vbc	infinite	V
ITF	transit time dependency on Ic	0	A
PTF	excess phase @ $1/(2\pi TF)$ Hz	0	°C
TR	ideal reverse transit time	0	s
EG	band-gap voltage (barrier height)	1.11	eV

Model Parameters		Default Values
XTB	forward and reverse beta temp coefficient	0
XTI (PT)	IS temperature effect exponent	3
KF	flicker noise coefficient	0
AF	flicker noise exponent	1

The bipolar transistor, as shown in Fig. D.5, is modeled as an intrinsic transistor with ohmic resistance RC in series with the collector, a variable resistance Rb in series with the base, and an ohmic resistance RE in series with the emitter. The substrate node is optional, defaulting to ground unless otherwise specified.

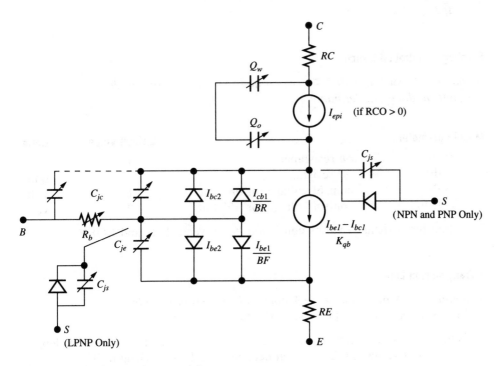

Fig. D.5 Bipolar transistor model.

R Resistor

R<*name*><*+node*><*−node*>[*model name*]<*value*>

Model Parameters		Default Values	Units
R	resistance multiplier	1	
TC1	linear temperature coefficient	0	°C^{-1}
TC2	quadratic temperature coefficient	0	°C^{-2}
TCE	exponential temperature coefficient	0	%/°C

If [*model name*] is included and TCE is not specified, then the resistance is

$<value> \cdot R(1 + TC1(T-T_{nom}) + TC2(T-T_{nom})^2)$

If [*model name*] is included and TCE is specified, then the resistance is

$<value> \cdot R \cdot 1.01^{TCE(T-Tnom)}.$

T_{nom} is the nominal temperature.

Noise is calculated assuming a 1-Hz bandwidth. The resistor generates thermal noise with this power density:

$$i^2 = \frac{R}{4kT}$$

S Voltage-Controlled Switch

S*<name><+switch node><-switch node><+controlling node>*
<-cont. node><model name>

Model Parameters		Default Value	Units
RON	*on* resistance	1	ohm
ROFF	*off* resistance	1E6	ohms
VON	control voltage for *on* state	1	volt
VOFF	control voltage for *off* state	0	volt

Note that the resistance varies continuously between RON and ROFF.

T Transmission Line

T*<name><+A port node><-A port node><+B port node>*
*<-B port node>*Z0=*<>*[TD=*<>*][F=*<>*[NL=*<>*]]

Z0 is the characteristic impedance, *F* is frequency, and *NL* is relative wavelength (with a default value of 0.25 (*F* then becomes the 1/4-wavelength *f*).

The transmission line, as shown in Fig. D.6, is modeled as a bidirectional delay line with two ports. The port A is shown on the left with nodes *1* and *2*; the port B is shown on the right with nodes *3* and *4*.

V Independent Voltage Source

V*<name><+node><-node>*[[DC]*<>*][AC*<magnitude>*[*phase*]][*transient*]

If [*transient*] is specified, it must be one of these:

EXP*<>*,PULSE*<>*,PWL*<>*,SFFM*<>*,or SIN*<>*

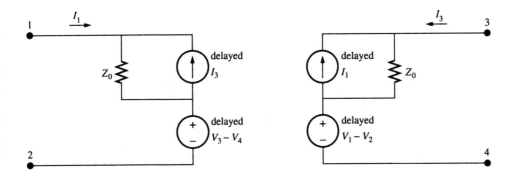

Fig. D.6 Transmission-line model.

W Current-Controlled Switch

W$<$*name*$><$+*switch node*$><$−*switch node*$>$
$<$*controlling V device name*$><$*model*$>$

Model Parameters		Default Value	Units
RON	*on* resistance	1	ohm
ROFF	*off* resistance	1E6	ohms
ION	control current for *on* state	0.001	ampere
IOFF	control current for *off* state	0	ampere

Note that the resistance varies continuously between RON and ROFF.

X Subcircuit Call

X$<$*name*$>$[*node*]*$<$*subcircuit name*$>$

There must be the same number of nodes in the call as in the subcircuit definition.

Sample Standard Device Library

This is a reduced version of MicroSim's standard parts libraries. Some components from several types of component libraries have been included here. The MicroSim library included with the production version of PSpice includes over 3500 analog devices and over 1500 digital devices.

It takes time for PSpice to scan a library file. To speed this up, PSpice creates an index file, called <filename>.IND. The index file is re-created whenever PSpice senses that the library file has changed.

The following is a summary of parts in this library:

Part Name	Part Type
Q2N2222A	NPN bipolar transistor
Q2N2907A	PNP bipolar transistor
Q2N3904	NPN bipolar transistor
Q2N3906	PNP bipolar transistor
D1N750	zener diode
MV2201	voltage variable capacitance diode
D1N4148	switching diode
MBD101	switching diode
J2N3819	N-channel junction field effect transistor
J2N4393	N-channel junction field effect transistor
LM324	linear operational amplifier
UA741	linear operational amplifier
LM111	voltage comparator
K3019PL_3C8	ferroxcube pot magnetic core

Part Name	Part Type
KRM8PL_3C8	ferroxcube pot magnetic core
K502T300_3C8	ferroxcube pot magnetic core
IRF150	N-type power MOS field effect transistor
IRF9140	P-type power MOS field effect transistor
7402	TTL digital 2-input NOR gate
7404	TTL digital inverter
7405	TTL digital inverter, open collector
7414	TTL digital inverter, schmidt trigger
7474	TTL digital D-type flip-flop
74107	TTL digital JK-type flip-flop
74393	TTL digital 4-bit binary counter
A4N25	optocoupler
2N1595	silicon controlled rectifier
2N5444	triac

LIBRARY OF BIPOLAR TRANSISTOR MODEL PARAMETERS

The parameters in this model library were derived from the data sheets for each part. Each part was characterized using the Parts option. Devices can also be characterized without Parts as follows:

NE, NC	Normally set to 4.
BF, ISE, IKF	These are adjusted to give the nominal beta vs. collector current curve. BF controls the mid-range beta. ISE/IS controls the low-current roll-off. IKF controls the high-current rolloff.
ISC	Set to ISE.
IS, RB, RE, RC	These are adjusted to give the nominal V_{BE} vs. I_C and V_{CE} vs. I_C curves in saturation. IS controls the low-current value of V_{BE}. RB+RE controls the rise of V_{BE} with I_C. RE+RC controls the rise of V_{CE} with I_C. RC is normally set to 0.
VAF	Using the voltages specified on the data sheet VAF is set to give the nominal output impedance (RO on the .OP printout) on the data sheet.
CJC, CJE	Using the voltages specified on the data sheet C_{JC} and C_{JE} are set to give the nominal input and output capacitances (CPI and CMU on the .OP printout; C_{ibo} and C_{obo} on the data sheet).
TF	Using the voltages and currents specified on the data sheet for FT, TF is adjusted to produce the nominal value of FT on the .OP printout.
TR	Using the rise and fall time circuits on the data sheet, TR (and if necessary TF) are adjusted to give a transient analysis which shows the nominal values of the turn-on delay, rise time, storage time, and fall time.

KF, AF   These parameters are only set if the data sheet has a spec for noise. Then, AF is set to 1 and KF is set to produce a total noise at the collector which is greater than the generator noise at the collector by the rated number of decibels.

```
.model Q2N2222A NPN(Is=14.34f Xti=3 Eg=1.11 Vaf=74.03 Bf=255.9 Ne=1.307
+               Ise=14.34f Ikf=.2847 Xtb=1.5 Br=6.092 Nc=2 Isc=0 Ikr=0 Rc=1
+               Cjc=7.306p Mjc=.3416 Vjc=.75 Fc=.5 Cje=22.01p Mje=.377 Vje=.75
+               Tr=46.91n Tf=411.1p Itf=.6 Vtf=1.7 Xtf=3 Rb=10)
                National  pid=19     case=TO18
                88-09-07 bam     creation
```

```
.model Q2N2907A PNP(Is=650.6E-18 Xti=3 Eg=1.11 Vaf=115.7 Bf=231.7 Ne=1.829
+               Ise=54.81f Ikf=1.079 Xtb=1.5 Br=3.563 Nc=2 Isc=0 Ikr=0 Rc=.715
+               Cjc=14.76p Mjc=.5383 Vjc=.75 Fc=.5 Cje=19.82p Mje=.3357 Vje=.75
+               Tr=111.3n Tf=603.7p Itf=.65 Vtf=5 Xtf=1.7 Rb=10)
                National  pid=63     case=TO18
                88-09-09 bam     creation
```

```
.model Q2N3904       NPN(Is=6.734f Xti=3 Eg=1.11 Vaf=74.03 Bf=416.4 Ne=1.259
                Ise=6.734f Ikf=66.78m Xtb=1.5 Br=.7371 Nc=2 Isc=0 Ikr=0 Rc=1
                Cjc=3.638p Mjc=.3085 Vjc=.75 Fc=.5 Cje=4.493p Mje=.2593 Vje=.75
+               Tr=239.5n Tf=301.2p Itf=.4 Vtf=4 Xtf=2 Rb=10)
                National  pid=23     case=TO92
```

```
.model Q2N3906   PNP(Is=1.41f Xti=3 Eg=1.11 Vaf=18.7 Bf=180.7 Ne=1.5 Ise=0
+          Ikf=80m Xtb=1.5 Br=4.977 Nc=2 Isc=0 Ikr=0 Rc=2.5 Cjc=9.728p
+          Mjc=.5776 Vjc=.75 Fc=.5 Cje=8.063p Mje=.3677 Vje=.75 Tr=33.42n
+          Tf=179.3p Itf=.4 Vtf=4 Xtf=6 Rb=10)
           National  pid=66     case=TO92
           88-09-09 bam     creation
```

LIBRARY OF DIODE MODEL PARAMETERS

This is a reduced version of MicroSim's diode model library. The parameters in this model library were derived from the data sheets for each part. Most parts were characterized using the Parts option. Devices can also be characterized without Parts as follows:

IS   nominal leakage current

RS   for zener diodes: nominal small-signal impedance at specified operating current

IB   for zener diodes: set to nominal leakage current

IBV   for zener diodes: at specified operating current IBV is adjusted to give the rated zener voltage

Zener Diodes

"A" suffix zeners have the same parameters (e.g., 1N750A has the same parameters as 1N750).

```
.model D1N750   D(Is=880.5E-18 Rs=.25 Ikf=0 N=1 Xti=3 Eg=1.11 Cjo=175p M=.5516
+          Vj=.75 Fc=.5 Isr=1.859n Nr=2 Bv=4.7 Ibv=20.245m Nbv=1.6989
+          Ibvl=1.9556m Nbvl=14.976 Tbvl=-21.277u)
           Motorola     pid=1N750     case=DO-35
           89-9-18 gjg
           Vz = 4.7 @ 20mA, Zz = 300 @ 1mA, Zz = 12.5 @ 5mA, Zz = 2.6 @ 20mA
```

Voltage-Variable Capacitance Diodes

The parameters in this model library were derived from the data sheets for each part. Each part was characterized using the Parts option.

```
.model MV2201   D(Is=1.365p Rs=1 Ikf=0 N=1 Xti=3 Eg=1.11 Cjo=14.93p M=.4261
+          Vj=.75 Fc=.5 Isr=16.02p Nr=2 Bv=25 Ibv=10u)
           Motorola     pid=MV2201     case=182-03
           88-09-22 bam     creation
```

Switching Diodes

```
.model D1N4148   D(Is=0.1p Rs=16 CJO=2p Tt=12n Bv=100 Ibv=0.1p)
           85-??-??     Original library
```

```
.model MBD101   D(Is=192.1p Rs=.1 Ikf=0 N=1 Xti=3 Eg=1.11 Cjo=893.8f M=98.29m
+          Vj=.75 Fc=.5 Isr=16.91n Nr=2 Bv=5 Ibv=10u)
           Motorola     pid=MBD101     case=182-03
           88-09-22 bam     creation
```

LIBRARY OF JUNCTION FIELD—EFFECT TRANSISTOR (JFET) MODEL PARAMETERS

This is a reduced version of MicroSim's JFET model library. The parameters in this model library were derived from the data sheets for each part. Each part was characterized using the Parts option.

```
.model J2N3819   NJF(Beta=1.304m Betatce=-.5 Rd=1 Rs=1 Lambda=2.25m Vto=-3
+          Vtotc=-2.5m Is=33.57f Isr=322.4f N=1 Nr=2 Xti=3 Alpha=311.7
+          Vk=243.6 Cgd=1.6p M=.3622 Pb=1 Fc=.5 Cgs=2.414p Kf=9.882E-18
+          Af=1)
           National     pid=50     case=TO92
           88-08-01 rmn     BVmin=25
```

```
.model J2N4393   NJF(Beta=9.109m Betatce=-.5 Rd=1 Rs=1 Lambda=6m Vto=-1.422
+          Vtotc=-2.5m Is=205.2f Isr=1.988p N=1 Nr=2 Xti=3 Alpha=20.98u
+          Vk=123.7 Cgd=4.57p M=.4069 Pb=1 Fc=.5 Cgs=4.06p Kf=123E-18
+          Af=1)
           National     pid=51     case=TO18
           88-07-13 bam     BVmin=40
```

LIBRARY OF LINEAR IC DEFINITIONS

This is a reduced version of MicroSim's linear subcircuit library. The parameters in the opamp library were derived from the data sheets for each part. the macro-model used is similar to the one described in:

"Macromodeling of Integrated Circuit Operational Amplifiers" by
Graeme Boyle, Barry Cohn, Donald Pederson, and James Solomon,
IEEE Journal of Solid-State Circuits, Vol. SC-9, no. 6, DEC. 1974.

Differences from the reference (above) occur in the output limiting stage which was modified to reduce internally generated currents associated with output voltage limiting, as well as short-circuit current limiting.

The opamps are modeled at room temperature and do not track changes with temperature. This library file contains models for nominal, not worst case, devices.

```
connections:        non-inverting input
                    |   inverting input
                    |   |   positive power supply
                    |   |   |   negative power supply
                    |   |   |   |   output
                    |   |   |   |   |
. subckt LM324      1   2   3   4   5

    c1    11 12 2.887E-12
    c2     6  7 30.00E-12
    dc     5 53 dx
    de    54  5 dx
    dlp   90 91 dx
    dln   92 90 dx
    dp     4  3 dx
    egnd  99  0 poly(2)  (3,0)  (4,0)  0 .5 .5
    fb     7 99 poly(5) vb vc ve vlp vln 0 21.22E6 -20E6 20E6 20E6 -20E6
    ga     6  0 11 12 188.5E-6
    gcm    0  6 10 99 59.61E-9
    iee    3 10 dc 15.09E-6
    hlim  90  0 vlim 1K
    q1    11  2 13 qx
    q2    12  1 14 qx
    r2     6  9 100.0E3
    rc1    4 11 5.305E3
    rc2    4 12 5.305E3
    re1   13 10 1.845E3
    re2   14 10 1.845E3
    ree   10 99 13.25E6
    ro1    8  5 50
    ro2    7 99 25
    rp     3  4 9.082E3
    vb     9  0 dc 0
    vc     3 53 dc 1.500
    ve    54  4 dc 0
    vlim   7  8 dc 0
    vlp   91  0 dc 40
    vln    0 92 dc 40
. model dx D(Is=800.0E-18 Rs=1)
. model qx PNP(Is=800.0E-18 Bf=166.7)
. ends
```

connections:
```
                    non-inverting input
                    |   inverting input
                    |   |   positive power supply
                    |   |   |   negative power supply
                    |   |   |   |   output
                    |   |   |   |   |
.subckt uA741       1   2   3   4   5
```

```
   c1    11  12  8.661E-12
   c2     6   7  30.00E-12
   dc     5  53  dx
   de    54   5  dx
   dlp   90  91  dx
   dln   92  90  dx
   dp     4   3  dx
   egnd  99   0  poly(2) (3,0) (4,0) 0 .5 .5
   fb     7  99  poly(5) vb vc ve vlp vln 0 10.61E6 -10E6 10E6 10E6 -10E6
   ga     6   0  11 12 188.5E-6
   gcm    0   6  10 99 5.961E-9
   iee   10   4  dc 15.16E-6
   hlim  90   0  vlim 1K
   q1    11   2  13 qx
   q2    12   1  14 qx
   r2     6   9  100.0E3
   rc1    3  11  5.305E3
   rc2    3  12  5.305E3
   re1   13  10  1.836E3
   re2   14  10  1.836E3
   ree   10  99  13.19E6
   ro1    8   5  50
   ro2    7  99  100
   rp     3   4  18.16E3
   vb     9   0  dc 0
   vc     3  53  dc 1
   ve    54   4  dc 1
   vlim   7   8  dc 0
   vlp   91   0  dc 40
   vln    0  92  dc 40
.model dx D(Is=800.0E-18 Rs=1)
.model qx NPN(Is=800.0E-18 Bf=93.75)
.ends
```

VOLTAGE COMPARATORS

The parameters in this comparator library were derived from data sheets for each part. The macromodel used was developed by MicroSim Corporation and is produced by the Parts option to PSpice. The comparators are modeled at room temperature. The macromodel does not track changes with temperature. This library file contains models for nominal, not worst case, devices.

```
connections:        non-inverting input
                    |   inverting input
                    |   |   positive power supply
                    |   |   |   negative power supply
                    |   |   |   |   open collector output
                    |   |   |   |   |   output ground
                    |   |   |   |   |   |
. subckt LM111      1   2   3   4   5   6


    f1      9   3  v1  1
    iee     3   7  dc  100.0E-6
    vi1    21   1  dc  .45
    vi2    22   2  dc  .45
    q1      9  21   7  qin
    q2      8  22   7  qin
    q3      9   8   4  qmo
    q4      8   8   4  qmi
. model qin PNP(Is=800.0E-18 Bf=833.3)
. model qmi NPN(Is=800.0E-18 Bf=1002)
. model qmo NPN(Is=800.0E-18 Bf=1000 Cjc=1E-15 Tr=118.8E-9)
    e1     10   6   9   4  1
    v1     10  11  dc  0
    q5      5  11   6  qoc
. model qoc NPN(Is=800.0E-18 Bf=34.49E3 Cjc=1E-15 Tf=364.6E-12
Tr=79.34E-9)
    dp      4   3  dx
    rp      3   4  6.122E3
. model dx   D(Is=800.0E-18 Rs=1)
. ends
```

LIBRARY OF MAGNETIC CORE MODEL PARAMETERS

This is a reduced version of MicroSim's magnetic core library.

The parameters in this model library were derived from the data sheets for each core. The Jiles-Atherton magnetics model is described in:
"Theory of Ferromagnetic Hysteresis," by D. C. Jiles and D. L. Atherton, *Journal of Magnetism and Magnetic Materials,* vol. 61 (1986), pp. 48–60.
Model parameters for ferrite material (Ferroxcube 3C8) were obtained by trial simulations, using the *B-H* curves from the manufacturer's catalog. Then the library was compiled from the data sheets for each core geometry. Notice that only the geometric values change once a material is characterized.

Example use: K2 L2 .99 K1409PL_3C8

Notes:

1. Using a *K* device (formerly only for mutual coupling) with a model reference changes the meaning of the *L* device: the inductance value becomes the number of turns for the winding.

2. *K* devices can "get away" with specifying only one inductor, as in the example above, to simulate power inductors.

Example circuit file:

Demonstration of power inductor *B-H* curve

To view results with Probe (*B-H* curve):
  1) Add Trace for *B*(K1)
  2) set X-axis variable to *H*(K1)

Probe x-axis unit is Oersted
Probe y-axis unit is Gauss

```
.trans .1 4
igen0 0 1 sin(0 .1amp 1Hz 0)  ; Generator: starts with 0.1 amp sinewave, then
igen1 0 1 sin(0 .1amp 1Hz 1)  ;    +0.1 amps, starting at 1 second
igen2 0 1 sin(0 .2amp 1Hz 2)  ;    +0.2 amps, starting at 2 seconds
igen3 0 1 sin(0 .8amp 1Hz 3)  ;    +0.4 amps, starting at 3 seconds
RL 1 0 1ohm              ; generator source resistance
L1 1 0 20               ; inductor with 20 turns
K1 L1 .9999 K528T500_3C8 ; Ferroxcube torroid core
.model K528T500_3C8 Core(MS=420E3 ALPHA=2E-5 A=26 K=18 C=1.05
+            AREA=1.17 PATH=8.49)
.options it15=0
.probe
.end
```

Ferroxcube Pot Cores: 3C8 Material

```
.model K3019PL_3C8    Core(Ms=420E3 Alpha=2E-5 A=26 K=18 C=1.05
+            Area=1.38 Path=4.52)
```

Ferroxcube Square Cores: 3C8 Material

```
.model KRM8PL_3C8    Core(Ms=420E3 Alpha=2E-5 A=26 K=18 C=1.05
+            Area=.630 Path=3.84)
```

Ferroxcube Toroid Cores: 3C8 Material

```
.model K502T300_3C8    Core(Ms=420E3 Alpha=2E-5 A=26 K=18 C=1.05
+            Area=.371 Path=7.32)
```

LIBRARY OF MOSFET MODEL PARAMETERS (FOR "POWER" MOSFET DEVICES)

This is a reduced version of MicroSim's power MOSFET model library. The parameters in this model library were derived from the data sheets for each part. Each part was characterized using the Parts option.

Device can also be characterized without Parts as follows:

LEVEL	Set to 3 (short-channel device).
TOX	Determined from gate ratings.
L, LD, W, WD	Assume L=2u. Calculate from input capacitance.
XJ, NSUB	Assume usual technology.
IS, RD, RB	Determined from "source-drain diode forward voltage" specification or curve (Idr vs. Vsd).
RS	Determine from Rds(on) specification.
RDS	Calculated from Idss specification or curves.
VTO, UO, THETA	Determined from "output characteristics" curve family (Ids vs. Vds, stepped Vgs).
ETA, VMAX, CBS	Set for null effect.
CBD, PB, MJ	Determined from "capacitance vs. Vds" curves.
RG	Calculate from rise/fall time specification or curves.
CGSO, CGDO	Determined from gate-charge, turn-on/off delay and rise time specifications.

Note: When specifying the instance of a device in your circuit file:

BE SURE to have the source and bulk nodes connected together, as this is the way the real device is constructed.

DO NOT include values for L, W, AD, AS, PD, PS, NRD, or NDS. The PSpice default values for these parameters are taken into account in the library model statements. Of course, you should NOT reset the default values using the .OPTIONS statement, either.

Example use: M17 15 23 7 7 IRF150

The "power" MOSFET device models benefit from relatively complete specification of static and dynamic characteristics by their manufacturers. The following effects are modeled:

DC transfer curves in forward operation

gate drive characteristics and switching delay

"on" resistance

reverse-mode "body-diode" operation

The factors not modeled include:

maximum ratings (e.g., high-voltage breakdown)

safe operating area (e.g., power dissipation)

latch-up

noise

For high-current switching applications, we advise that you include series inductance elements, for the source and drain, in your circuit file. In doing so, voltage spikes due to *di/dt* will be modeled. According to the 1985 *International Rectifier Data Book*, the following case styles have lead inductance values of:

TO-204 (modified TO-3)   source = 12.5nH   drain = 5.0nH
TO-220   source = 7.5nH   drain = 3.5–4.5nH

```
.model IRF150    NMOS(Level=3 Gamma=0 Delta=0 Eta=0 Theta=0 Kappa=0 Vmax=0 Xj=0
+        Tox=100n Uo=600 Phi=.6 Rs=1.624m Kp=20.53u W=.3 L=2u Vto=2.831
+          Rd=1.031m Rds=444.4K Cbd=3.229n Pb=.8 Mj=.5 Fc=.5 Cgso=9.027n
+        Cgdo=1.679n Rg=13.89 Is=194E-18 N=1 Tt=288n)
        Int'l Rectifier    pid=IRFC150  case=TO3
        88-08-25 bam       creation
```

```
.model IRF9140    PMOS(Level=3 Gamma=0 Delta=0 Eta=0 Theta=0 Kappa=0 Vmax=0 Xj=0
+        Tox=100n Uo=300 Phi=.6 Rs=70.6m Kp=10.15u W=1.9 L=2u Vto=-3.67
+        Rd=60.66m Rds=444.4K Cbd=2.141n Pb=.8 Mj=.5 Fc=.5 Cgso=877.2p
+        Cgdo=369.3p Rg=.811 Is=52.23E-18 N=2 Tt=140n)
        Int'l Rectifier    pid=IRFC9140    case=TO3
        88-08-25 bam       creation
```

LIBRARY OF DIGITAL LOGIC

This is a reduced version of MicroSim's Digital components library.

The parameters in this model library were derived from:

The TTL Data Book, Texas Instruments, 1985, vol. 2

Each device is modeled by a subcircuit. The interface pins of the subcircuit have the same name as the pin labels in the data book. The general order is inputs followed by outputs, but on the more complex devices you will have to look at the subcircuit definition. The word "BAR" is appended to inverted inputs or outputs. There are two optional power supply pins for each digital subcircuit. You do not need to specify these if you are using a 5-V supply with analog and digital ground connected. If you use another power supply configuration, then the pins should be connected to that supply.

The timing characteristics from the data book are included in the models, with all data sheet effects modeled, unless noted in this file.

If a device contains multiple, independent, identical functions, only one is contained in the subcircuit. (For example, the 7400 contains four two-input NAND gates, but there is only one in the 7400 subckt.)

The subcircuit name is the part name. Only the 74 series (not the 54 series) is included in the library, except for a few parts which are made only in the 54 series (e.g., 54L00).

7402 Quadruple 2-Input Positive-NOR Gates

The TTL Data Book Texas Instruments, 1985, vol. 2
    tdn   06/23/89   Update interface and model names

```
.subckt 7402  A B Y
+    optional: DPWR=$G_DPWR DGND=$G_DGND
+    params: MNTYMXDLY=0 IO_LEVEL=0
U1 nor(2) DPWR DGND
+    A B   Y
+    D_02 IO_STD MNTYMXDLY={MNTYMXDLY} IO_LEVEL={IO_LEVEL}
.ends

.model D_02 ugate (
+    tplhty=12ns      tplhmx=22ns
+    tphlty=8ns tphlmx=15ns
+    )
```

7404 Hex Inverters

The TTL Data Book, Texas Instruments, 1985, vol. 2

tdn   06/23/89   Update interface and model names

```
.subckt 7404  A Y
+    optional: DPWR=$G_DPWR DGND=$G_DGND
+    params: MNTYMXDLY=0 IO_LEVEL=0
U1 inv DPWR DGND
+    A   Y
+    D_04 IO_STD MNTYMXDLY={MNTYMXDLY} IO_LEVEL={IO_LEVEL}
.ends

.model D_04 ugate (
+    tplhty=12ns      tplhmx=22ns
+    tphlty=8ns tphlmx=15ns
+    )
```

7405 Hex Inverters with Open-Collector Outputs

The TTL Data Book, Texas Instruments, 1985, vol. 2

tdn   06/23/89   Update interface and model names

```
.subckt 7405  A Y
+    optional: DPWR=$G_DPWR DGND=$G_DGND
+    params: MNTYMXDLY=0 IO_LEVEL=0
U1 inv DPWR DGND
+    A   Y
+    D_05 IO_STD_OC MNTYMXDLY={MNTYMXDLY} IO_LEVEL={IO_LEVEL}
.ends

.model D_05 ugate (
+    tplhty=40ns      tplhmx=55ns
+    tphlty=8ns tphlmx=15ns
+    )
```

7414 Hex Schmitt-Trigger Inverters

The TTL Data Book, Texas Instruments, 1985, vol. 2

tdn   06/23/89   Update interface and model names

```
.subckt 7414  A Y
+    optional: DPWR=$G_DPWR DGND=$G_DGND
+    params: MNTYMXDLY=0 IO_LEVEL=0
```

Note: These devices are modeled as simple inverters. Hysteresis is modeled in the A to D interface.

```
U1 inv DPWR DGND
+    A   Y
+    D_14 IO_STD_ST MNTYMXDLY={MNTYMXDLY} IO_LEVEL={IO_LEVEL}
.ends

.model D_14 ugate (
+    tplhty=15ns      tplhmx=22ns
+    tphlty=15ns      tphlmx=22ns
+    )
```

7474 Dual D-Type Positive-Edge-Triggered Flip-Flops with Preset and Clear

The TTL Data Book, Texas Instruments, 1985, vol. 2

tdn   06/28/89   Update interface and model names

```
.subckt 7474   1CLRBAR 1D 1CLK 1PREBAR 1Q 1QBAR
+    optional: DPWR=$G_DPWR DGND=$G_DGND
+    params: MNTYMXDLY=0 IO_LEVEL=0
UFF1 dff(1) DPWR DGND
+    1PREBAR 1CLRBAR 1CLK   1D    1Q 1QBAR
+    D_74 IO_STD MNTYMXDLY={MNTYMXDLY} IO_LEVEL={IO_LEVEL}
.ends

.model D_74 ueff (
+    twpclmn=30 ns      twclklmn=37ns
+    twclkhmn=30ns      tsudclkmn=20ns
+    thdclkmn=5ns       tppcqlhmx=25ns
+    tppcqhlmx=40ns     tpclkqlhty=14ns
+    tpclkqlhmx=25ns tpclkqhlty=20ns
+    tpclkqhlmx=40ns
+    )
```

74107 Dual J-K Flip-Flops with Clear

The TTL Data Book, Texas Instruments, 1985, vol. 2

tdn   06/29/89   Update interface and model names

```
.subckt 74107   CLK CLRBAR J K Q QBAR
+    optional: DPWR=$G_DPWR DGND=$G_DGND
+    params: MNTYMXDLY=0 IO_LEVEL=0
UIBUF bufa(3) DPWR DGND
+    CLRBAR J K    CLRBAR_BUF J_BUF K_BUF
+    D0_GATE IO_STD IO_LEVEL={IO_LEVEL}
U2BUF buf DPWR DGND
+    CLK    CLK_BUF
+    D_107_4 IO_STD MNTYMXDLY={MNTYMXDLY} IO_LEVEL={IO_LEVEL}
U1 srff(1) DPWR DGND
+    $D_HI CLRBAR_BUF CLK_BUF   W1 W2   Y YB
+    D_107_1 IO_STD MNTYMXDLY={MNTYMXDLY}
U2 srff(1) DPWR DGND
+    $D_HI CLRBAR_BUF CLKBAR    Y YB    QBUF QBAR_BUF
+    D_107_2 IO_STD MNTYMXDLY={MNTYMXDLY}
U3 inva(3) DPWR DGND
+    CLK_BUF J_BUF K_BUF    CLKBAR JB KB
+    D0_GATE IO_STD
```

```
U4 ao(3,2) DPWR DGND
+      J_BUF K_BUF QBAR_BUFD J_BUF KB $D_HI      W1
+      D_107_3 IO_STD MNTYMXDLY={MNTYMXDLY}
U5 ao(3,2) DPWR DGND
+      J_BUF K_BUF QBUFD JB K_BUF $D_HI      W2
+      D_107_3 IO_STD MNTYMXDLY={MNTYMXDLY}
UBUF bufa(4) DPWR DGND
+      QBUF QBAR_BUF QBUF QBAR_BUF    Q QBAR QBUFD QBAR_BUFD
+      D_107_3 IO_STD MNTYMXDLY={MNTYMXDLY} IO_LEVEL={IO_LEVEL}
.ends


.model D_107_1 ugff (
+      twghmx=19ns        twghty=19ns
+      twpclmx=47ns       twpclty=27ns
+      )
.model D_107_2 ugff (
+      tppcqlhty=10ns     tppcqlhmx=19ns
+      tppcqhlty=19ns     tppcqhlmx=34ns
+      tpgqlhty=10ns      tpgqlhmx=19ns
+      tpgqhlty=19ns      tpgqhlmx=34ns
+      twghmx=20ns        twghty=20ns
+      twpclmx=47ns       twpclty=47ns
+      )
.model D_107_3 ugate (
+      tplhty=6ns tplhmx=6ns
+      tphlty=6ns tphlmx=6ns
+      )
.model D_107_4 ugate (
+      tplhmn=6ns tplhmx=6ns
+      )
```

74393 Dual 4-Bit Binary Counter with Individual Clocks

The TTL Data Book, Texas Instruments, 1985, vol. 2

atl   7/18/89   Update interface and model names

```
.subckt 74393  A CLR QA QB QC QD
+      optional: DPWR=$G_DPWR DGND=$G_DGND
+      params: MNTYMXDLY=0 IO_LEVEL=0
UINV inv DPWR DGND
+      CLR    CLRBAR
+      D0_GATE IO_STD IO_LEVEL={IO_LEVEl}
U1 jkff(1) DPWR DGND
+      $D_HI CLRBAR A    $D_HI $D_HI    QA_BUF $D_NC
+      D_393_1 IO_STD MNTYMXDLY={MNTYMXDLY} IO_LEVEL={IO_LEVEL}
U2 jkff(1) DPWR DGND
+      $D_HI CLRBAR QA_BUF    $D_HI $D_HI    QB_BUF $D_NC
+      D_393_2 IO_STD MNTYMXDLY={MNTYMXDLY}
U3 jkff(1) DPWR DGND
+      $D_HI CLRBAR QB_BUF    $D_HI $D_HI    QC_BUF $D_NC
+      D_393_2 IO_STD MNTYMXDLY={MNTYMXDLY}
U4 jkff(1) DPWR DGND
+      $D_HI CLRBAR QC_BUF    $D_HI $D_HI    QD_BUF $D_NC
+      D_393_3 IO_STD MNTYMXDLY={MNTYMXDLY}
UBUFF bufa(4) DPWR DGND
+      QA_BUF QB_BUF QC_BUF QD_BUF    QA QB QC QD
+      D_393_4 IO_STD MNTYMXDLY={MNTYMXDLY} IO_LEVEL={IO_LEVEL}
.ends
```

```
.model D_393_1 ueff  (
+    tppcqhlty=18ns    tppcqhlmx=33ns
+    tpclkqlhty=6ns    tpclkqlhmx=14ns
+    tpclkqhlty=7ns    tpclkqhlmx=14ns
+    twclkhmn=20ns     twclklmn=20ns
+    twpclmn=20ns      tsudclkmn=25ns
+    )
.model D_393_2 ueff ()
.model D_393_3 ueff (
+    tpclkqlhty=27ns  tpclkqlhmx=40ns
+    tpclkqhlty=27ns  tpclkqhlmx=40ns
+    )
.model D_393_4 ugate (
+    tphlty=6ns  tphlmx=6ns
+    tplhty=6ns  tplhmx=6ns
+    )
```

AtoD and DtoA SUBCIRCUITS

The subcircuits in this library are used to convert analog signals into digital signals (AtoD) and digital signals into analog signals (DtoA). The PSpice Digital Simulation Option creases "X" devices which reference these subcircuits whenever it needs to convert a digital or analog signal. The user usually will not need to use these subcircuits directly. However, if you need to add new AtoD or DtoA subcircuits, the interface nodes must be in the following order and have the following parameters:

AtoD:

```
.subckt <name> <analog-node> <dig-node> <dig-pwr> <dig-gnd>
+ params: CAPACITANCE=0
```

DtoA:

```
.subckt <name> <dig-node> <analog-node> <dig-pwr> <dig-gnd>
+ params: DRVL=0 DRVH=0 CAPACITANCE=0
```

I/O Models

I/O models specify the names of the AtoD and DtoA subcircuits that PSpice must use to convert analog signals to digital signals or vice versa. (I/O models also describe driving and loading characteristics.) Up to four of each AtoD and DtoA subcircuit names may be specified in an I/O model, using parameters AtoD1 through AtoD4, and DtoA1 through DtoA4. The subcircuit that PSpice actually uses depends on the value of the IO_LEVEL parameters in a subcircuit reference.

As implemented in this library, the levels have the following definitions:
IO_LEVEL definition:

1   Basic (simple) model with X, R, and F between VIL max and VIH min (AtoD)

2   Basic (simple) model without intermediate X value

3   Elaborate model with X between VIL max and VIH min (AtoD)
4   Elaborate model without intermediate X, R, and F value

The Elaborate has a more accurate *I-V* curve, including clamping diodes, but since it has more devices, it can take longer to simulate when it is used.

For example, to specify the basic interface without an intermediate X value, you would use:

```
X1 in out 74LS04 PARAMS: IO_LEVEL=2
```

If the IO_LEVEL is not specified for a device, the default IO_LEVEL is used. The default level is controlled by the .OPTION parameter DIGIOLVL, which defaults to 1.

Digital Power Supply

PSpice automatically creates one instance of this subcircuit if any AtoD or DtoA interfaces are created. PSpice always uses node *0* as the required analog reference node "GND." The digital power and ground nodes default to global nodes named $G_DPWR and $G_DGND, which are used throughout the digital libraries. The default output is 5.0 V.

To create your own power supply, simply create an instance of this subcircuit, using your own digital power and ground node names, and the desired voltage. For example:

```
XMYPOWER 0 MY_PWR MY_GND DIGIFPWR params: VOLTAGE=3.5V

.subckt DIGIFPWR AGND
+       optional: DPWR=$G_DPWR DGND=$G_DGND
+       params:   VOLTAGE=5.0v REFERENCE=0v

VDPWR DPWR DGND    {VOLTAGE}
R1      DPWR AGND    1MEG
VDGND DGND REF      {REFERENCE}
R2      REF  AGND    1E-6
R3      DGND AGND    1MEG
.ends
```

STIMULUS DEVICE MODELS AND SUBCIRCUITS

Stimulus I/O Models

```
.model IO_STM uio (
+    drvh=0      drv1=0
+    DtoA1="DtoA_STM" DtoA2="DtoA_STM"
+    DtoA3="DtoA_STM" DtoA4="DtoA_STM"
+    )
.model IO_STM_OC uio (
+    drvh=1MEG  drv1=0
+    DtoA1="DtoA_STM_OC"  DtoA2="DtoA_STM_OC"
+    DtoA3="DtoA_STM_OC"  DtoA4="DtoA_STM_OC"
+    )
```

Stimulus DtoA Subcircuit

```
.subckt DtoA_STM D A   DPWR DGND
+      params: DRVL=0 DRVH=0 CAPACITANCE=0

N1   A DGND DPWR DINSTM DGTLNET=D IO_STM
C1   A 0 {CAPACITANCE+0.1pF}
.ends
```

Stimulus Open Collector DtoA Subcircuit

```
.subckt DtoA_STM_OC   D A   DPWR DGND
+      params: DRVL=0 DRVH=0 CAPACITANCE=0

N1   A DGND DPWR DINSTM=OC DGTLNET=D IO_STM_OC
C1   A 0 {CAPACITANCE+0.1pF}
.ends
```

Stimulus Digital Input/Output Models

We use 1/2 ohm and a 500 ps transition time, on the assumption that this will be a "strong" signal source with a "fast" switching time in most systems that use this library. Change the tsw's and/or the rlow and rhi values if these don't work for your system.

```
.model DINSTM dinput (
+      s0name="0" s0tsw=0.5ns   s0rlo=.5    s0rhi=1k
+      s1name="1" s1tsw=0.5ns   s1rlo=1k    s1rhi=.5
+      s2name="X" s2tsw=0.5ns   s2rlo=0.429    s2rhi=1.16 ; .313ohm, 1.35v
+      s3name="R" s3tsw=0.5ns   s3rlo=0.429    s3rhi=1.16 ; .313ohm, 1.35v
+      s4name="F" s4tsw=0.5ns   s4rlo=0.429    s4rhi=1.16 ; .313ohm, 1.35v
+      s5name="Z" s5tsw=0.5ns   s5rlo=1MEG s5rhi=1MEG
+      )
.model DINSTM_OC dinput (
+      s0name="0" s0tsw=0.5ns   s0rlo=.5   s0rhi=1k
+      s1name="1" s1tsw=0.5ns   s1rlo=1MEG s1rhi=1MEG
+      s2name="X" s2tsw=0.5ns   s2rlo=0.429    s2rhi=1.16 ; .313ohm, 1.35v
+      s3name="R" s3tsw=0.5ns   s3rlo=0.429    s3rhi=1.16 ; .313ohm, 1.35v
+      s4name="F" s4tsw=0.5ns   s4rlo=0.429    s4rhi=1.16 ; .313ohm, 1.35v
+      s5name="Z" s5tsw=0.5ns   s5rlo=1MEG     s5rhi=1MEG
+      )
```

Zero-Delay Models

Zero-Delay Gate Model

```
.model D0_GATE ugate ()
```

Zero-Delay Tristate Gate Model

```
.model D0_TGATE utgate ()
```

Zero-Delay Edge-Triggered Flip-Flop Model

```
.model D0_EFF ueff ()
```

Zero-Delay Gated Flip-Flop Model

```
.model D0_GFF ugff ()
```

74/54 Family (Standard TTL)

7400 I/O Models

```
.model IO_STD uio (
+      drvh=96.4  drvl=104
+      AtoD1="AtoD_STD" AtoD2="AtoD_STD_NX"
+      AtoD3="AtoD_STD_E"      AtoD4="AtoD_STD_NXE"
+      DtoA1="DtoA_STD" DtoA2="DtoA_STD"
+      DtoA3="DtoA_STD" DtoA4="DtoA_STD"
+      )
.model IO_STD_ST uio (
+      drvh=96.4 drvl=104
+      AtoD1="AtoD_STD_ST"  AtoD2="AtoD_STD_ST"
+      AtoD3="AtoD_STD_ST_E" AtoD4="AtoD_STD_ST_E"
+      DtoA1="DtoA_STD" DtoA2="DtoA_STD"
+      DtoA3="DtoA_STD" DtoA4="DtoA_STD"
+      )
.model IO_STD_OC uio (
+      drvh=1MEG drvl=104
+      AtoD1="AtoD_STD" AtoD2="AtoD_STD_NX"
+      AtoD3="AtoD_STD_E"      AtoD4="AtoD_STD_NXE"
+      DtoA1="DtoA_STD_OC"  DtoA2="DtoA_STD_OC"
+      DtoA3="DtoA_STD_OC"  DtoA4="DtoA_STD_OC"
+      )
```

7400 Standard AtoD Subcircuits
Simple models:

```
.subckt AtoD_STD   A D   DPWR DGND params: CAPACITANCE=0

O0   A DGND DO74 DGTLNET=D IO_STD
C1   A 0  {CAPACITANCE+0.1pF}
.ends

.subckt AtoD_STD_NX   A D   DPWR DGND params: CAPACITANCE=0

O0   A DGND DO74_NX DGTLNET=D IO_STD
C1   A 0  {CAPACITANCE+0.1pF}
.ends
```

Elaborate models:

```
.subckt AtoD_STD_E   A D   DPWR DGND params: CAPACITANCE=0

O0   A DGND DO74 DGTLNET=D IO_STD
C1   A 0  {CAPACITANCE+0.1pF}
D0    DGND a     D74CLMP
D1    1     2    D74
D2     2         DGND D74
R1   DPWR 3     4k
Q1     1    3      A     0    Q74 ; substrait should be DGND
.ends

.subckt AtoD_STD_NXE   A D   DPWR DGND params: CAPACITANCE=0
```

```
O0    A  DGND  DO74_NX  DGTLNET=D  IO_STD
C1    A  0  {CAPACITANCE+0.1pF}
D0    DGND  a       D74CLMP
D1    1     2       D74
D2       2          DGND D74
R1    DPWR 3       4k
Q1       1       3       A       0       Q74 ; substrait should be DGND
.ends
```

7400 Schmidt Trigger AtoD Subcircuits
Simple model:

```
.subckt AtoD_STD_ST   A D   DPWR DGND params: CAPACITANCE=0

O0    A  DGND  DO74_ST  DGTLNET=D  IO_STD
C1    A  0  {CAPACITANCE+0.1pF}
.ends
```

Elaborate model:

```
.subckt AtoD_STD_ST_E   A D   DPWR DGND params: CAPACITANCE=0

O0    A  DGND  DO74_ST  DGTLNET=D  IO_STD
C1    A  0  {CAPACITANCE+0.1pF}
D0    DGND  a       D74CLMP
D1    1     2       D74
D2       2          DGND D74
R1    DPWR 3       4k
Q1       1       3       A       0       Q74
.ends
```

7400 Standard DtoA Subcircuit

```
.subckt DtoA_STD   D A   DPWR DGND params: DRVL=0 DRVH=0 CAPACITANCE=0

N1    A  DGND  DPWR  DIN74  DGTLNET=D  IO_STD
C1    A  0  {CAPACITANCE+0.1pF}
.ends
```

7400 Open Collector DtoA Subcircuit

```
.subckt DtoA_STD_OC   D A   DPWR DGND params: DRVL=0 DRVH=0 CAPACITANCE=0

N1    A  DGND  DPWR  DIN74_OC  DGTLNET=D  IO_STD_OC
C1    A  0  {CAPACITANCE+0.1pF}
.ends
```

7400 Digital Input/Output Models

```
.model DIN74 dinput (
+      s0name="0" s0tsw=3.5ns    s0rlo=7.13 s0rhi=389 ; 7ohm,    0.09v
+      s1name="1" s1tsw=5.5ns    s1rlo=467 s1rhi=200 ; 140ohm,   3.5v
+      s2name="X" s2tsw=3.5ns    s2rlo=42.9 s2rhi=116 ; 31.3ohm, 1.35v
+      s3name="R" s3tsw=3.5ns    s3rlo=42.9 s3rhi=116 ; 31.3ohm, 1.35v
+      s4name="F" s4tsw=3.5ns    s4rlo=42.9 s4rhi=116 ; 31.3ohm, 1.35v
+      s5name="Z" s5tsw=3.5ns    s5rlo=200K s5rhi=200K
+      )
```

```
.model DIN74_OC dinput (
+     s0name="0" s0tsw=3.5ns    s0rlo=7.13 s0rhi=389 ; 7ohm,     0.09v
+     s1name="1" s1tsw=5.5ns    s1rlo=200K s1rhi=200K
+     s2name="X" s2tsw=3.5ns    s2rlo=42.9 s2rhi=116 ; 31.3ohm, 1.35v
+     s3name="R" s3tsw=3.5ns    s3rlo=42.9 s3rhi=116 ; 31.3ohm, 1.35v
+     s4name="F" s4tsw=3.5ns    s4rlo=42.9 s4rhi=116 ; 31.3ohm, 1.35v
+     s5name="Z" s5tsw=5.5ns    s5rlo=200K s5rhi=200K
+     )

.model DO74 doutput (
+     s0name="X" s0vlo=0.8  s0vhi=2.0
+     s1name="0" s1vlo=-1.5 s1vhi=0.8
+     s2name="R" s2vlo=0.8  s2vhi=1.4
+     s3name="R" s3vlo=1.3  s3vhi=2.0
+     s4name="X" s4vlo=0.8  s4vhi=2.0
+     s5name="1" s5vlo=2.0  s5vhi=7.0
+     s6name="F" s6vlo=1.3  s6vhi=2.0
+     s7name="F" s7vlo=0.8  s7vhi=1.4
+     )

.model DO74_NX doutput (
+     s0name="0" s0vlo=-1.5 s0vhi=1.35
+     s2name="1" s2vlo=1.35 s2vhi=7.0
+     )

.model DO74_ST doutput (
+     s0name="0" s0vlo=-1.5 s0vhi=1.7
+     s1name="1" s1vlo=0.9  s1vhi=7.0
+     )

.model D74 d (
+     is=1e-16  rs=25 cjo=2pf
+     )

.model D74CLMP d (
+     is=1e-15  rs=2 cjo=2pf
+     )

.model Q74 npn (
+     ise=1e-16  isc=4e-16
+     bf=49 br=.03
+     cje=1pf    cjc=.5pf
+     cjs=3pf    vje=0.9v
+     vjc=0.8v   vjs=0.7v
+     mje=0.5    mjc=0.33
+     mjs=0.33   tf=0.2ns
+     tr=10ns    rb=50
+     rc=20
+     )
```

LIBRARY OF OPTOCOUPLER MODELS

The parameters in this model library were derived from the data sheets for each part.

```
.model 4N25
```

```
6-pin DIP: pin      #1  #2    #4    #5    #6
                    |   |     |     |     |
 .subckt A4N25      pin1 pin2 pin4  pin5  pin6         params: rel_CTR=1
                    Motorola            pid=4N25
                    88-01-04 pwt
                    88-01-18 pwt       rework Cje approximation
```

The data sheet used for this model is from Motorola; it was the most complete for dc and switching parameters, and it was easy to find the component IR-LED and phototransistor as separate devices for further specifications.

```
d_MainLED      pin1 pin2 MainLED
d_PhotoLED     pin1 1        PhotoLED .001
v_PhotoLED     1 pin2        0

f_TempComp     0 2          v_PhotoLED 1
r_TempComp     2 0          TempComp {rel_CTR}

g_BaseSrc      5 6 2 0         .9
q_PhotoBJT     5 6 4        PhotoBJT
r_C        5 pin5          .1
r_B        6 pin6          .1
r_E        4 pin4          .1
```

Since active devices dominate pin-to-pin capacitance on each "side" of the optocoupler, isolation is modeled by identical capacitances and resistances linked to a common point; this gives isolation of 0.5 pF and 1E+11 ohms.

```
c_1        pin1 7          .4p
r_1        pin1 7          .12T
c_2        pin2 7          .4p
r_2        pin2 7          .12T
c_4        pin4 7          .4p
r_4        pin4 7          .12T
c_5        pin5 7          .4p
r_5        pin5 7          .12T
c_6        pin6 7          .4p
r_6        pin6 7          .12T
```

Similar to Motorola MLED15.

```
.model MainLED  D(Is=1.1p Rs=.66 Ikf=30m N=1.9 Xti=3 Cjo=40p M=.34 Vj=.75
+          Isr=30n Nr=3.8 Bv=6 Ibv=100u Tt=.5u)
```

Models photon generation: same as MainLED except on ac effects, no breakdown.

```
.model PhotoLED D(Is=1.1p Rs=.66 Ikf=30m N=1.9 Xti=3 Cjo=0 M=.34 Vj=.75
+          Isr=30n Nr=3.8 Bv=0 Tt=0)
```

Temperature compensation for system: 1.38x @ $-55'C$, .54x @ $+100'C$, all @ 10mA

Note: The photo BJT has its own temperature corrections, which must be kept as the transistor is electrically available.

```
.model TempComp RES(R=1 Tc1=-11.27m Tc2=43.46u)
```

Similar to Motorola MDR3050. $H_{fe} = 325$ @ $I_C = 550uA$, $V_{CE} = 5V$
   Use beta variation (w/Parts) to model change in current-transfer ratio (CTR).
   Hand adjust reverse beta (Br) to match saturation characteristics.
   Set Isc to model dark current.
   Hand adjust C_{jc} to match fall time @ $I_C = 10mA$ (which yields rise time, too).
   Hand adjust reverse transit-time (Tr) to match storage time @ $I_C = 10mA$.
   Delay time set by LED *I-V* and *C-V* characteristics; set C_{je} to 25% of C_{jc}, inspection of phototransistor chip layouts show that the emitter area is 20%-25% that of the collector area. The same layouts show that base resistance is made negligible by design; also, the operating currents are small. Hand adjust forward transit-time (Tf) to match MDR3050 pulse data. Check against 4N25 frequency response (Figs. 11, 12).

```
.model PhotoBJT NPN(Is=10f Xti=3 Vaf=60
+         Bf=400 Ne=3.75 Ise=580p Ikf=.26 Xtb=1.5
+         Br=.04 Nc=2    Isc=3.5n
+         Cjc=10p Mjc=.3333 Vjc=.75 Tr=88u
+         Cje=2.5p Mje=.3333 Vje=.75 Tf=1.5n)
.ends
```

.model 4N25A
6-pin DIP: pin #1   #2   #4   #5   #6
              | | | | |

```
.subckt A4N25A  pin1 pin2 pin4 pin5 pin6
          88-01-05 pwt
```
Same as 4N25 (UL recognized).

```
  x1 pin1 pin2 pin4 pin5 pin6 A4N25
.ends
```

LIBRARY OF THYRISTOR (SCR AND TRIAC) MODELS

Library of SCR models

Note: This library requires the Analog Behavioral Modeling option available with PSpice. A model developed without Behavioral Modeling was found to be very slow and not very robust.

   This macromodel uses a controlled switch as the basis SCR structure. In all cases, the designer should use the manufacturer's data book for actual part selection.

   The required parameters were derived from data sheet (Motorola) information on each part. When available, only "typical" parameters are used (except for Idrm which is always a "max" value). if a "typical" parameter is not available, a "min" or "max" value may be used in which case a comment is made in the library.

   The SCRs are modeled at room temperature and do not track changes with temperature. Note that Vdrm is specified by the manufacturer as valid over a temperature range. Also, in nearly all cases, *dVdt* and *Toff* are specified by the manufacturer at approximately 100° C. This results in a model that is somewhat "conservative" for a room temperature model.

The parameter *dVdt* (when available from the data sheet) is used to model the Critical Rate of Rise of Off-State Voltage. If not specified, *dVdt* is defaulted to 1000 V/microsecond. A side effect of this model is that the turn-on current, I_{ON}, is determined by *Vtm/(Ih Vdrm)*. *Vtm* is also used as the holding voltage.

```
.SUBCKT Scr anode gate cathode PARAMS:
+ Vdrm=400v      Vrrm=400v       Idrm=10u
+ Ih=6ma         dVdt=5e7
+ Igt=5ma        Vgt=0.7v
+ Vtm=1.7v       Itm=24+ Ton=1u          Toff=15u
```

Where:

$Vdrm \Rightarrow$ Forward breakover voltage

$Vrrm \Rightarrow$ Reverse breakdown voltage

$Idrm \Rightarrow$ Peak blocking current

$Ih \quad \Rightarrow$ Holding current

$dVdt \Rightarrow$ Critical value for *dV/dt* triggering

$Igt \quad \Rightarrow$ Gate trigger current

$Vgt \quad \Rightarrow$ Gate trigger voltage

$Vtm \quad \Rightarrow$ On-state voltage

$Itm \quad \Rightarrow$ On-state current

$Ton \quad \Rightarrow$ Turn-on time

$Toff \Rightarrow$ Turn-off time

Main conduction path

```
Scr       anode    anode0   control 0      Vswitch ; controlled switch
Dak1      anode0   anode2   Dakfwd  OFF             ; SCR is initially off
Dka       cathode  anode0   Dkarev  OFF
VIak      anode2   cathode                          ; current sensor
```

dVdt Turn-on

```
Emon      dvdt0    0        TABLE {v(anode,cathode)} (0 0)  (2000 2000)
CdVdt     dvdt0    dvdt1    100pfd                  ; displacement current
Rdlay     dvdt1    dvdt2    1k
VdVdt     dvdt2    cathode DC 0.0
EdVdt     condvdt  0        TABLE {i(vdVdt)-100p dvdt}  (0 0 )  (.1m 10)
RdVdt     condvdt  0        1meg
```

Gate

```
Rseries   gate     gate1    {(Vgt-0.65)/Igt}
Rshunt    gate1    gate2    {0.65/Igt}
Dgkjf     gate1    gate2    Dgk
VIgf      gate2    cathode                          ; current sensor
```

Gate Turn-on

```
Egate1    gate4    0      TABLE {i(Vigf)-0.95 Igt} (0 0)  (1m 10)
Rgate1    gate4    0      1meg
Egon1     congate 0       TABLE {v(gate4) v(anode,cathode)} (0 0)  (10 10)
Rgon1     congate 0       1meg
```

Main Turn-on

```
EItot     Itot     0      TABLE {i(VIak)+5E-5 i(VIgf)/Igt} (0 0) (2000
2000)
RItot     Itot     0      1meg
Eprod     prod     0      TABLE {v(anode,cathode) v(Itot)} (0 0) (1 1)
Rprod     prod     0      1meg
Elin      conmain 0       TABLE
+         {10 (v(prod) - (Vtm Ih))/(Vtm Ih)} (0 0) (2 10)
Rlin      conmain 0       1meg
```

Turn-on/Turn-off control

```
Eonoff    contot   0      TABLE
+         {v(congate)+v(conmain){v(condvdt)} (0 0)  (10 10)
```

Turn-on/Turn-off delays

```
Rton      contot  dlay1   825
Dton      dlay1   control Delay
Rtoff     contot  dlay2   {290 Toff/Ton}
Dtoff     control dlay2   Delay
Cton      control 0       {Ton/454}
```

Reverse breakdown

```
Dbreak    anode   break1  Dbreak
Dbreak2   cathode break1  Dseries
```

Controlled Switch Model

```
.MODEL Vswitch vswitch
+  Ron = {(Vtm-0.7)/Itm}, Roff = {Vdrm Vdrm/(Vtm Ih)},
+  Von = 5.0,             Voff = 1.5)
```

Diodes

```
.MODEL  Dgk      D      (Is=1E-16 Cjo=50pf Rs=5)
.MODEL  Dseries  D      (Is=1E-14)
.MODEL  Delay    D      (Is=1E-12 Cjo=5pf  Rs=0.01)
.MODEL  Dkarev   D      (Is=1E-10 Cjo=5pf  Rs=0.01)
.MODEL  Dakfwd   D      (Is=4E-11 Cjo=5pf)
.MODEL  Dbreak   D      (Ibv=1E-7 Bv={1.1 Vrrm} Cjo=5pf Rs=0.5)
```

Allow the gate to float if required.

```
Rfloat  gate    cathode 1e10
```

```
.ENDS
```

```
.SUBCKT 2N1595        anode gate cathode
```

"Typical" parameters

```
X1 anode gate cathode Scr PARAMS:
+ Vdrm=50v    Vrrm=50v    Ih=5ma     Vtm=1.1v    Itm=1
+ dVdt=1e9    Igt=2ma     Vgt=.7v    Ton=0.8u    Toff=10u
+ Idrm=10u
```

90-5-18 Motorola DL137, Rev 2, 3/89

```
.ENDS
```

Library of Triac Models

Note: This library requires the Analog Behavioral Modeling option available with PSpice.

This macromodel uses two controlled switches as the basic triac structure. The model was developed to provide firing in all four quadrants. It should be noted, however, that the library contains parts that the manufacturer has guaranteed will fire in 4 quadrants, 3 quadrants, or 2 quadrants. Therefore, the designer should always use the manufacturer's data book for part selection.

The required parameters were derived from data sheet (Motorola) information on each part. When available, only "typical" parameters are used (except for Idrm which is always a "max" value). If a "typical" parameter is not available, a "min" or "max" value may be used in which case a comment is made in the library.

The triacs are modeled at room temperature and do not track changes with temperature. Note that *Vdrm* is specified by the manufacturer as valid over a temperature range. Also, in nearly all cases, *dVdt* is specified by the manufacturer at approximately 100°C. This results in a model that is somewhat "conservative" for a room temperature model.

The parameter *dVdt* (when available from the data sheet) is used to model the Critical Rate of Rise of Off-State Voltage. If not specified, *dVdt* is defaulted to 1000 V/μs. The Critical Rate of Rise of Commutation Voltage is not modeled.

It is generally good practice to use an *RC* snubber network across the triac to limit the commutating *dVdt* to a value below the maximum allowable rating (see manufacturer's data sheet and application notes). Also, note that the turn-off time is assumed to be zero.

```
.SUBCKT Triac MT2 gate MT1 PARAMS:
+ Vdrm=400v    Idrm=10u
+ Ih=6ma       dVdt=50e6
+ Igt=20ma     Vgt=0.9v
+ Vtm=1.3v     Itm=17
+ Ton=1.5u
```

Where:

$Vdrm \Rightarrow$ Forward breakover voltage

$Idrm \Rightarrow$ Peak blocking current

$Ih \quad \Rightarrow$ Holding current [MT2(+)]

$dVdt \Rightarrow$ Critical value for dV/dt triggering

Igt $\Rightarrow$ Gate trigger current [MT2(+),G(−)]
Vgt $\Rightarrow$ Gate trigger voltage [MT2(+),G(−)]
Vtm $\Rightarrow$ On-state voltage
Itm $\Rightarrow$ On-state current
Ton $\Rightarrow$ Turn-on time

Main conduction path

```
Striac   MT2      MT20     cntrol    0            Vswitch ; controlled switch
Dak1     MT20     MT22     Dak       OFF                  ; triac is initially off
VIak     MT22     MT1                                     ; current sensor
Striacr  MT2      MT23     cntrolr   0            Vswitch ; controlled switch
Dka1     MT21     MT23     Dak       OFF                  ; triac is initially off
VIka     MT1      MT21                                    ; reverse current sense
```

dVdt Turn-on

```
Emon     dvdt0    0        TABLE {ABS (V(MT2,MT1))} (0 0) (2000 2000)
CdVdt    dvdt0    dvdt1    100pfd                 ; displacement current
Rdlay    dvdt1    dvdt2    1k
VdVdt    dvdt2    MT1      DC 0.0
EdVdt    condvdt  0        TABLE {i(vdVdt)-100p dVdt} (0 0 ) (.1m 10)
RdVdt    condvdt  0        1meg
```

Gate

```
Rseries  gate     gate1    {(Vgt-0.65)/Igt}
Rshunt   gate1    gate2    {0.65/Igt}
Dgkf     gate1    gate2    Dgk
Dgkr     gate2    gate1    Dgk
VIgf     gate2    MT1      DC 0.0                 ; current sensor
```

Gate Turn-on

```
Egate    congate  0        TABLE {(ABS(i(VIgf))-0.95 Igt)} (0 0) (1m 10)
Rgate    congate  0        1meg
```

Holding current, holding voltage (Quadrant I)

```
Emain1   main1    0        TABLE {i(VIak)-Ih+5e-3 i(VIgf)/Igt} (0 0) (.1m 1)
Rmain1   main1    0        1meg
Emain2   main2    0        TABLE {v(MT2,MT1)-(Ih Vtm/Itm)} (0 0) (.1m 1)
Rmain2   main2    0        1meg
Emain3   cnhold   0        TABLE {v(main1,0) v(main2,0)} (0 0 (1 10)
Rmain3   cnhold   0        1meg
```

Holding current, holding voltage (Quadrant III)

```
Emain1r  main1r   0        TABLE {i(VIka)-Ih-5e-3 i(VIgf)/Igt} (0 0) (.1m 1)
Rmain1r  main1r   0        1meg
Emain2r  main2r   0        TABLE {v(MT1,MT2)-(Ih Vtm/Itm)} (0 0) (.1m 1)
Rmain2r  main2r   0        1meg
Emain3r  cnholdr  0        TABLE {v(main1r,0) v(main2r,0)} (0 0 (1 10)
Rmain3r  cnholdr  0        1meg
```

Main

```
Emain4    main4    0              table {(1.0-ABS(i(VIgf))/Igt)} (0 0)  (1 1)
Rmain4    main4    0              1meg
Emain5    cnmain   0              table {v(mt2,mt1)-1.05 Vdrm v(main4)} (0 0)  (1 10)
Remain5   cnmain   0              1meg
Emain5r   cnmainr  0              table {v(mt1,mt2)-1.05 Vdrm v(main4)} (0 0)  (1 10)
Rmain5r   cnmainr  0              1meg
```

Turn-on/Turn-off control (Quadrant I)

```
Eonoff    contot   0              TABLE
+                  {v(cnmain)+v(congate)+v(cnhold)+v(condvdt)} (0 0)  (10 10)
```

Turn-on/Turn-off delays (Quadrant I)

```
Rton      contot   dlay1          825
Dton      dlay1    cntrol         Delay
Rtoff     contot   dlay2          {2.9E-3/Ton}
Dtoff     cntrol   dlay2          Delay
Cton      cntrol   0              {Ton/454}
```

Turn-on/Turn-off control (Quadrant III)

```
Eonoffr   contotr  0              TABLE
+                  {v(cnmainr)+v(congate)+v(cnholdr)+v(condvdt)} (0 0)  (10 10)
```

Turn-on/Turn-off delays (Quadrant III)

```
Rtonr     contotr  dlayr1         825
Dtonr     dlayr1   cntrolr        Delay
Rtoffr    contotr  dlayr2         {2.9E-3/Ton}
Dtoffr    cntrolr  dlayr2         Delay
Ctonr     cntrolr  0              {Ton454}
```

Controlled Switch Model

```
.MODEL Vswitch vswitch
+ (Ron = { (Vtm-0.7)/Itm}, Roff = {1.75E-3 Vdrm/Idrm},
+   Von = 5.0,             Voff = 1.5)
```

Diodes

```
.MODEL  Dgk    D       (Is=1E-16 Cjo=50pf Rs=5)
.MODEL  Delay  D       (Is=1E-12 Cjo=5pf  Rs=0.01)
.MODEL  Dak    D       (Is=4E-11 Cjo=5pf)
```

Allow the gate to float if required.

```
Rfloat  gate   MT1 1e10
.ENDS

.SUBCKT 2N5444        MT2 gate MT1
```

Min and Max parameters

```
X1 MT2 gate MT1 Triac PARAMS:
+ Vdrm=200v   Idrm=10u      Ih=70ma      dVdt=50e6   Ton=1u
+ Igt=70ma    Vgt=2.0v      Vtm=1.65v    Itm=56
```

90-5-18 Motorola   DL137, Rev 2, 3/89

```
. ENDS
```

Index